D0142682

The William Lowell Putnam Mathematical Competition 1985–2000

Problems, Solutions, and Commentary

Ravi Vakil's photo on p. 337 is courtesy of Gabrielle Vogel.

Library of Congress Catalog Card Number 2002107972
ISBN 0-88385-807-X
Printed in the United States of America
Current Printing (last digit):
10 9 8 7 6 5 4 3 2

The William Lowell Putnam Mathematical Competition 1985–2000

Problems, Solutions, and Commentary

Kiran S. Kedlaya
University of California, Berkeley

Bjorn Poonen
University of California, Berkeley

Ravi Vakil
Stanford University

Published and distributed by
The Mathematical Association of America

MAA PROBLEM BOOKS SERIES

Problem Books is a series of the Mathematical Association of America consisting of collections of problems and solutions from annual mathematical competitions; compilations of problems (including unsolved problems) specific to particular branches of mathematics; books on the art and practice of problem solving, etc.

The Inquisitive Problem Solver, Paul Vaderlind, Richard K. Guy, and Loren L. Larson

Mathematical Olympiads 1998–1999: Problems and Solutions From Around the World, edited by Titu Andreescu and Zuming Feng

Mathematical Olympiads 1999–2000: Problems and Solutions From Around the World, edited by Titu Andreescu and Zuming Feng

The William Lowell Putnam Mathematical Competition 1985–2000: Problems, Solutions, and Commentary, Kiran S. Kedlaya, Bjorn Poonen, Ravi Vakil

USA and International Mathematical Olympiads 2000, edited by Titu Andreescu and Zuming Feng

USA and International Mathematical Olympiads 2001, edited by Titu Andreescu and Zuming Feng

MAA Service Center
P. O. Box 91112
Washington, DC 20090-1112
1-800-331-1622 fax: 1-301-206-9789
www.maa.org

Dedicated to the Putnam contestants

Introduction

This book is the third collection of William Lowell Putnam Mathematical Competition problems and solutions, following [PutnamI] and [PutnamII]. As the subtitle indicates, the goals of our volume differ somewhat from those of the earlier volumes.

Many grand ideas of mathematics are best first understood through simple problems, with the inessential details stripped away. When developing new theory, research mathematicians often turn to toy[†] problems as a means of getting a foothold. For this reason, Putnam problems and solutions should be considered not in isolation, but instead in the context of important mathematical themes. Many of the best problems contain kernels of sophisticated ideas, or are connected to some of the most important research done today. We have tried to emphasize the organic nature of mathematics, by highlighting the connections of problems and solutions to other problems, to the curriculum, and to more advanced topics. A quick glance at the index will make clear the wide range of powerful ideas connected to these problems. For example, Putnam problems connect to the Generalized Riemann Hypothesis (1988B1) and the Weil Conjectures (1991B5 and 1998B6).

1 Structure of this book

The first section contains the problems, as they originally appeared in the competition, but annotated to clarify occasional infelicities of wording. We have included a list of the Questions Committee with each competition, and we note here that in addition Loren Larson has served as an *ex officio* member of the committee for nearly the entire period covered by this book. Next is a section containing a brief hint for each problem. The hints may often be more mystifying than enlightening. Nonetheless, we hope that they encourage readers to spend more time wrestling with a problem before turning to the solution section.

The heart of this book is in the solutions. For each problem, we include every solution we know, eliminating solutions only if they are essentially equivalent to one already given, or clearly inferior to one already given. Putnam problems are usually constructed so that they admit a solution involving nothing more than calculus, linear algebra, and a bit of real analysis and abstract algebra; hence we always

[†] A "toy" problem does not necessarily mean an easy problem. Rather, it means a relatively tractable problem where a key issue has been isolated, and all extraneous detail has been stripped away.

include one solution requiring no more background than this. On the other hand, as mentioned above, the problems often relate to deep and beautiful mathematical ideas, and concealing these ideas makes some solutions look like isolated tricks; therefore where germane we mention additional problems solvable by similar methods, alternate solutions possibly involving more advanced concepts, and further remarks relating the problem to the mathematical literature. Our alternate solutions are sometimes more terse than the first one. The top of each solution includes the score distribution of the top contestants: see page 51. When we write "see 1997A6," we mean "see the solution(s) to 1997A6 and the surrounding material."

After the solutions comes a list of the winning individuals and teams. This includes one-line summaries of the winners' histories, when known to us. Finally, we reprint an article by Joseph A. Gallian, "Putnam Trivia for the Nineties," and an article by Bruce Reznick, "Some Thoughts on Writing for the Putnam."

2 The Putnam Competition over the years

The competition literature states: "The competition began in 1938, and was designed to stimulate a healthy rivalry in mathematical studies in the colleges and universities of the United States and Canada. It exists because Mr. William Lowell Putnam had a profound conviction in the value of organized team competition in regular college studies. Mr. Putnam, a member of the Harvard class of 1882, wrote an article for the December 1921 issue of the *Harvard Graduates' Magazine* in which he described the merits of an intercollegiate competition. To establish such a competition, his widow, Elizabeth Lowell Putnam, in 1927 created a trust fund known as the William Lowell Putnam Intercollegiate Memorial Fund. The first competition supported by this fund was in the field of English and a few years later a second experimental competition was held, this time in mathematics between two institutions. It was not until after Mrs. Putnam's death in 1935 that the examination assumed its present form and was placed under the administration of the Mathematical Association of America."

Since 1962, the competition has consisted of twelve problems, usually numbered A1 through A6 and B1 through B6, given in two sessions of three hours each on the first Saturday in December. For more information about the history of the Putnam Competition, see the articles of Garrett Birkhoff and L. E. Bush in [PutnamI].

The competition is open to regularly enrolled undergraduates in the U.S. and Canada who have not yet received a college degree. No individual may participate in the competition more than four times. Each college or university with at least three participants names a team of three individuals. But the team must be chosen *before* the competition, so schools often fail to select their highest three scores; indeed, some schools are notorious for this. Also, the team rank is determined by the sum of the ranks of the team members, so one team member having a bad day can greatly lower the team rank. These two factors add an element of uncertainty to the team competition.

Prizes are awarded to the mathematics departments of the institutions with the five winning teams, and to the team members. The five highest ranking individuals are designated Putnam Fellows; prizes are awarded to these individuals and to each of the next twenty highest ranking contestants. One of the Putnam Fellows is also awarded the William Lowell Putnam Prize Scholarship at Harvard. Also, in some years, beginning in 1992, the Elizabeth Lowell Putnam Prize has been awarded to a woman whose performance has been deemed particularly meritorious. The winners of this prize are listed in the "Individual Results" section. The purpose of the Putnam Competition is not only to select a handful of prize winners, however; it is also to provide a stimulating challenge to all the contestants.

The nature of the problems has evolved. A few of the changes reflect changing emphases in the discipline of mathematics itself: for example, there are no more problems on Newtonian mechanics, and the number of problems involving extended algebraic manipulations has decreased. Other changes seem more stylistic: problems from recent decades often admit relatively short solutions, and are never open-ended.

The career paths of recent Putnam winners promise to differ in some ways from those of their predecessors recorded in [PutnamI]. Although it is hard to discern patterns among recent winners since many are still in school, it seems that fewer are becoming pure mathematicians than in the past. Most still pursue a Ph.D. in mathematics or some other science, but many then go into finance or cryptography, or begin other technology-related careers. It is also true that some earlier winners have switched from pure mathematics to other fields. For instance, David Mumford, a Putnam Fellow in 1955 and 1956 who later won a Fields Medal for his work in algebraic geometry, has been working in computer vision since the 1980s.

3 Advice to the student reader

The first lesson of the Putnam is: don't be intimidated. Some of the problems relate to complex mathematical ideas, but all can be solved using only the topics in a typical undergraduate mathematics curriculum, admittedly combined in clever ways. By working on these problems and afterwards studying their solutions, you will gain insight into beautiful aspects of mathematics beyond what you may have seen before.

Be patient when working on a problem. Learning comes more from struggling with problems than from solving them. If after some time, you are still stuck on a problem, see if the hint will help, and sleep on it before giving up. Most students, when they first encounter Putnam problems, do not solve more than a few, if any at all, because they give up too quickly. Also keep in mind that problem-solving becomes easier with experience; it is not a function of cleverness alone.

Be patient with the solutions as well. Mathematics is meant to be read slowly and carefully. If there are some steps in a solution that you do not follow, try discussing it with a knowledgeable friend or instructor. Most research mathematicians do the same when they are stuck (which is most of the time); the best mathematics research is almost never done in isolation, but rather in dialogue with other mathematicians, and in consultation of their publications. When you read the solutions, you will often find interesting side remarks and related problems to think about, as well as connections to other beautiful parts of mathematics, both elementary and advanced. Maybe you will create new problems that are not in this book. We hope that you follow up on the ideas that interest you most.

Year	Cut-off score for			
	Median	Top ~ 200	Honorable Mention	Putnam Fellow
1985	2	37	66	91
1986	19	33	51	81
1987	1	26	49	88
1988	16	40	65	110
1989	0	29	50	77
1990	2	28	50	77
1991	11	40	62	93
1992	2	32	53	92
1993	10	29	41	60
1994	3	28	47	87
1995	8	35	52	85
1996	3	26	43	76
1997	1	25	42	69
1998	10	42	69	98
1999	0	21	45	69
2000	0	21	43	90

TABLE 1. Score cut-offs

4 Scoring

Scores in the competition tend to be very low. The questions are difficult and the grading is strict, with little partial credit awarded. Students who solve one question and write it up perfectly do better than those with partial ideas for a number of problems.

Each of the twelve problems is graded on a basis of 0 to 10 points, so the maximum possible score is 120. Table 1 shows the scores required in each of the years covered in this volume to reach the median, the top 200, Honorable Mention, and the rank of Putnam Fellow (top five, or sometimes six in case of a tie). Keep in mind that the contestants are self-selected from among the brightest in two countries. As you can see from Table 1, solving a single problem should be considered a success. In particular, the Putnam is not a "test" with passing and failing grades; instead it is an open-ended challenge, a competition between you and the problems.

Along with each solution in this book, we include the score distribution of the top 200 or so contestants on that problem: see page 51. This may be used as a rough indicator of the difficulty of a problem, but of course, different individuals may find different problems difficult, depending on background. The problems with highest scores were 1988A1 and 1988B1, and the problems with the lowest scores were 1999B4 and 1999B5. When an easier problem was accidentally placed toward the end of the competition, the scores tended to be surprisingly low. We suspect that this is because contestants expected the problem to be more difficult than it actually was.

5 Some basic notation

The following definitions are standard in modern mathematics, so we use them throughout this book:

$$\mathbb{Z} = \text{the ring of integers} = \{\dots, -2, -1, 0, 1, 2, \dots\}$$
$$\mathbb{Q} = \text{the field of rational numbers} = \{\, m/n : m, n \in \mathbb{Z}, n \neq 0 \,\}$$
$$\mathbb{R} = \text{the field of real numbers}$$
$$\mathbb{C} = \text{the field of complex numbers} = \{\, a + bi : a, b \in \mathbb{R} \,\}, \text{ where } i = \sqrt{-1}$$
$$\mathbb{F}_q = \text{the finite field of } q \text{ elements.}$$

The cardinality of a set S is denoted $\#S$ or sometimes $|S|$. If $a, b \in \mathbb{Z}$, then "$a \mid b$" means that a divides b, that is, that there exists $k \in \mathbb{Z}$ such that $b = ka$. Similarly, "$a \nmid b$" means that a does not divide b. The set of positive real numbers is denoted by $\mathbb{R}^+$.

We use the notation $\ln x$ for the natural logarithm function, even though in higher mathematics the synonym $\log x$ is more frequently used. It is tacitly assumed that the base of the logarithm, if unspecified, equals $e = 2.71828\dots$. If logarithms to the base 10 are intended, it is better to write $\log_{10} x$. More generally, $\log_a x = (\log x)/(\log a)$ denotes logarithm to the base a. In computer science, the notation $\lg n$ is sometimes used as an abbreviation for $\log_2 n$. (In number theory, when p is a prime number, $\log_p x$ sometimes also denotes the p-adic logarithm function [Kob, p. 87], a function with similar properties but defined on nonzero p-adic numbers instead of positive real numbers. But this book will have no need for this p-adic function.)

Rings for us are associative and have a multiplicative unit 1. If R is a ring, then $R[x]$ denotes the ring of all polynomials

$$a_n x^n + a_{n-1} x^{n-1} + \cdots + a_1 x + a_0$$

where n is any nonnegative integer, and $a_0, a_1, \dots, a_n \in R$. Also, $R[[x]]$ denotes the ring of formal power series

$$a_0 + a_1 x + a_2 x^2 + \cdots$$

where the a_i belong to R.

If R is a ring and $n \geq 1$, $M_n(R)$ denotes the set of $n \times n$ matrices with coefficients in R, and $\mathrm{GL}_n(R)$ denotes the subset of matrices $A \in M_n(R)$ that have an inverse in $M_n(R)$. When R is a field, a matrix $A \in M_n(R)$ has such an inverse if and only if its determinant $\det(A)$ is nonzero; more generally, for any commutative ring, A has such an inverse if and only if $\det(A)$ is a unit of R. (The reason to insist that the determinant be a unit, and not just nonzero, is that it makes $\mathrm{GL}_n(R)$ a group under multiplication.) For instance, $\mathrm{GL}_2(\mathbb{Z})$ is the set of matrices $\begin{pmatrix} a & b \\ c & d \end{pmatrix}$ with $a, b, c, d \in \mathbb{Z}$ and $ad - bc = \pm 1$.

6 Acknowledgements

We are grateful to the many individuals who have shared ideas with us. Much of our material is adapted from the annual articles in the *American Mathematical Monthly* and *Mathematics Magazine*, by Alexanderson, Klosinski, and Larson. Many additional solutions were taken from the web, especially from annual postings of Dave Rusin to the `sci.math` newsgroup, and from postings in recent years of Manjul Bhargava, Kiran Kedlaya, and Lenny Ng at the website

`http://www.unl.edu/amc`

hosted by American Mathematics Competitions; hopefully these postings will continue in future years. We thank Gabriel Carroll, Sabin Cautis, Keith Conrad, Ioana Dumitriu, J.P. Grossman, Doug Jungreis, Andrew Kresch, Abhinav Kumar, Greg Kuperberg, Russ Mann, Lenny Ng, Naoki Sato, Dave Savitt, Hoeteck Wee, and Eric Wepsic, who read parts of this book and contributed many suggestions and ideas that were incorporated into the text, often without attribution. We thank Jerry Alexanderson, Loren Larson, and Roger Nelsen for detailed and helpful comments on the entire manuscript. We thank Pramod Achar, Art Benjamin, George Bergman, Mira Bernstein, Anders Buch, Robert Burckel, Ernie Croot, Charles Fefferman, Donald Sarason, Jun Song, Bernd Sturmfels, Mark van Raamsdonk, and Balint Virag for additional comments, and for suggesting references. We thank Joe Gallian and Bruce Reznick for permission to reprint their articles [G2] and [Re4].

We thank also the members of the Questions Committee in the years covered in this volume: Bruce Reznick, Richard P. Stanley, Harold M. Stark, Abraham P. Hillman, Gerald A. Heuer, Paul R. Halmos, Kenneth A. Stolarsky, George E. Andrews, George T. Gilbert, Eugene Luks, Fan Chung, Mark I. Krusemeyer, Richard K. Guy, Michael J. Larsen, David J. Wright, Steven G. Krantz, Andrew J. Granville, and Carl Pomerance. Loren Larson has served as an *ex officio* member of the committee for nearly the entire period covered by this book. Finally, we thank Don Albers, Elaine Pedreira, Martha Pennigar, Beverly Ruedi, and the other staff at the Mathematical Association of America for their assistance and support throughout this project.

Kiran S. Kedlaya
Bjorn Poonen
Ravi Vakil

Berkeley / Palo Alto
Fall 2001

Contents

Introduction **vii**
1 Structure of this book . . . vii
2 The Putnam Competition over the years . . . viii
3 Advice to the student reader . . . ix
4 Scoring . . . x
5 Some basic notation . . . xi
6 Acknowledgements . . . xi

Problems **1**

Hints **35**

Solutions **51**
The Forty-Sixth Competition (1985) . . . 53
The Forty-Seventh Competition (1986) . . . 65
The Forty-Eighth Competition (1987) . . . 76
The Forty-Ninth Competition (1988) . . . 88
The Fiftieth Competition (1989) . . . 101
The Fifty-First Competition (1990) . . . 116
The Fifty-Second Competition (1991) . . . 135
The Fifty-Third Competition (1992) . . . 154
The Fifty-Fourth Competition (1993) . . . 171
The Fifty-Fifth Competition (1994) . . . 191
The Fifty-Sixth Competition (1995) . . . 204
The Fifty-Seventh Competition (1996) . . . 217
The Fifty-Eighth Competition (1997) . . . 232
The Fifty-Ninth Competition (1998) . . . 250
The Sixtieth Competition (1999) . . . 262
The Sixty-First Competition (2000) . . . 278

Results **295**
Individual Results . . . 295
Team Results . . . 301

Putnam Trivia for the Nineties
by Joseph A. Gallian **307**
Answers 321

Some Thoughts on Writing for the Putnam
by Bruce Reznick **311**

Bibliography **323**

Index **333**

About the Authors **337**

Problems

The Forty-Sixth William Lowell Putnam Mathematical Competition
December 7, 1985

Questions Committee: Bruce Reznick, Richard P. Stanley, and Harold M. Stark

See page 35 for hints.

A1. Determine, with proof, the number of ordered triples (A_1, A_2, A_3) of sets which have the property that

(i) $A_1 \cup A_2 \cup A_3 = \{1, 2, 3, 4, 5, 6, 7, 8, 9, 10\}$, and

(ii) $A_1 \cap A_2 \cap A_3 = \emptyset$,

where $\emptyset$ denotes the empty set. Express the answer in the form $2^a 3^b 5^c 7^d$, where a, b, c, and d are nonnegative integers. (page 53)

A2. Let T be an acute triangle. Inscribe a pair R, S of rectangles in T as shown:

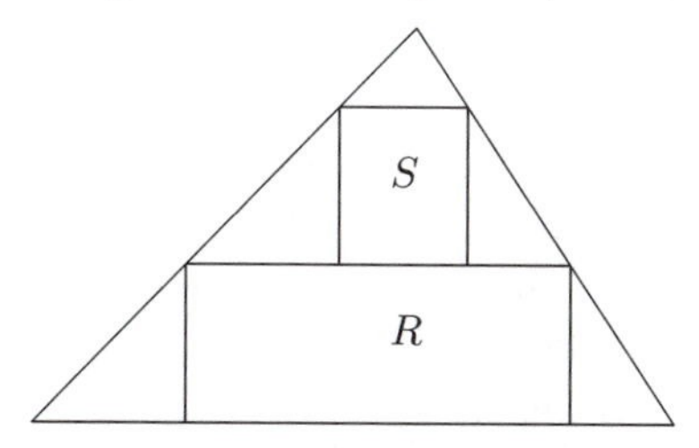

Let $A(X)$ denote the area of polygon X. Find the maximum value, or show that no maximum exists, of $\frac{A(R)+A(S)}{A(T)}$, where T ranges over all triangles and R, S over all rectangles as above. (page 54)

A3. Let d be a real number. For each integer $m \geq 0$, define a sequence $\{a_m(j)\}$, $j = 0, 1, 2, \ldots$ by the condition

$$a_m(0) = d/2^m, \quad \text{and} \quad a_m(j+1) = (a_m(j))^2 + 2a_m(j), \quad j \geq 0.$$

Evaluate $\lim_{n\to\infty} a_n(n)$. (page 56)

A4. Define a sequence $\{a_i\}$ by $a_1 = 3$ and $a_{i+1} = 3^{a_i}$ for $i \geq 1$. Which integers between 00 and 99 inclusive occur as the last two digits in the decimal expansion of infinitely many a_i? (page 57)

A5. Let $I_m = \int_0^{2\pi} \cos(x)\cos(2x)\cdots\cos(mx)\,dx$. For which integers m, $1 \leq m \leq 10$, is $I_m \neq 0$? (page 58)

A6. If $p(x) = a_0 + a_1 x + \cdots + a_m x^m$ is a polynomial with real coefficients a_i, then set

$$\Gamma(p(x)) = a_0^2 + a_1^2 + \cdots + a_m^2.$$

Let $f(x) = 3x^2 + 7x + 2$. Find, with proof, a polynomial $g(x)$ with real coefficients such that

(i) $g(0) = 1$, and

(ii) $\Gamma(f(x)^n) = \Gamma(g(x)^n)$

for every integer $n \geq 1$. (page 59)

B1. Let k be the smallest positive integer with the following property:

There are distinct integers m_1, m_2, m_3, m_4, m_5 such that the polynomial

$$p(x) = (x - m_1)(x - m_2)(x - m_3)(x - m_4)(x - m_5)$$

has exactly k nonzero coefficients.

Find, with proof, a set of integers m_1, m_2, m_3, m_4, m_5 for which this minimum k is achieved. (page 60)

B2. Define polynomials $f_n(x)$ for $n \geq 0$ by $f_0(x) = 1$, $f_n(0) = 0$ for $n \geq 1$, and

$$\frac{d}{dx}\left(f_{n+1}(x)\right) = (n+1) f_n(x+1)$$

for $n \geq 0$. Find, with proof, the explicit factorization of $f_{100}(1)$ into powers of distinct primes. (page 61)

B3. Let

$$\begin{array}{cccc} a_{1,1} & a_{1,2} & a_{1,3} & \cdots \\ a_{2,1} & a_{2,2} & a_{2,3} & \cdots \\ a_{3,1} & a_{3,2} & a_{3,3} & \cdots \\ \vdots & \vdots & \vdots & \ddots \end{array}$$

be a doubly infinite array of positive integers, and suppose each positive integer appears exactly eight times in the array. Prove that $a_{m,n} > mn$ for some pair of positive integers (m, n). (page 61)

B4. Let C be the unit circle $x^2 + y^2 = 1$. A point p is chosen randomly on the circumference of C and another point q is chosen randomly from the interior of C (these points are chosen independently and uniformly over their domains). Let R be the rectangle with sides parallel to the x- and y-axes with diagonal pq. What is the probability that no point of R lies outside of C? (page 62)

B5. Evaluate $\int_0^\infty t^{-1/2} e^{-1985(t+t^{-1})}\,dt$. You may assume that $\int_{-\infty}^\infty e^{-x^2}\,dx = \sqrt{\pi}$. (page 62)

B6. Let G be a finite set of real $n \times n$ matrices $\{M_i\}$, $1 \leq i \leq r$, which form a group under matrix multiplication. Suppose that $\sum_{i=1}^{r} \operatorname{tr}(M_i) = 0$, where $\operatorname{tr}(A)$ denotes the trace of the matrix A. Prove that $\sum_{i=1}^{r} M_i$ is the $n \times n$ zero matrix. (page 63)

The Forty-Seventh William Lowell Putnam Mathematical Competition
December 6, 1986

Questions Committee: Richard P. Stanley,
Harold M. Stark, and Abraham P. Hillman
See page 36 for hints.

A1. Find, with explanation, the maximum value of $f(x) = x^3 - 3x$ on the set of all real numbers x satisfying $x^4 + 36 \leq 13x^2$. (page 65)

A2. What is the units (i.e., rightmost) digit of $\left\lfloor \frac{10^{20000}}{10^{100}+3} \right\rfloor$? Here $\lfloor x \rfloor$ is the greatest integer $\leq x$. (page 65)

A3. Evaluate $\sum_{n=0}^{\infty} \operatorname{Arccot}(n^2+n+1)$, where $\operatorname{Arccot} t$ for $t \geq 0$ denotes the number θ in the interval $0 < \theta \leq \pi/2$ with $\cot \theta = t$. (page 65)

A4. A *transversal* of an $n \times n$ matrix A consists of n entries of A, no two in the same row or column. Let $f(n)$ be the number of $n \times n$ matrices A satisfying the following two conditions:

(a) Each entry $\alpha_{i,j}$ of A is in the set $\{-1, 0, 1\}$.

(b) The sum of the n entries of a transversal is the same for all transversals of A.

An example of such a matrix A is

$$A = \begin{pmatrix} -1 & 0 & -1 \\ 0 & 1 & 0 \\ 0 & 1 & 0 \end{pmatrix}.$$

Determine with proof a formula for $f(n)$ of the form

$$f(n) = a_1 b_1^n + a_2 b_2^n + a_3 b_3^n + a_4,$$

where the a_i's and b_i's are rational numbers. (page 67)

A5. Suppose $f_1(x), f_2(x), \ldots, f_n(x)$ are functions of n real variables $x = (x_1, \ldots, x_n)$ with continuous second-order partial derivatives everywhere on $\mathbb{R}^n$. Suppose further that there are constants c_{ij} such that

$$\frac{\partial f_i}{\partial x_j} - \frac{\partial f_j}{\partial x_i} = c_{ij}$$

for all i and j, $1 \leq i \leq n$, $1 \leq j \leq n$. Prove that there is a function $g(x)$ on $\mathbb{R}^n$ such that $f_i + \partial g / \partial x_i$ is linear for all i, $1 \leq i \leq n$. (A linear function is one of the form

$$a_0 + a_1 x_1 + a_2 x_2 + \cdots + a_n x_n.)$$

(page 68)

A6. Let $a_1, a_2, \ldots, a_n$ be real numbers, and let $b_1, b_2, \ldots, b_n$ be distinct positive integers. Suppose there is a polynomial $f(x)$ satisfying the identity

$$(1-x)^n f(x) = 1 + \sum_{i=1}^{n} a_i x^{b_i}.$$

Find a simple expression (not involving any sums) for $f(1)$ in terms of $b_1, b_2, \ldots, b_n$ and n (but independent of $a_1, a_2, \ldots, a_n$). (page 69)

B1. Inscribe a rectangle of base b and height h and an isosceles triangle of base b in a circle of radius one as shown. For what value of h do the rectangle and triangle have the same area?

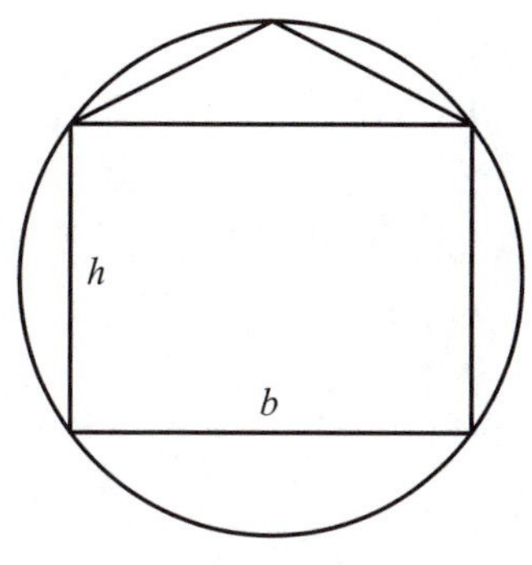

(page 70)

B2. Prove that there are only a finite number of possibilities for the ordered triple $T = (x - y, y - z, z - x)$, where x, y, and z are complex numbers satisfying the simultaneous equations

$$x(x-1) + 2yz = y(y-1) + 2zx = z(z-1) + 2xy,$$

and list all such triples T. (page 71)

B3. Let Γ consist of all polynomials in x with integer coefficients. For f and g in Γ and m a positive integer, let $f \equiv g \pmod{m}$ mean that every coefficient of $f - g$ is an integral multiple of m. Let n and p be positive integers with p prime. Given that f, g, h, r, and s are in Γ with $rf + sg \equiv 1 \pmod{p}$ and $fg \equiv h \pmod{p}$, prove that there exist F and G in Γ with $F \equiv f \pmod{p}$, $G \equiv g \pmod{p}$, and $FG \equiv h \pmod{p^n}$. (page 71)

B4. For a positive real number r, let $G(r)$ be the minimum value of $\left|r - \sqrt{m^2 + 2n^2}\right|$ for all integers m and n. Prove or disprove the assertion that $\lim_{r\to\infty} G(r)$ exists and equals 0. (page 72)

B5. Let $f(x, y, z) = x^2 + y^2 + z^2 + xyz$. Let $p(x, y, z)$, $q(x, y, z)$, $r(x, y, z)$ be polynomials with real coefficients satisfying

$$f(p(x, y, z), q(x, y, z), r(x, y, z)) = f(x, y, z).$$

Prove or disprove the assertion that the sequence p, q, r consists of some permutation of $\pm x$, $\pm y$, $\pm z$, where the number of minus signs is 0 or 2. (page 73)

B6. Suppose A, B, C, D are $n \times n$ matrices with entries in a field F, satisfying the conditions that AB^t and CD^t are symmetric and $AD^t - BC^t = I$. Here I is the $n \times n$ identity matrix, and if M is an $n \times n$ matrix, M^t is the transpose of M. Prove that $A^tD - C^tB = I$. (page 74)

The Forty-Eighth William Lowell Putnam Mathematical Competition
December 5, 1987

Questions Committee: Harold M. Stark, Abraham P. Hillman, and Gerald A. Heuer

See page 37 for hints.

A1. Curves A, B, C, and D are defined in the plane as follows:†

$$A = \left\{ (x,y) : x^2 - y^2 = \frac{x}{x^2+y^2} \right\},$$

$$B = \left\{ (x,y) : 2xy + \frac{y}{x^2+y^2} = 3 \right\},$$

$$C = \{ (x,y) : x^3 - 3xy^2 + 3y = 1 \},$$

$$D = \{ (x,y) : 3x^2y - 3x - y^3 = 0 \}.$$

Prove that $A \cap B = C \cap D$. (page 76)

A2. The sequence of digits

$$123456789101112131415161718192021\ldots$$

is obtained by writing the positive integers in order. If the 10^nth digit in this sequence occurs in the part of the sequence in which the m-digit numbers are placed, define $f(n)$ to be m. For example, $f(2) = 2$ because the 100th digit enters the sequence in the placement of the two-digit integer 55. Find, with proof, $f(1987)$. (page 76)

A3. For all real x, the real-valued function $y = f(x)$ satisfies

$$y'' - 2y' + y = 2e^x.$$

(a) If $f(x) > 0$ for all real x, must $f'(x) > 0$ for all real x? Explain.

(b) If $f'(x) > 0$ for all real x, must $f(x) > 0$ for all real x? Explain. (page 78)

A4. Let P be a polynomial, with real coefficients, in three variables and F be a function of two variables such that

$$P(ux, uy, uz) = u^2 F(y - x, z - x) \quad \text{for all real } x, y, z, u,$$

and such that $P(1,0,0) = 4$, $P(0,1,0) = 5$, and $P(0,0,1) = 6$. Also let A, B, C be complex numbers with $P(A,B,C) = 0$ and $|B - A| = 10$. Find $|C - A|$. (page 78)

A5. Let

$$\vec{G}(x,y) = \left(\frac{-y}{x^2+4y^2}, \frac{x}{x^2+4y^2}, 0 \right).$$

Prove or disprove that there is a vector-valued function

$$\vec{F}(x,y,z) = (M(x,y,z), N(x,y,z), P(x,y,z))$$

with the following properties:

---

† The equations defining A and B are indeterminate at $(0,0)$. The point $(0,0)$ belongs to neither.

(i) M, N, P have continuous partial derivatives for all $(x, y, z) \neq (0,0,0)$;

(ii) Curl $\vec{F} = \vec{0}$ for all $(x, y, z) \neq (0,0,0)$;

(iii) $\vec{F}(x, y, 0) = \vec{G}(x, y)$.

(page 79)

A6. For each positive integer n, let $a(n)$ be the number of zeros in the base 3 representation of n. For which positive real numbers x does the series

$$\sum_{n=1}^{\infty} \frac{x^{a(n)}}{n^3}$$

converge? (page 79)

B1. Evaluate

$$\int_2^4 \frac{\sqrt{\ln(9-x)}\,dx}{\sqrt{\ln(9-x)} + \sqrt{\ln(x+3)}}.$$

(page 80)

B2. Let r, s, and t be integers with $0 \le r$, $0 \le s$, and $r + s \le t$. Prove that

$$\frac{\binom{s}{0}}{\binom{t}{r}} + \frac{\binom{s}{1}}{\binom{t}{r+1}} + \cdots + \frac{\binom{s}{s}}{\binom{t}{r+s}} = \frac{t+1}{(t+1-s)\binom{t-s}{r}}.$$

(Note: $\binom{n}{k}$ denotes the binomial coefficient $\frac{n(n-1)\cdots(n+1-k)}{k(k-1)\cdots 3\cdot 2\cdot 1}$.) (page 81)

B3. Let F be a field in which $1 + 1 \neq 0$. Show that the set of solutions to the equation $x^2 + y^2 = 1$ with x and y in F is given by $(x, y) = (1, 0)$ and

$$(x, y) = \left(\frac{r^2-1}{r^2+1}, \frac{2r}{r^2+1}\right),$$

where r runs through the elements of F such that $r^2 \neq -1$. (page 83)

B4. Let $(x_1, y_1) = (0.8, 0.6)$ and let $x_{n+1} = x_n \cos y_n - y_n \sin y_n$ and $y_{n+1} = x_n \sin y_n + y_n \cos y_n$ for $n = 1, 2, 3, \ldots$. For each of $\lim_{n\to\infty} x_n$ and $\lim_{n\to\infty} y_n$, prove that the limit exists and find it or prove that the limit does not exist. (page 85)

B5. Let O_n be the n-dimensional vector $(0, 0, \ldots, 0)$. Let M be a $2n \times n$ matrix of complex numbers such that whenever $(z_1, z_2, \ldots, z_{2n})M = O_n$, with complex z_i, not all zero, then at least one of the z_i is not real. Prove that for arbitrary real numbers $r_1, r_2, \ldots, r_{2n}$, there are complex numbers $w_1, w_2, \ldots, w_n$ such that

$$\operatorname{Re}\left[M\begin{pmatrix} w_1 \\ \vdots \\ w_n \end{pmatrix}\right] = \begin{pmatrix} r_1 \\ \vdots \\ r_{2n} \end{pmatrix}.$$

(Note: if C is a matrix of complex numbers, $\operatorname{Re}(C)$ is the matrix whose entries are the real parts of the entries of C.) (page 85)

B6. Let F be the field of p^2 elements where p is an odd prime. Suppose S is a set of $(p^2-1)/2$ distinct nonzero elements of F with the property that for each $a \neq 0$ in F, exactly one of a and $-a$ is in S. Let N be the number of elements in the intersection $S \cap \{2a : a \in S\}$. Prove that N is even. (page 86)

The Forty-Ninth William Lowell Putnam Mathematical Competition December 3, 1988

Questions Committee: Abraham P. Hillman, Gerald A. Heuer, and Paul R. Halmos

See page 38 for hints.

A1. Let R be the region consisting of the points (x, y) of the cartesian plane satisfying both $|x| - |y| \leq 1$ and $|y| \leq 1$. Sketch the region R and find its area.

(page 88)

A2. A not uncommon calculus mistake is to believe that the product rule for derivatives says that $(fg)' = f'g'$. If $f(x) = e^{x^2}$, determine, with proof, whether there exists an open interval (a, b) and a nonzero function g defined on (a, b) such that this wrong product rule is true for x in (a, b).

(page 88)

A3. Determine, with proof, the set of real numbers x for which

$$\sum_{n=1}^{\infty}\left(\frac{1}{n}\csc\frac{1}{n}-1\right)^{x}$$

converges.

(page 89)

A4.

(a) If every point of the plane is painted one of three colors, do there necessarily exist two points of the same color exactly one inch apart?

(b) What if "three" is replaced by "nine"?

Justify your answers.

(page 90)

A5. Prove that there exists a *unique* function f from the set $\mathbb{R}^+$ of positive real numbers to $\mathbb{R}^+$ such that

$$f(f(x)) = 6x - f(x) \qquad \text{and} \qquad f(x) > 0 \qquad \text{for all } x > 0.$$

(page 92)

A6. If a linear transformation A on an n-dimensional vector space has $n + 1$ eigenvectors such that any n of them are linearly independent, does it follow that A is a scalar multiple of the identity? Prove your answer.

(page 93)

B1. A *composite* (positive integer) is a product ab with a and b not necessarily distinct integers in $\{2, 3, 4, \dots\}$. Show that every composite is expressible as $xy + xz + yz + 1$, with x, y, and z positive integers.

(page 94)

B2. Prove or disprove: if x and y are real numbers with $y \geq 0$ and $y(y+1) \leq (x+1)^2$, then $y(y-1) \leq x^2$.

(page 95)

B3. For every n in the set $\mathbb{Z}^+ = \{1, 2, \dots\}$ of positive integers, let r_n be the minimum value of $|c - d\sqrt{3}|$ for all nonnegative integers c and d with $c + d = n$. Find, with proof, the smallest positive real number g with $r_n \leq g$ for all $n \in \mathbb{Z}^+$.

(page 96)

B4. Prove that if $\sum_{n=1}^{\infty} a_n$ is a convergent series of positive real numbers, then so is $\sum_{n=1}^{\infty} (a_n)^{n/(n+1)}$.
(page 97)

B5. For positive integers n, let $\mathbf{M}_n$ be the $2n+1$ by $2n+1$ skew-symmetric matrix for which each entry in the first n subdiagonals below the main diagonal is 1 and each of the remaining entries below the main diagonal is -1. Find, with proof, the rank of $\mathbf{M}_n$. (According to one definition, the rank of a matrix is the largest k such that there is a $k \times k$ submatrix with nonzero determinant.)

One may note that

$$\mathbf{M}_1 = \begin{pmatrix} 0 & -1 & 1 \\ 1 & 0 & -1 \\ -1 & 1 & 0 \end{pmatrix} \quad \text{and} \quad \mathbf{M}_2 = \begin{pmatrix} 0 & -1 & -1 & 1 & 1 \\ 1 & 0 & -1 & -1 & 1 \\ 1 & 1 & 0 & -1 & -1 \\ -1 & 1 & 1 & 0 & -1 \\ -1 & -1 & 1 & 1 & 0 \end{pmatrix}.$$

(page 97)

B6. Prove that there exist an infinite number of ordered pairs (a, b) of integers such that for every positive integer t the number $at + b$ is a triangular number if and only if t is a triangular number. (The triangular numbers are the $t_n = n(n+1)/2$ with n in $\{0, 1, 2, \ldots\}$.)
(page 100)

The Fiftieth William Lowell Putnam Mathematical Competition December 2, 1989

Questions Committee: Gerald A. Heuer, Paul R. Halmos, and Kenneth A. Stolarsky

See page 39 for hints.

A1. How many primes among the positive integers, written as usual in base 10, are such that their digits are alternating 1's and 0's, beginning and ending with 1? (page 101)

A2. Evaluate $\int_0^a \int_0^b e^{\max\{b^2x^2, a^2y^2\}}\, dy\, dx$, where a and b are positive. (page 101)

A3. Prove that if

$$11z^{10} + 10iz^9 + 10iz - 11 = 0,$$

then $|z| = 1$. (Here z is a complex number and $i^2 = -1$.) (page 101)

A4. If α is an irrational number, $0 < \alpha < 1$, is there a finite game with an honest coin such that the probability of one player winning the game is α? (An honest coin is one for which the probability of heads and the probability of tails are both $1/2$. A game is finite if with probability 1 it must end in a finite number of moves.) (page 102)

A5. Let m be a positive integer and let $\mathcal{G}$ be a regular $(2m+1)$-gon inscribed in the unit circle. Show that there is a positive constant A, independent of m, with the following property. For any point p inside $\mathcal{G}$ there are two distinct vertices v_1 and v_2 of $\mathcal{G}$ such that

$$\big|\, |p - v_1| - |p - v_2| \,\big| < \frac{1}{m} - \frac{A}{m^3}.$$

Here $|s - t|$ denotes the distance between the points s and t. (page 103)

A6. Let $\alpha = 1 + a_1x + a_2x^2 + \cdots$ be a formal power series with coefficients in the field of two elements. Let

$$a_n = \begin{cases} 1 & \text{if every block of zeros in the binary expansion of } n \\ & \qquad \text{has an even number of zeros in the block,} \\ \\ 0 & \text{otherwise.} \end{cases}$$

(For example, $a_{36} = 1$ because $36 = 100100_2$, and $a_{20} = 0$ because $20 = 10100_2$.) Prove that $\alpha^3 + x\alpha + 1 = 0$. (page 107)

B1. A dart, thrown at random, hits a square target. Assuming that any two parts of the target of equal area are equally likely to be hit, find the probability that the point hit is nearer to the center than to any edge. Express your answer in the form $(a\sqrt{b} + c)/d$, where a, b, c, d are positive integers. (page 108)

B2. Let S be a nonempty set with an associative operation that is left and right cancellative ($xy = xz$ implies $y = z$, and $yx = zx$ implies $y = z$). Assume that for every a in S the set $\{\, a^n : n = 1, 2, 3, \ldots \}$ is finite. Must S be a group? (page 109)

B3. Let f be a function on $[0, \infty)$, differentiable and satisfying

$$f'(x) = -3f(x) + 6f(2x)$$

for $x > 0$. Assume that $|f(x)| \leq e^{-\sqrt{x}}$ for $x \geq 0$ (so that $f(x)$ tends rapidly to 0 as x increases). For n a nonnegative integer, define

$$\mu_n = \int_0^\infty x^n f(x)\, dx$$

(sometimes called the nth moment of f).

a. Express μ_n in terms of μ_0.

b. Prove that the sequence $\{\mu_n 3^n/n!\}$ always converges, and that the limit is 0 only if $\mu_0 = 0$. (page 109)

B4. Can a countably infinite set have an uncountable collection of nonempty subsets such that the intersection of any two of them is finite? (page 111)

B5. Label the vertices of a trapezoid T (quadrilateral with two parallel sides) inscribed in the unit circle as A, B, C, D so that AB is parallel to CD and A, B, C, D are in counterclockwise order. Let s_1, s_2, and d denote the lengths of the line segments AB, CD, and OE, where E is the point of intersection of the diagonals of T, and O is the center of the circle. Determine the least upper bound of $(s_1 - s_2)/d$ over all such T for which $d \neq 0$, and describe all cases, if any, in which it is attained. (page 112)

B6. Let $(x_1, x_2, \ldots, x_n)$ be a point chosen at random from the n-dimensional region defined by $0 < x_1 < x_2 < \cdots < x_n < 1$. Let f be a continuous function on $[0, 1]$ with $f(1) = 0$. Set $x_0 = 0$ and $x_{n+1} = 1$. Show that the expected value of the Riemann sum

$$\sum_{i=0}^{n} (x_{i+1} - x_i) f(x_{i+1})$$

is $\int_0^1 f(t)P(t)\, dt$, where P is a polynomial of degree n, independent of f, with $0 \leq P(t) \leq 1$ for $0 \leq t \leq 1$. (page 113)

The Fifty-First William Lowell Putnam Mathematical Competition
December 1, 1990

Questions Committee: Paul R. Halmos,
Kenneth A. Stolarsky, and George E. Andrews
See page 40 for hints.

A1. Let

$$T_0 = 2, \;\; T_1 = 3, \;\; T_2 = 6,$$

and for $n \geq 3$,

$$T_n = (n+4)T_{n-1} - 4nT_{n-2} + (4n-8)T_{n-3}.$$

The first few terms are

$$2, \; 3, \; 6, \; 14, \; 40, \; 152, \; 784, \; 5168, \; 40576, \; 363392.$$

Find, with proof, a formula for T_n of the form $T_n = A_n + B_n$, where (A_n) and (B_n) are well-known sequences. (page 116)

A2. Is $\sqrt{2}$ the limit of a sequence of numbers of the form $\sqrt[3]{n} - \sqrt[3]{m}$, $(n, m = 0, 1, 2, \ldots)$? (page 117)

A3. Prove that any convex pentagon whose vertices (no three of which are collinear) have integer coordinates must have area $\geq 5/2$. (page 118)

A4. Consider a paper punch that can be centered at any point of the plane and that, when operated, removes from the plane precisely those points whose distance from the center is irrational. How many punches are needed to remove every point? (page 120)

A5. If $\mathbf{A}$ and $\mathbf{B}$ are square matrices of the same size such that $\mathbf{ABAB} = \mathbf{0}$, does it follow that $\mathbf{BABA} = \mathbf{0}$? (page 121)

A6. If X is a finite set, let $|X|$ denote the number of elements in X. Call an ordered pair (S, T) of subsets of $\{1, 2, \ldots, n\}$ *admissible* if $s > |T|$ for each $s \in S$, and $t > |S|$ for each $t \in T$. How many admissible ordered pairs of subsets of $\{1, 2, \ldots, 10\}$ are there? Prove your answer. (page 123)

B1. Find all real-valued continuously differentiable functions f on the real line such that for all x

$$(f(x))^2 = \int_0^x \left((f(t))^2 + (f'(t))^2\right) dt + 1990.$$

(page 124)

B2. Prove that for $|x| < 1$, $|z| > 1$,

$$1 + \sum_{j=1}^{\infty} (1 + x^j) \frac{(1-z)(1-zx)(1-zx^2)\cdots(1-zx^{j-1})}{(z-x)(z-x^2)(z-x^3)\cdots(z-x^j)} = 0.$$

(page 125)

B3. Let S be a set of 2×2 integer matrices whose entries a_{ij} (1) are all squares of integers, and, (2) satisfy $a_{ij} \leq 200$. Show that if S has more than 50387 ($= 15^4 - 15^2 - 15 + 2$) elements, then it has two elements that commute. (page 125)

B4. Let G be a finite group of order n generated by a and b. Prove or disprove: there is a sequence

$$g_1, g_2, g_3, \ldots, g_{2n}$$

such that

(1) every element of G occurs exactly twice, and

(2) g_{i+1} equals $g_i a$ or $g_i b$, for $i = 1, 2, \ldots, 2n$. (Interpret g_{2n+1} as g_1.)

(page 126)

B5. Is there an infinite sequence $a_0, a_1, a_2, \ldots$ of nonzero real numbers such that for $n = 1, 2, 3, \ldots$ the polynomial

$$p_n(x) = a_0 + a_1 x + a_2 x^2 + \cdots + a_n x^n$$

has exactly n distinct real roots? (page 127)

B6. Let S be a nonempty closed bounded convex set in the plane. Let K be a line and t a positive number. Let L_1 and L_2 be support lines for S parallel to K, and let $\overline{L}$ be the line parallel to K and midway between L_1 and L_2. Let $B_S(K,t)$ be the band of points whose distance from $\overline{L}$ is at most $(t/2)w$, where w is the distance between L_1 and L_2. What is the smallest t such that

$$S \cap \bigcap_K B_S(K,t) \neq \emptyset$$

for all S? (K runs over all lines in the plane.)

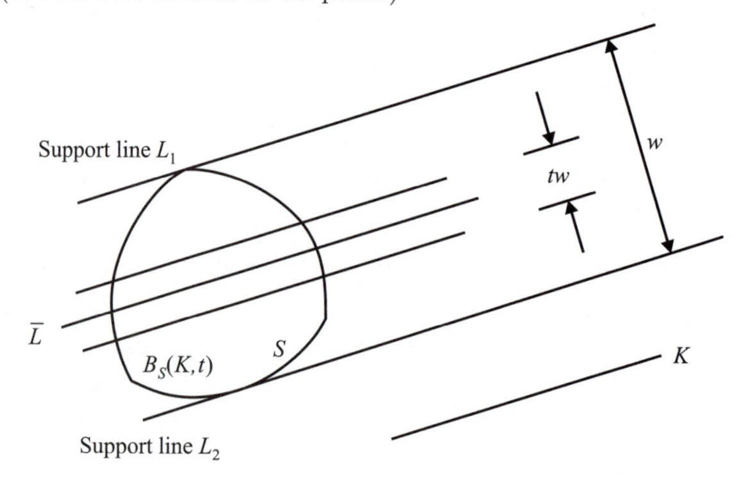

(page 128)

The Fifty-Second William Lowell Putnam Mathematical Competition
December 7, 1991

Questions Committee: Kenneth A. Stolarsky,
George E. Andrews, and George T. Gilbert
See page 41 for hints.

A1. A 2×3 rectangle has vertices at $(0,0)$, $(2,0)$, $(0,3)$, and $(2,3)$. It rotates 90° clockwise about the point $(2,0)$. It then rotates 90° clockwise about the point $(5,0)$, then 90° clockwise about the point $(7,0)$, and finally, 90° clockwise about the point $(10,0)$. (The side originally on the x-axis is now back on the x-axis.) Find the area of the region above the x-axis and below the curve traced out by the point whose initial position is $(1,1)$. (page 135)

A2. Let $\mathbf{A}$ and $\mathbf{B}$ be different $n \times n$ matrices with real entries. If $\mathbf{A}^3 = \mathbf{B}^3$ and $\mathbf{A}^2\mathbf{B} = \mathbf{B}^2\mathbf{A}$, can $\mathbf{A}^2 + \mathbf{B}^2$ be invertible? (page 135)

A3. Find all real polynomials $p(x)$ of degree $n \geq 2$ for which there exist real numbers $r_1 < r_2 < \cdots < r_n$ such that

(i) $p(r_i) = 0, \qquad i = 1, 2, \ldots, n$, and

(ii) $p'\left(\frac{r_i + r_{i+1}}{2}\right) = 0, \qquad i = 1, 2, \ldots, n-1$,

where $p'(x)$ denotes the derivative of $p(x)$. (page 135)

A4. Does there exist an infinite sequence of closed discs $D_1, D_2, D_3, \ldots$ in the plane, with centers $c_1, c_2, c_3, \ldots$, respectively, such that

(i) the c_i have no limit point in the finite plane,

(ii) the sum of the areas of the D_i is finite, and

(iii) every line in the plane intersects at least one of the D_i?

(page 137)

A5. Find the maximum value of

$$\int_0^y \sqrt{x^4 + (y - y^2)^2}\, dx$$

for $0 \leq y \leq 1$. (page 138)

A6. Let $A(n)$ denote the number of sums of positive integers $a_1 + a_2 + \cdots + a_r$ which add up to n with $a_1 > a_2 + a_3$, $a_2 > a_3 + a_4$, $\ldots$, $a_{r-2} > a_{r-1} + a_r$, $a_{r-1} > a_r$. Let $B(n)$ denote the number of $b_1 + b_2 + \cdots + b_s$ which add up to n, with

(i) $b_1 \geq b_2 \geq \cdots \geq b_s$,

(ii) each b_i is in the sequence $1, 2, 4, \ldots, g_j, \ldots$ defined by $g_1 = 1$, $g_2 = 2$, and $g_j = g_{j-1} + g_{j-2} + 1$, and

(iii) if $b_1 = g_k$ then every element in $\{1, 2, 4, \ldots, g_k\}$ appears at least once as a b_i.

Prove that $A(n) = B(n)$ for each $n \geq 1$.

(For example, $A(7) = 5$ because the relevant sums are 7, $6+1$, $5+2$, $4+3$, $4+2+1$, and $B(7) = 5$ because the relevant sums are $4+2+1$, $2+2+2+1$, $2+2+1+1+1$, $2+1+1+1+1+1$, $1+1+1+1+1+1+1$.) (page 139)

B1. For each integer $n \geq 0$, let $S(n) = n - m^2$, where m is the greatest integer with $m^2 \leq n$. Define a sequence $(a_k)_{k=0}^{\infty}$ by $a_0 = A$ and $a_{k+1} = a_k + S(a_k)$ for $k \geq 0$. For what positive integers A is this sequence eventually constant? (page 141)

B2. Suppose f and g are nonconstant, differentiable, real-valued functions on $\mathbb{R}$. Furthermore, suppose that for each pair of real numbers x and y,

$$f(x+y) = f(x)f(y) - g(x)g(y),$$

$$g(x+y) = f(x)g(y) + g(x)f(y).$$

If $f'(0) = 0$, prove that $(f(x))^2 + (g(x))^2 = 1$ for all x. (page 142)

B3. Does there exist a real number L such that, if m and n are integers greater than L, then an $m \times n$ rectangle may be expressed as a union of 4×6 and 5×7 rectangles, any two of which intersect at most along their boundaries? (page 143)

B4. Suppose p is an odd prime. Prove that

$$\sum_{j=0}^{p} \binom{p}{j}\binom{p+j}{j} \equiv 2^p + 1 \pmod{p^2}.$$

(page 145)

B5. Let p be an odd prime and let $\mathbb{Z}_p$ denote[†] (the field of) integers modulo p. How many elements are in the set

$$\{x^2 : x \in \mathbb{Z}_p\} \cap \{y^2 + 1 : y \in \mathbb{Z}_p\}?$$

(page 148)

B6. Let a and b be positive numbers. Find the largest number c, in terms of a and b, such that

$$a^x b^{1-x} \leq a\frac{\sinh ux}{\sinh u} + b\frac{\sinh u(1-x)}{\sinh u}$$

for all u with $0 < |u| \leq c$ and for all x, $0 < x < 1$. (Note: $\sinh u = (e^u - e^{-u})/2$.) (page 151)

[†] This notation is becoming nonstandard in current mathematics; see the warning preceding the solution.

The Fifty-Third William Lowell Putnam Mathematical Competition December 5, 1992

Questions Committee: George E. Andrews, George T. Gilbert, and Eugene Luks

See page 42 for hints.

A1. Prove that $f(n) = 1 - n$ is the only integer-valued function defined on the integers that satisfies the following conditions:

(i) $f(f(n)) = n$, for all integers n;

(ii) $f(f(n+2)+2) = n$ for all integers n;

(iii) $f(0) = 1$. (page 154)

A2. Define $C(\alpha)$ to be the coefficient of x^{1992} in the power series expansion about $x = 0$ of $(1+x)^{\alpha}$. Evaluate

$$\int_0^1 C(-y-1)\left(\frac{1}{y+1} + \frac{1}{y+2} + \frac{1}{y+3} + \cdots + \frac{1}{y+1992}\right) dy.$$

(page 154)

A3. For a given positive integer m, find all triples (n, x, y) of positive integers, with n relatively prime to m, which satisfy $(x^2 + y^2)^m = (xy)^n$. (page 154)

A4. Let f be an infinitely differentiable real-valued function defined on the real numbers. If

$$f\left(\frac{1}{n}\right) = \frac{n^2}{n^2+1}, \qquad n = 1, 2, 3, \ldots,$$

compute the values of the derivatives $f^{(k)}(0)$, $k = 1, 2, 3, \ldots$. (page 155)

A5. For each positive integer n, let

$$a_n = \begin{cases} 0 & \text{if the number of 1's in the binary representation of } n \text{ is even,} \\ 1 & \text{if the number of 1's in the binary representation of } n \text{ is odd.} \end{cases}$$

Show that there do not exist positive integers k and m such that

$$a_{k+j} = a_{k+m+j} = a_{k+2m+j},$$

for $0 \le j \le m - 1$. (page 156)

A6. Four points are chosen at random on the surface of a sphere. What is the probability that the center of the sphere lies inside the tetrahedron whose vertices are at the four points? (It is understood that each point is independently chosen relative to a uniform distribution on the sphere.) (page 159)

B1. Let S be a set of n distinct real numbers. Let A_S be the set of numbers that occur as averages of two distinct elements of S. For a given $n \ge 2$, what is the smallest possible number of elements in A_S? (page 160)

B2. For nonnegative integers n and k, define $Q(n,k)$ to be the coefficient of x^k in the expansion of $(1+x+x^2+x^3)^n$. Prove that

$$Q(n,k)=\sum_{j=0}^{k}\binom{n}{j}\binom{n}{k-2j},$$

where $\binom{a}{b}$ is the standard binomial coefficient. (Reminder: For integers a and b with $a \geq 0$, $\binom{a}{b}=\frac{a!}{b!(a-b)!}$ for $0 \leq b \leq a$, with $\binom{a}{b}=0$ otherwise.) (page 161)

B3. For any pair (x,y) of real numbers, a sequence $(a_n(x,y))_{n\geq 0}$ is defined as follows:

$$a_0(x,y)=x,$$

$$a_{n+1}(x,y)=\frac{(a_n(x,y))^2+y^2}{2}, \qquad \text{for } n \geq 0.$$

Find the area of the region $\{\,(x,y)|(a_n(x,y))_{n\geq 0} \text{ converges}\,\}$. (page 161)

B4. Let $p(x)$ be a nonzero polynomial of degree less than 1992 having no nonconstant factor in common with x^3-x. Let

$$\frac{d^{1992}}{dx^{1992}}\left(\frac{p(x)}{x^3-x}\right)=\frac{f(x)}{g(x)}$$

for polynomials $f(x)$ and $g(x)$. Find the smallest possible degree of $f(x)$. (page 163)

B5. Let D_n denote the value of the $(n-1)\times(n-1)$ determinant

$$\begin{vmatrix} 3 & 1 & 1 & 1 & \cdots & 1 \\ 1 & 4 & 1 & 1 & \cdots & 1 \\ 1 & 1 & 5 & 1 & \cdots & 1 \\ 1 & 1 & 1 & 6 & \cdots & 1 \\ \vdots & \vdots & \vdots & \vdots & \ddots & \vdots \\ 1 & 1 & 1 & 1 & \cdots & n+1 \end{vmatrix}.$$

Is the set $\{D_n/n!\}_{n\geq 2}$ bounded? (page 164)

B6. Let $\mathcal{M}$ be a set of real $n\times n$ matrices such that

(i) $I \in \mathcal{M}$, where I is the $n\times n$ identity matrix;

(ii) if $A \in \mathcal{M}$ and $B \in \mathcal{M}$, then either $AB \in \mathcal{M}$ or $-AB \in \mathcal{M}$, but not both;

(iii) if $A \in \mathcal{M}$ and $B \in \mathcal{M}$, then either $AB = BA$ or $AB=-BA$;

(iv) if $A \in \mathcal{M}$ and $A \neq I$, there is at least one $B \in \mathcal{M}$ such that $AB=-BA$.

Prove that $\mathcal{M}$ contains at most n^2 matrices. (page 166)

The Fifty-Fourth William Lowell Putnam Mathematical Competition December 4, 1993

Questions Committee: George T. Gilbert, Eugene Luks, and Fan Chung

See page 43 for hints.

A1. The horizontal line $y = c$ intersects the curve $y = 2x - 3x^3$ in the first quadrant as in the figure. Find c so that the areas of the two shaded regions are equal.

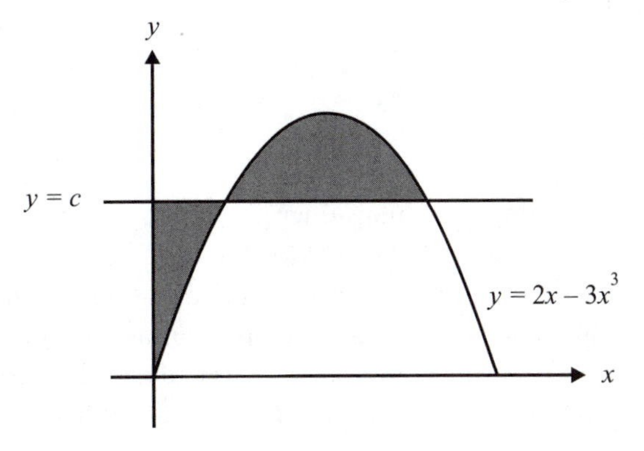

(page 171)

A2. Let $(x_n)_{n \geq 0}$ be a sequence of nonzero real numbers such that

$$x_n^2 - x_{n-1}x_{n+1} = 1 \text{ for } n = 1, 2, 3, \ldots.$$

Prove there exists a real number a such that $x_{n+1} = ax_n - x_{n-1}$ for all $n \geq 1$.

(page 171)

A3. Let $\mathcal{P}_n$ be the set of subsets of $\{1, 2, \ldots, n\}$. Let $c(n, m)$ be the number of functions $f : \mathcal{P}_n \to \{1, 2, \ldots, m\}$ such that $f(A \cap B) = \min\{f(A), f(B)\}$. Prove that

$$c(n, m) = \sum_{j=1}^{m} j^n.$$

(page 173)

A4. Let $x_1, x_2, \ldots, x_{19}$ be positive integers each of which is less than or equal to 93. Let $y_1, y_2, \ldots, y_{93}$ be positive integers each of which is less than or equal to 19. Prove that there exists a (nonempty) sum of some x_i's equal to a sum of some y_j's.

(page 174)

A5. Show that

$$\int_{-100}^{-10} \left(\frac{x^2 - x}{x^3 - 3x + 1} \right)^2 dx + \int_{\frac{1}{101}}^{\frac{1}{11}} \left(\frac{x^2 - x}{x^3 - 3x + 1} \right)^2 dx + \int_{\frac{101}{100}}^{\frac{11}{10}} \left(\frac{x^2 - x}{x^3 - 3x + 1} \right)^2 dx$$

is a rational number.

(page 175)

A6. The infinite sequence of 2's and 3's

$$2, 3, 3, 2, 3, 3, 3, 2, 3, 3, 3, 2, 3, 3, 2, 3, 3, 3, 2, 3, 3, 3, 2, 3, 3, 3, 2, 3, 3, 2, 3, 3, 3, 2, \ldots$$

has the property that, if one forms a second sequence that records the number of 3's between successive 2's, the result is identical to the given sequence. Show that there exists a real number r such that, for any n, the nth term of the sequence is 2 if and only if $n = 1 + \lfloor rm \rfloor$ for some nonnegative integer m. (Note: $\lfloor x \rfloor$ denotes the largest integer less than or equal to x.) (page 178)

B1. Find the smallest positive integer n such that for every integer m, with $0 < m < 1993$, there exists an integer k for which

$$\frac{m}{1993} < \frac{k}{n} < \frac{m+1}{1994}.$$

(page 180)

B2. Consider the following game played with a deck of $2n$ cards numbered from 1 to $2n$. The deck is randomly shuffled and n cards are dealt to each of two players, A and B. Beginning with A, the players take turns discarding one of their remaining cards and announcing its number. The game ends as soon as the sum of the numbers on the discarded cards is divisible by $2n + 1$. The last person to discard wins the game. Assuming optimal strategy by both A and B, what is the probability that A wins? (page 182)

B3. Two real numbers x and y are chosen at random in the interval (0,1) with respect to the uniform distribution. What is the probability that the closest integer to x/y is even? Express the answer in the form $r + s\pi$, where r and s are rational numbers. (page 182)

B4. The function $K(x, y)$ is positive and continuous for $0 \le x \le 1, 0 \le y \le 1$, and the functions $f(x)$ and $g(x)$ are positive and continuous for $0 \le x \le 1$. Suppose that for all x, $0 \le x \le 1$,

$$\int_0^1 f(y)K(x,y)\,dy = g(x) \qquad \text{and} \qquad \int_0^1 g(y)K(x,y)\,dy = f(x).$$

Show that $f(x) = g(x)$ for $0 \le x \le 1$. (page 184)

B5. Show there do not exist four points in the Euclidean plane such that the pairwise distances between the points are all odd integers. (page 185)

B6. Let S be a set of three, not necessarily distinct, positive integers. Show that one can transform S into a set containing 0 by a finite number of applications of the following rule: Select two of the three integers, say x and y, where $x \le y$ and replace them with $2x$ and $y - x$. (page 188)

The Fifty-Fifth William Lowell Putnam Mathematical Competition
December 3, 1994

Questions Committee: Eugene Luks, Fan Chung, and Mark I. Krusemeyer
See page 44 for hints.

A1. Suppose that a sequence a_1, a_2, a_3, ... satisfies $0 < a_n \leq a_{2n} + a_{2n+1}$ for all $n \geq 1$. Prove that the series $\sum_{n=1}^{\infty} a_n$ diverges. (page 191)

A2. Let A be the area of the region in the first quadrant bounded by the line $y = \frac{1}{2}x$, the x-axis, and the ellipse $\frac{1}{9}x^2 + y^2 = 1$. Find the positive number m such that A is equal to the area of the region in the first quadrant bounded by the line $y = mx$, the y-axis, and the ellipse $\frac{1}{9}x^2 + y^2 = 1$. (page 191)

A3. Show that if the points of an isosceles right triangle of side length 1 are each colored with one of four colors, then there must be two points of the same color which are at least a distance $2 - \sqrt{2}$ apart. (page 192)

A4. Let A and B be 2×2 matrices with integer entries such that A, $A+B$, $A+2B$, $A+3B$, and $A+4B$ are all invertible matrices whose inverses have integer entries. Show that $A+5B$ is invertible and that its inverse has integer entries. (page 193)

A5. Let $(r_n)_{n\geq 0}$ be a sequence of positive real numbers such that $\lim_{n\to\infty} r_n = 0$. Let S be the set of numbers representable as a sum

$$r_{i_1} + r_{i_2} + \cdots + r_{i_{1994}},$$

with $i_1 < i_2 < \cdots < i_{1994}$. Show that every nonempty interval (a, b) contains a nonempty subinterval (c, d) that does not intersect S. (page 194)

A6. Let f_1, f_2, ... , f_{10} be bijections of the set of integers such that for each integer n, there is some composition $f_{i_1} \circ f_{i_2} \circ \cdots \circ f_{i_m}$ of these functions (allowing repetitions) which maps 0 to n. Consider the set of 1024 functions

$$\mathcal{F} = \{f_1^{e_1} \circ f_2^{e_2} \circ \cdots \circ f_{10}^{e_{10}}\},$$

$e_i = 0$ or 1 for $1 \leq i \leq 10$. (f_i^0 is the identity function and $f_i^1 = f_i$.) Show that if A is any nonempty finite set of integers, then at most 512 of the functions in $\mathcal{F}$ map A to itself. (page 195)

B1. Find all positive integers that are within 250 of exactly 15 perfect squares. (page 196)

B2. For which real numbers c is there a straight line that intersects the curve

$$y = x^4 + 9x^3 + cx^2 + 9x + 4$$

in four distinct points? (page 196)

B3. Find the set of all real numbers k with the following property: For any positive, differentiable function f that satisfies $f'(x) > f(x)$ for all x, there is some number N such that $f(x) > e^{kx}$ for all $x > N$. (page 198)

B4. For $n \geq 1$, let d_n be the greatest common divisor of the entries of $A^n - I$, where

$$A = \begin{pmatrix} 3 & 2 \\ 4 & 3 \end{pmatrix} \text{ and } I = \begin{pmatrix} 1 & 0 \\ 0 & 1 \end{pmatrix}.$$

Show that $\lim_{n\to\infty} d_n = \infty$. (page 198)

B5. For any real number α, define the function $f_\alpha(x) = \lfloor \alpha x \rfloor$. Let n be a positive integer. Show that there exists an α such that for $1 \leq k \leq n$,†

$$f_\alpha^k(n^2) = n^2 - k = f_{\alpha^k}(n^2).$$

(page 200)

B6. For any integer a, set

$$n_a = 101a - 100 \cdot 2^a.$$

Show that for $0 \leq a, b, c, d \leq 99$, $n_a + n_b \equiv n_c + n_d \pmod{10100}$ implies $\{a, b\} = \{c, d\}$. (page 202)

† Here $f_\alpha^k(n^2) = f_\alpha(\cdots(f_\alpha(n^2))\cdots)$, where f_α is applied k times to n^2.

The Fifty-Sixth William Lowell Putnam Mathematical Competition December 2, 1995

Questions Committee: Fan Chung, Mark I. Krusemeyer, and Richard K. Guy

See page 45 for hints.

A1. Let S be a set of real numbers which is closed under multiplication (that is, if a and b are in S, then so is ab). Let T and U be disjoint subsets of S whose union is S. Given that the product of any *three* (not necessarily distinct) elements of T is in T and that the product of any three elements of U is in U, show that at least one of the two subsets T, U is closed under multiplication. (page 204)

A2. For what pairs (a, b) of positive real numbers does the improper integral

$$\int_b^\infty \left(\sqrt{\sqrt{x+a}-\sqrt{x}}-\sqrt{\sqrt{x}-\sqrt{x-b}}\right) dx$$

converge? (page 204)

A3. The number $d_1d_2\dots d_9$ has nine (not necessarily distinct) decimal digits. The number $e_1e_2\dots e_9$ is such that each of the nine 9-digit numbers formed by replacing just one of the digits d_i in $d_1d_2\dots d_9$ by the corresponding digit e_i $(1 \le i \le 9)$ is divisible by 7. The number $f_1f_2\dots f_9$ is related to $e_1e_2\dots e_9$ is the same way: that is, each of the nine numbers formed by replacing one of the e_i by the corresponding f_i is divisible by 7. Show that, for each i, $d_i - f_i$ is divisible by 7. [For example, if $d_1d_2\dots d_9 = 199501996$, then e_6 may be 2 or 9, since 199502996 and 199509996 are multiples of 7.] (page 205)

A4. Suppose we have a necklace of n beads. Each bead is labelled with an integer and the sum of all these labels is $n-1$. Prove that we can cut the necklace to form a string whose consecutive labels $x_1, x_2, \dots, x_n$ satisfy

$$\sum_{i=1}^{k} x_i \le k-1 \qquad \text{for} \quad k = 1, 2, \dots, n.$$

(page 205)

A5. Let $x_1, x_2, \dots, x_n$ be differentiable (real-valued) functions of a single variable t which satisfy

$$\begin{aligned} \frac{dx_1}{dt} &= a_{11}x_1 + a_{12}x_2 + \cdots + a_{1n}x_n \\ \frac{dx_2}{dt} &= a_{21}x_1 + a_{22}x_2 + \cdots + a_{2n}x_n \\ &\vdots \\ \frac{dx_n}{dt} &= a_{n1}x_1 + a_{n2}x_2 + \cdots + a_{nn}x_n \end{aligned}$$

for some constants $a_{ij} > 0$. Suppose that for all i, $x_i(t) \to 0$ as $t \to \infty$. Are the functions $x_1, x_2, \dots, x_n$ necessarily linearly dependent? (page 206)

A6. Suppose that each of n people writes down the numbers 1, 2, 3 in random order in one column of a $3 \times n$ matrix, with all orders equally likely and with the orders for different columns independent of each other. Let the row sums a, b, c of the resulting matrix be rearranged (if necessary) so that $a \leq b \leq c$. Show that for some $n \geq 1995$, it is at least four times as likely that both $b = a + 1$ and $c = a + 2$ as that $a = b = c$.
(page 207)

B1. For a partition π of $\{1, 2, 3, 4, 5, 6, 7, 8, 9\}$, let $\pi(x)$ be the number of elements in the part containing x. Prove that for any two partitions π and π', there are two distinct numbers x and y in $\{1, 2, 3, 4, 5, 6, 7, 8, 9\}$ such that $\pi(x) = \pi(y)$ and $\pi'(x) = \pi'(y)$. [A *partition* of a set S is a collection of disjoint subsets (parts) whose union is S.]
(page 209)

B2. An ellipse, whose semi-axes have lengths a and b, rolls without slipping on the curve $y = c \sin\left(\frac{x}{a}\right)$. How are a, b, c related, given that the ellipse completes one revolution when it traverses one period of the curve?
(page 209)

B3. To each positive integer with n^2 decimal digits, we associate the determinant of the matrix obtained by writing the digits in order across the rows. For example, for $n = 2$, to the integer 8617 we associate $\det \begin{pmatrix} 8 & 6 \\ 1 & 7 \end{pmatrix} = 50$. Find, as a function of n, the sum of all the determinants associated with n^2-digit integers. (Leading digits are assumed to be nonzero; for example, for $n = 2$, there are 9000 determinants.)
(page 211)

B4. Evaluate

$$\sqrt[8]{2207 - \cfrac{1}{2207 - \cfrac{1}{2207 - \cdots}}}.$$

Express your answer in the form $\frac{a+b\sqrt{c}}{d}$, where a, b, c, d are integers.
(page 211)

B5. A game starts with four heaps of beans, containing 3, 4, 5 and 6 beans. The two players move alternately. A move consists of taking **either**

(a) one bean from a heap, provided at least two beans are left behind in that heap, **or**

(b) a complete heap of two or three beans.

The player who takes the last heap wins. To win the game, do you want to move first or second? Give a winning strategy.
(page 212)

B6. For a positive real number α, define

$$S(\alpha) = \{ \lfloor n\alpha \rfloor : n = 1, 2, 3, \dots \}.$$

Prove that $\{1, 2, 3, \dots\}$ cannot be expressed as the disjoint union of three sets $S(\alpha), S(\beta)$ and $S(\gamma)$.
(page 214)

The Fifty-Seventh William Lowell Putnam Mathematical Competition December 7, 1996

Questions Committee: Mark I. Krusemeyer, Richard K. Guy, and Michael J. Larsen
See page 46 for hints.

A1. Find the least number A such that for any two squares of combined area 1, a rectangle of area A exists such that the two squares can be packed in the rectangle (without the interiors of the squares overlapping). You may assume that the sides of the squares will be parallel to the sides of the rectangle. (page 217)

A2. Let C_1 and C_2 be circles whose centers are 10 units apart, and whose radii are 1 and 3. Find, with proof, the locus of all points M for which there exists points X on C_1 and Y on C_2 such that M is the midpoint of the line segment XY. (page 218)

A3. Suppose that each of 20 students has made a choice of anywhere from 0 to 6 courses from a total of 6 courses offered. Prove or disprove: there are 5 students and 2 courses such that all 5 have chosen both courses or all 5 have chosen neither course. (page 218)

A4. Let S be a set of ordered triples (a, b, c) of distinct elements of a finite set A. Suppose that

(1) $(a, b, c) \in S$ if and only if $(b, c, a) \in S$;

(2) $(a, b, c) \in S$ if and only if $(c, b, a) \notin S$ [for a, b, c distinct];

(3) (a, b, c) and (c, d, a) are both in S if and only if (b, c, d) and (d, a, b) are both in S.

Prove that there exists a one-to-one function g from A to $\mathbb{R}$ such that $g(a) < g(b) < g(c)$ implies $(a, b, c) \in S$. (page 219)

A5. If p is a prime number greater than 3 and $k = \lfloor 2p/3 \rfloor$, prove that the sum

$$\binom{p}{1} + \binom{p}{2} + \cdots + \binom{p}{k}$$

of binomial coefficients is divisible by p^2. (page 220)

A6. Let $c \geq 0$ be a constant. Give a complete description, with proof, of the set of all continuous functions $f : \mathbb{R} \to \mathbb{R}$ such that $f(x) = f(x^2 + c)$ for all $x \in \mathbb{R}$. (page 220)

B1. Define a *selfish* set to be a set which has its own cardinality (number of elements) as an element. Find, with proof, the number of subsets of $\{1, 2, \ldots, n\}$ which are *minimal* selfish sets, that is, selfish sets none of whose proper subsets is selfish. (page 222)

B2. Show that for every positive integer n,

$$\left(\frac{2n-1}{e}\right)^{\frac{2n-1}{2}} < 1 \cdot 3 \cdot 5 \cdots (2n-1) < \left(\frac{2n+1}{e}\right)^{\frac{2n+1}{2}}.$$

(page 224)

B3. Given that $\{x_1, x_2, \ldots, x_n\} = \{1, 2, \ldots, n\}$, find, with proof, the largest possible value, as a function of n (with $n \geq 2$), of

$$x_1x_2 + x_2x_3 + \cdots + x_{n-1}x_n + x_nx_1.$$

(page 225)

B4. For any square matrix A, we can define $\sin A$ by the usual power series:

$$\sin A = \sum_{n=0}^{\infty} \frac{(-1)^n}{(2n+1)!} A^{2n+1}.$$

Prove or disprove: there exists a 2×2 matrix A with real entries such that

$$\sin A = \begin{pmatrix} 1 & 1996 \\ 0 & 1 \end{pmatrix}.$$

(page 227)

B5. Given a finite string S of symbols X and O, we write $\Delta(S)$ for the number of X's in S minus the number of O's. For example, $\Delta(XOOXOOX) = -1$. We call a string S *balanced* if every substring T of (consecutive symbols of) S has $-2 \leq \Delta(T) \leq 2$. Thus, $XOOXOOX$ is not balanced, since it contains the substring $OOXOO$. Find, with proof, the number of balanced strings of length n. (page 229)

B6. Let $(a_1, b_1), (a_2, b_2), \ldots, (a_n, b_n)$ be the vertices of a convex polygon which contains the origin in its interior. Prove that there exist positive real numbers x and y such that

$$(a_1, b_1)x^{a_1}y^{b_1} + (a_2, b_2)x^{a_2}y^{b_2} + \cdots + (a_n, b_n)x^{a_n}y^{b_n} = (0, 0).$$

(page 230)

The Fifty-Eighth William Lowell Putnam Mathematical Competition
December 6, 1997

Questions Committee: Richard K. Guy, Michael J. Larsen, and David J. Wright

See page 47 for hints.

A1. A rectangle, $HOMF$, has sides $HO = 11$ and $OM = 5$. A triangle ABC has H as the intersection of the altitudes, O the center of the circumscribed circle, M the midpoint of BC, and F the foot of the altitude from A. What is the length of BC?

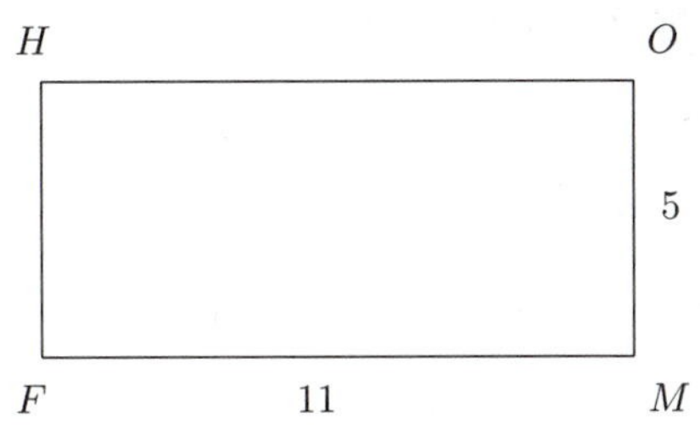

(page 232)

A2. Players $1, 2, 3, \ldots, n$ are seated around a table, and each has a single penny. Player 1 passes a penny to Player 2, who then passes two pennies to Player 3. Player 3 then passes one penny to Player 4, who passes two pennies to Player 5, and so on, players alternately passing one penny or two to the next player who still has some pennies. A player who runs out of pennies drops out of the game and leaves the table. Find an infinite set of numbers n for which some player ends up with all n pennies.

(page 233)

A3. Evaluate

$$\int_0^\infty \left(x - \frac{x^3}{2} + \frac{x^5}{2\cdot 4} - \frac{x^7}{2\cdot 4\cdot 6} + \cdots\right)\left(1 + \frac{x^2}{2^2} + \frac{x^4}{2^2\cdot 4^2} + \frac{x^6}{2^2\cdot 4^2\cdot 6^2} + \cdots\right) dx.$$

(page 234)

A4. Let G be a group with identity e and $\phi : G \to G$ a function such that

$$\phi(g_1)\phi(g_2)\phi(g_3) = \phi(h_1)\phi(h_2)\phi(h_3)$$

whenever $g_1g_2g_3 = e = h_1h_2h_3$. Prove that there exists an element $a \in G$ such that $\psi(x) = a\phi(x)$ is a homomorphism (that is, $\psi(xy) = \psi(x)\psi(y)$ for all $x, y \in G$).

(page 237)

A5. Let N_n denote the number of ordered n-tuples of positive integers $(a_1, a_2, \ldots, a_n)$ such that $1/a_1 + 1/a_2 + \cdots + 1/a_n = 1$. Determine whether N_{10} is even or odd.

(page 237)

A6. For a positive integer n and any real number c, define x_k recursively by $x_0 = 0$, $x_1 = 1$, and for $k \geq 0$,

$$x_{k+2} = \frac{cx_{k+1} - (n-k)x_k}{k+1}.$$

Fix n and then take c to be the largest value for which $x_{n+1} = 0$. Find x_k in terms of n and k, $1 \leq k \leq n$. (page 238)

B1. Let $\{x\}$ denote the distance between the real number x and the nearest integer. For each positive integer n, evaluate

$$S_n = \sum_{m=1}^{6n-1} \min\left(\left\{\frac{m}{6n}\right\}, \left\{\frac{m}{3n}\right\}\right).$$

(Here $\min(a, b)$ denotes the minimum of a and b.) (page 242)

B2. Let f be a twice-differentiable real-valued function satisfying

$$f(x) + f''(x) = -xg(x)f'(x),$$

where $g(x) \geq 0$ for all real x. Prove that $|f(x)|$ is bounded. (page 243)

B3. For each positive integer n, write the sum $\sum_{m=1}^{n} \frac{1}{m}$ in the form $\frac{p_n}{q_n}$, where p_n and q_n are relatively prime positive integers. Determine all n such that 5 does not divide q_n. (page 244)

B4. Let $a_{m,n}$ denote the coefficient of x^n in the expansion of $(1 + x + x^2)^m$. Prove that for all integers $k \geq 0$,

$$0 \leq \sum_{i=0}^{\lfloor \frac{2k}{3} \rfloor} (-1)^i a_{k-i,i} \leq 1.$$

(page 246)

B5. Prove that for $n \geq 2$,

$$\left.2^{2^{\cdot^{\cdot^{\cdot^{2}}}}}\right\}n \equiv \left.2^{2^{\cdot^{\cdot^{\cdot^{2}}}}}\right\}n-1 \pmod{n}.$$

(page 247)

B6. The dissection of the 3–4–5 triangle shown below has diameter $5/2$.

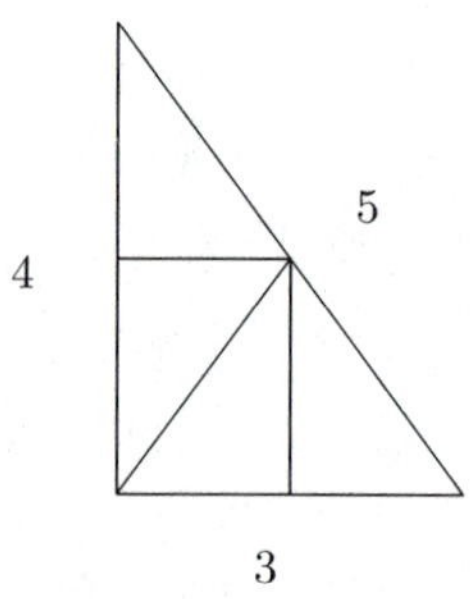

Find the least diameter of a dissection of this triangle into four parts. (The diameter of a dissection is the least upper bound of the distances between pairs of points belonging to the same part.) (page 248)

The Fifty-Ninth William Lowell Putnam Mathematical Competition
December 5, 1998

Questions Committee: Michael J. Larsen, David J. Wright, and Steven G. Krantz

See page 48 for hints.

A1. A right circular cone has base of radius 1 and height 3. A cube is inscribed in the cone so that one face of the cube is contained in the base of the cone. What is the side-length of the cube? (page 250)

A2. Let s be any arc of the unit circle lying entirely in the first quadrant. Let A be the area of the region lying below s and above the x-axis and let B be the area of the region lying to the right of the y-axis and to the left of s. Prove that $A+B$ depends only on the arc length, and not on the position, of s. (page 250)

A3. Let f be a real function on the real line with continuous third derivative. Prove that there exists a point a such that

$$f(a) \cdot f'(a) \cdot f''(a) \cdot f'''(a) \geq 0.$$

(page 251)

A4. Let $A_1 = 0$ and $A_2 = 1$. For $n > 2$, the number A_n is defined by concatenating the decimal expansions of A_{n-1} and A_{n-2} from left to right. For example $A_3 = A_2A_1 = 10$, $A_4 = A_3A_2 = 101$, $A_5 = A_4A_3 = 10110$, and so forth. Determine all n such that 11 divides A_n. (page 252)

A5. Let $\mathcal{F}$ be a finite collection of open discs in $\mathbb{R}^2$ whose union contains a set $E \subseteq \mathbb{R}^2$. Show that there is a pairwise disjoint subcollection $D_1, \ldots, D_n$ in $\mathcal{F}$ such that

$$E \subseteq \bigcup_{j=1}^{n} 3D_j.$$

Here, if D is the disc of radius r and center P, then $3D$ is the disc of radius $3r$ and center P. (page 252)

A6. Let A, B, C denote distinct points with integer coordinates in $\mathbb{R}^2$. Prove that if

$$(|AB| + |BC|)^2 < 8 \cdot [ABC] + 1$$

then A, B, C are three vertices of a square. Here $|XY|$ is the length of segment XY and $[ABC]$ is the area of triangle ABC. (page 253)

B1. Find the minimum value of

$$\frac{(x+1/x)^6 - (x^6+1/x^6) - 2}{(x+1/x)^3 + (x^3+1/x^3)}$$

for $x > 0$. (page 253)

B2. Given a point (a, b) with $0 < b < a$, determine the minimum perimeter of a triangle with one vertex at (a, b), one on the x-axis, and one on the line $y = x$. You may assume that a triangle of minimum perimeter exists. (page 254)

B3. Let H be the unit hemisphere $\{\,(x, y, z) : x^2 + y^2 + z^2 = 1, z \geq 0\,\}$, C the unit circle $\{\,(x, y, 0) : x^2 + y^2 = 1\,\}$, and P the regular pentagon inscribed in C. Determine the surface area of that portion of H lying over the planar region inside P, and write your answer in the form $A \sin \alpha + B \cos \beta$, where A, B, α, β are real numbers.

(page 255)

B4. Find necessary and sufficient conditions on positive integers m and n so that

$$\sum_{i=0}^{mn-1} (-1)^{\lfloor i/m \rfloor + \lfloor i/n \rfloor} = 0.$$

(page 256)

B5. Let N be the positive integer with 1998 decimal digits, all of them 1; that is,

$$N = 1111 \cdots 11.$$

Find the thousandth digit after the decimal point of $\sqrt{N}$. (page 257)

B6. Prove that, for any integers a, b, c, there exists a positive integer n such that $\sqrt{n^3 + an^2 + bn + c}$ is not an integer. (page 258)

The Sixtieth William Lowell Putnam Mathematical Competition December 4, 1999

Questions Committee: David J. Wright, Steven G. Krantz, and Andrew J. Granville

See page 49 for hints.

A1. Find polynomials $f(x)$, $g(x)$, and $h(x)$, if they exist, such that, for all x,

$$|f(x)| - |g(x)| + h(x) = \begin{cases} -1 & \text{if } x < -1 \\ 3x + 2 & \text{if } -1 \leq x \leq 0 \\ -2x + 2 & \text{if } x > 0. \end{cases}$$

(page 262)

A2. Let $p(x)$ be a polynomial that is nonnegative for all real x. Prove that for some k, there are polynomials $f_1(x), \ldots, f_k(x)$ such that

$$p(x) = \sum_{j=1}^{k} (f_j(x))^2.$$

(page 263)

A3. Consider the power series expansion

$$\frac{1}{1 - 2x - x^2} = \sum_{n=0}^{\infty} a_n x^n.$$

Prove that, for each integer $n \geq 0$, there is an integer m such that

$$a_n^2 + a_{n+1}^2 = a_m.$$

(page 264)

A4. Sum the series

$$\sum_{m=1}^{\infty} \sum_{n=1}^{\infty} \frac{m^2 n}{3^m (n 3^m + m 3^n)}.$$

(page 265)

A5. Prove that there is a constant C such that, if $p(x)$ is a polynomial of degree 1999, then

$$|p(0)| \leq C \int_{-1}^{1} |p(x)|\, dx.$$

(page 266)

A6. The sequence $(a_n)_{n \geq 1}$ is defined by $a_1 = 1, a_2 = 2, a_3 = 24$, and, for $n \geq 4$,

$$a_n = \frac{6a_{n-1}^2 a_{n-3} - 8a_{n-1}a_{n-2}^2}{a_{n-2}a_{n-3}}.$$

Show that, for all n, a_n is an integer multiple of n. (page 267)

B1. Right triangle ABC has right angle at C and $\angle BAC = \theta$; the point D is chosen on AB so that $|AC| = |AD| = 1$; the point E is chosen on BC so that $\angle CDE = \theta$. The perpendicular to BC at E meets AB at F. Evaluate $\lim_{\theta\to 0} |EF|$. [Here $|PQ|$ denotes the length of the line segment PQ.]

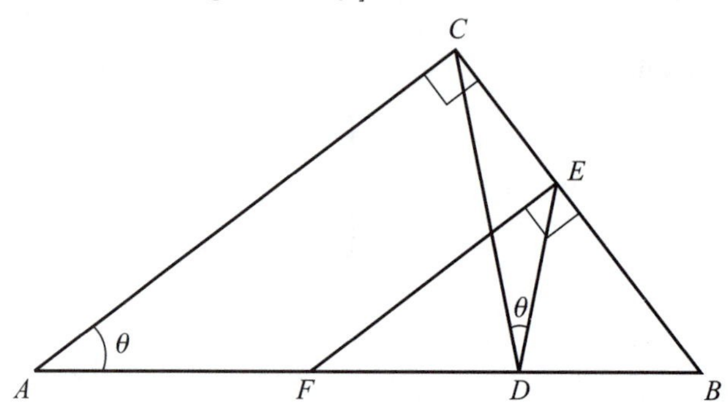

(page 268)

B2. Let $P(x)$ be a polynomial of degree n such that $P(x) = Q(x)P''(x)$, where $Q(x)$ is a quadratic polynomial and $P''(x)$ is the second derivative of $P(x)$. Show that if $P(x)$ has at least two distinct roots then it must have n distinct roots. [The roots may be either real or complex.] (page 269)

B3. Let $A = \{\, (x,y) : 0 \le x, y < 1\,\}$. For $(x,y) \in A$, let

$$S(x,y) = \sum_{\frac{1}{2} \le \frac{m}{n} \le 2} x^m y^n,$$

where the sum ranges over all pairs (m,n) of positive integers satisfying the indicated inequalities. Evaluate

$$\lim_{\substack{(x,y)\to(1,1)\\(x,y)\in A}} (1-xy^2)(1-x^2y)S(x,y).$$

(page 271)

B4. Let f be a real function with a continuous third derivative such that $f(x)$, $f'(x)$, $f''(x)$, $f'''(x)$ are positive for all x. Suppose that $f'''(x) \le f(x)$ for all x. Show that $f'(x) < 2f(x)$ for all x. (page 272)

B5. For an integer $n \ge 3$, let $\theta = 2\pi/n$. Evaluate the determinant of the $n\times n$ matrix $I+A$, where I is the $n\times n$ identity matrix and $A = (a_{jk})$ has entries $a_{jk} = \cos(j\theta + k\theta)$ for all j, k. (page 276)

B6. Let S be a finite set of integers, each greater than 1. Suppose that for each integer n there is some $s \in S$ such that $\gcd(s,n) = 1$ or $\gcd(s,n) = s$. Show that there exist $s, t \in S$ such that $\gcd(s,t)$ is prime. (page 277)

The Sixty-First William Lowell Putnam Mathematical Competition
December 2, 2000

Questions Committee: Steven G. Krantz, Andrew J. Granville,
Carl Pomerance, and Eugene Luks
See page 50 for hints.

A1. Let A be a positive real number. What are the possible values of $\sum_{j=0}^{\infty} x_j^2$, given that $x_0, x_1, \ldots$ are positive numbers for which $\sum_{j=0}^{\infty} x_j = A$? (page 278)

A2. Prove that there exist infinitely many integers n such that n, $n+1$, $n+2$ are each the sum of two squares of integers. [Example: $0 = 0^2 + 0^2$, $1 = 0^2 + 1^2$, and $2 = 1^2 + 1^2$.] (page 278)

A3. The octagon $P_1P_2P_3P_4P_5P_6P_7P_8$ is inscribed in a circle, with the vertices around the circumference in the given order. Given that the polygon $P_1P_3P_5P_7$ is a square of area 5 and the polygon $P_2P_4P_6P_8$ is a rectangle of area 4, find the maximum possible area of the octagon. (page 280)

A4. Show that the improper integral

$$\lim_{B\to\infty} \int_0^B \sin(x)\sin(x^2)\,dx$$

converges. (page 281)

A5. Three distinct points with integer coordinates lie in the plane on a circle of radius $r > 0$. Show that two of these points are separated by a distance of at least $r^{1/3}$. (page 285)

A6. Let $f(x)$ be a polynomial with integer coefficients. Define a sequence $a_0, a_1, \ldots$ of integers such that $a_0 = 0$ and $a_{n+1} = f(a_n)$ for all $n \geq 0$. Prove that if there exists a positive integer m for which $a_m = 0$ then either $a_1 = 0$ or $a_2 = 0$. (page 288)

B1. Let a_j, b_j, c_j be integers for $1 \leq j \leq N$. Assume, for each j, at least one of a_j, b_j, c_j is odd. Show that there exist integers r, s, t such that $ra_j + sb_j + tc_j$ is odd for at least $4N/7$ values of j, $1 \leq j \leq N$. (page 289)

B2. Prove that the expression

$$\frac{\gcd(m,n)}{n}\binom{n}{m}$$

is an integer for all pairs of integers $n \geq m \geq 1$. (page 290)

B3. Let $f(t) = \sum_{j=1}^{N} a_j \sin(2\pi jt)$, where each a_j is real and a_N is not equal to 0. Let N_k denote the number of zeros† (including multiplicities) of $\frac{d^kf}{dt^k}$. Prove that

$$N_0 \leq N_1 \leq N_2 \leq \cdots \text{ and } \lim_{k\to\infty} N_k = 2N.$$

(page 290)

---

† The proposers intended for N_k to count only the zeros in the interval $[0, 1)$.

B4. Let $f(x)$ be a continuous function such that $f(2x^2-1) = 2xf(x)$ for all x. Show that $f(x) = 0$ for $-1 \leq x \leq 1$. (page 292)

B5. Let S_0 be a finite set of positive integers. We define finite sets $S_1, S_2, \ldots$ of positive integers as follows: the integer a is in S_{n+1} if and only if exactly one of $a-1$ or a is in S_n. Show that there exist infinitely many integers N for which $S_N = S_0 \cup \{\, N + a : a \in S_0 \,\}$. (page 293)

B6. Let B be a set of more than $2^{n+1}/n$ distinct points with coordinates of the form $(\pm 1, \pm 1, \ldots, \pm 1)$ in n-dimensional space with $n \geq 3$. Show that there are three distinct points in B which are the vertices of an equilateral triangle. (page 294)

Hints

The Forty-Sixth William Lowell Putnam Mathematical Competition
December 7, 1985

A1. Interpret the problem as asking for the number of ways of placing the numbers 1 through 10 in a Venn diagram. Alternatively, identify triples of subsets of $\{1, 2, \ldots, 10\}$ with 10×3 matrices with entries in $\{0, 1\}$. The conditions in the problem correspond to conditions on the rows of such a matrix.

A2. Draw the altitude in the top subtriangle. Deleting R and S leaves six right triangles each of which is similar to the left or right part of T cut by its altitude; minimize the sum of their areas as a fraction of $A(T)$.

A3. Let $b_m(j) = a_m(j) + 1$.

A4. If 3 does not divide n, then $3^a \bmod n$ is determined by $a \bmod \phi(n)$, where $\phi(n)$ denotes the Euler ϕ-function.

A5. Substitute $\cos x = (e^{ix} + e^{-ix})/2$ everywhere (de Moivre's Theorem) and expand.

A6. If $p(x)p(x^{-1}) = q(x)q(x^{-1})$ as Laurent polynomials, equating coefficients of x^0 yields $\Gamma(p(x)) = \Gamma(q(x))$.

B1. A symmetric choice of the m_i attains $k = 3$. Polynomials with $k = 1$ and $k = 2$ cannot have distinct integer zeros.

B2. Factor the first few $f_n(x)$.

B3. If not, at least how many array entries would be less than or equal to a given integer k?

B4. The condition is satisfied if and only if the absolute values of the x- and y-coordinates of q are less than or equal to those of p, respectively.

B5. Substitute $u = 1/t$ and combine the resulting integral with the original integral.

B6. What is $\left(\sum_{i=1}^{r} M_i\right)^2$?

The Forty-Seventh William Lowell Putnam Mathematical Competition
December 6, 1986

A1. Factor $x^4 - 13x^2 + 36$.

A2. Expand in a geometric series.

A3. Simplify the first few partial sums. Alternatively, note that $\operatorname{Arccot}(a/b)$ is the argument of $a + bi$ for $a \geq 0$.

A4. Given the first row of such a matrix, what are the possibilities for the other rows? Count the matrices according to the set of distinct values appearing in the first row.

A5. Find n linear functions $h_i : \mathbb{R}^n \to \mathbb{R}$ such that

$$\frac{\partial h_i}{\partial x_j} - \frac{\partial h_j}{\partial x_i} = c_{ij}.$$

A6. Write down a differential equation satisfied by $F(t) = \sum_{i=1}^{n} a_i e^{b_i t}$.

B1. Express the altitude of the triangle in terms of h.

B2. Subtract and factor.

B3. Use induction.

B4. The assertion is true. Use a greedy algorithm: for each r, choose the largest m, and then the largest n, subject to the constraint $\sqrt{m^2 + 2n^2} \leq r$.

B5. It is false!

B6. Consider the $(2n) \times (2n)$ block matrix $X = \begin{pmatrix} A & B \\ C & D \end{pmatrix}$. Use the identities given to find X^{-1} as another block matrix involving transposes of A, B, C, and D.

The Forty-Eighth William Lowell Putnam Mathematical Competition
December 5, 1987

A1. Use complex numbers.

A2. Let $g(m)$ denote the total number of digits in the integers with m or fewer digits. Estimate $g(m)$ to guess the m such that $g(m-1) < 10^{1987} \leq g(m)$.

A3. Find the general solution to the differential equation explicitly.

A4. Show that $F(Y, Z) = P(0, Y, Z)$. Prove that $F(Y, Z)$ has the form $aY^2 + bYZ + cZ^2$ and solve for a, b, c.

A5. If $\vec{F}$ existed, Stokes' Theorem would imply that the line integral of $\vec{G}$ over a certain ellipse would vanish.

A6. Convergence is not affected if n^3 is replaced by 3^{3k} where 3^k is the greatest power of 3 less than or equal to n.

B1. Use the symmetry of the interval.

B2. Use induction on s.

B3. Express r as a rational function of $\frac{r^2-1}{r^2+1}$ and $\frac{2r}{r^2+1}$. Alternatively, intersect the circle with lines through $(1, 0)$.

B4. Use the trigonometric substitution $(x_n, y_n) = (\cos\theta_n, \sin\theta_n)$.

B5. Write $M = A + iB$, and express everything in terms of the $2n \times 2n$ real matrix $(A \ \ B)$.

B6. Compare the product of the elements of $\{2a : a \in S\}$ to the product of the elements of S in two different ways.

The Forty-Ninth William Lowell Putnam Mathematical Competition
December 3, 1988

A1. Graph the part of R in the first quadrant; then use symmetry.

A2. The differential equation satisfied by g admits a solution on some intervals, either by the existence and uniqueness theorem, or explicitly by separation of variables.

A3. Use Taylor series to estimate $\frac{1}{n}\csc\frac{1}{n} - 1$.

A4. (a) The answer is yes. Consider equilateral triangles, and triangles of side lengths $\sqrt{3}$, $\sqrt{3}$, 1.
(b) Use a chessboard coloring.

A5. For any x, the sequence

$$x,\ f(x),\ f(f(x)),\ f(f(f(x))),\ \ldots$$

is linear recursive.

A6. The trace of A is independent of choice of basis.

B1. Take $z = 1$.

B2. Reduce to the case $y > 1$, and obtain lower bounds for $|x+1|$ and then $|x|$.

B3. The $c - d\sqrt{3}$ for a fixed n form an arithmetic progression.

B4. Divide the terms according to whether $a_n \geq 1/2^{n+1}$.

B5. The eigenvectors of M are $(1, \zeta, \zeta^2, \ldots, \zeta^{2n})$ where $\zeta^{2n+1} = 1$.

B6. An integer t is a triangular number if and only if $8t + 1$ is a square.

The Fiftieth William Lowell Putnam Mathematical Competition December 2, 1989

A1. Note that $100^k - 1 = (10^k + 1)(10^k - 1)$.

A2. Divide the rectangle into two parts by the diagonal line $ay = bx$.

A3. Show that the fractional linear transformation $\frac{11-10iz}{11z+10i}$ interchanges $\{ z : |z| < 1 \}$ and $\{ z : |z| > 1 \} \cup \{\infty\}$.

A4. Let coin flips determine digits past the decimal point in the binary expansion of a real number.

A5. An interval of what length is needed to contain all the distances $|p - v|$?

A6. Prove $\alpha^4 + x\alpha^2 + \alpha = 0$.

B1. Assume that the dartboard has corners at $(\pm 1, \pm 1)$, and find the equations of the curves bounding one-eighth of the specified region.

B2. The answer is yes. Pick a, and find an identity among the powers of a.

B3. Integrate by parts.

B4. Every real number is a limit of a sequence of rational numbers.

B5. Assume that AB and CD are horizontal, with AB below CD. Let $y = mx + e$ be the equation of BD. Use coordinate geometry to show that the least upper bound is 2.

B6. Write down the answer as a sum of multivariable integrals, and change the order of integration within each term so that the variable at which f is evaluated is the last to be integrated over.

The Fifty-First William Lowell Putnam Mathematical Competition
December 1, 1990

A1. Evaluate $T_n - n!$ for the first few n.

A2. For any increasing sequence $\{a_n\}$ with $a_n \to \infty$ and $a_{n+1} - a_n \to 0$, the set $S = \{\, a_n - a_m : m, n \geq 1 \,\}$ is dense in $\mathbb{R}$.

A3. Use Pick's Theorem, and consider parity of coordinates.

A4. Use punches centered at $A = (-\alpha, 0)$, $B = (0,0)$, and $C = (\alpha, 0)$ where α^2 is irrational.

A5. Construct a counterexample where each of $\mathbf{A}$ and $\mathbf{B}$ maps each standard basis vector to another standard basis vector or to 0.

A6. For each n, the number of admissible ordered pairs of subsets of $\{1, 2, \ldots, n\}$ is a Fibonacci number.

B1. Differentiable functions $g(x)$ and $h(x)$ are equal if and only if $g(0) = h(0)$ and $g'(x) = h'(x)$ for all x.

B2. The partial sums factor completely.

B3. If not, then S contains at most one diagonal matrix, at most one multiple of $\begin{pmatrix} 1 & 1 \\ 1 & 1 \end{pmatrix}$, and at most one of $\begin{pmatrix} 1 & 1 \\ 0 & 1 \end{pmatrix}$ and $\begin{pmatrix} 1 & 4 \\ 0 & 1 \end{pmatrix}$.

B4. A connected directed graph in which each vertex has indegree 2 and outdegree 2 has a closed path traversing each arc once.

B5. Define the a_n inductively, with $|a_{n+1}| \ll |a_n|$. Alternatively, let $a_n = (-1)^n 10^{-n^2}$ and evaluate the polynomial at 1, 10^2, 10^4, $\ldots$.

B6. When $t \geq 1/3$, the intersection contains the centroid.

The Fifty-Second William Lowell Putnam Mathematical Competition
December 7, 1991

A1. The region is a disjoint union of triangles and quarter-circles.

A2. Show that $\mathbf{A}^2 + \mathbf{B}^2$ times something nonzero is zero. There is only one nonzero matrix in the problem statement.

A3. If $n > 2$, then

$$\frac{p'(x)}{p(x)} = \frac{1}{x - r_1} + \cdots + \frac{1}{x - r_n}$$

is positive at $(r_{n-1} + r_n)/2$.

A4. Cover the coordinate axes by discs D_i of radius a_i where $\sum a_i$ diverges and $\sum a_i^2$ converges.

A5. If $u, v \geq 0$, then $\sqrt{u^2 + v^2} \leq u + v$.

A6. Exhibit a bijection between the sets counted by $A(n)$ and $B(n)$. The inequalities $a_1 > a_2 + a_3$, etc., suggest subtracting g_r from a_1, g_{r-1} from a_2, g_{r-2} from a_3, and so on. This is on the right track, except that because $g_r = g_{r-1} + g_{r-2} + 1$, this operation may not preserve the strict inequalities. Modify this idea by subtracting something slightly different, so that strict inequality is preserved.

B1. If a_k is not a square, determine whether or not a_{k+1} can be a square.

B2. Let $h(x) = f(x) + ig(x)$. Alternatively, differentiate the given functional equations to show that $H(x) = f(x)^2 + g(x)^2$ is constant.

B3. If a and b are positive integers, there exists a number g such that every multiple of $\gcd(a, b)$ greater than g may be written in the form $ra + sb$, where r and s are nonnegative integers.

B4. The sum is the coefficient of x^p in $\sum_{j=0}^{p} \binom{p}{j} (1 + x)^{p+j}$.

B5. First count solutions to $x^2 = y^2 + 1$ by rewriting as $(x + y)(x - y) = 1$, and solving the system $x + y = r$, $x - y = r^{-1}$ for each nonzero r.

B6. Without loss of generality, $u > 0$ and $a \geq b$. Divide by b, set $r = a/b$, and set $v = e^u$. Guess a value of v for which equality holds.

The Fifty-Third William Lowell Putnam Mathematical Competition December 5, 1992

A1. Apply f to (ii), then use (i) on the left hand side.

A2. Try replacing 1992 with a smaller number such as 2 or 3, and look for a pattern. Alternatively, show that

$$C(-y-1)\left(\frac{1}{y+1}+\cdots+\frac{1}{y+1992}\right)=\frac{d}{dy}\left(\frac{(y+1)\cdots(y+1992)}{1992!}\right).$$

A3. Substitute $x = ad$ and $y = bd$ where $d = \gcd(x, y)$. No prime can divide a.

A4. Let $h(x) = f(x) - 1/(1 + x^2)$. Use Rolle's Theorem and continuity repeatedly to prove $h^{(n)}(0) = 0$ for all n.

A5. Suppose there were three identical blocks in a row. Look at such an example with minimal m. If m is odd, show that each block must consist of alternating 0's and 1's. If m is even, halve the even indices in the blocks to find a smaller example.

A6. For every configuration of four points, consider the 16 configurations obtainable by replacing some of the points by their opposites.

B1. Given a set S of n elements, what is the longest chain of pairs of distinct numbers of S such that each pair is obtained from the previous pair by replacing one of its elements by a larger element of S?

B2. Factor $1 + x + x^2 + x^3$.

B3. By symmetry, it suffices to consider the case where $x, y \geq 0$. For fixed (x, y), the sequence is obtained by iterating a quadratic polynomial (depending on y). If such a sequence converges, it must converge to a fixed point of the polynomial.

B4. Use partial fractions.

B5. Use row operations to make most entries zero, then use column operations to make the matrix upper or lower triangular.

B6. If $A \in \mathcal{M}$, then A^2 commutes with all elements of $\mathcal{M}$, so $A^2 = I$. Given any linear relation among the elements of $\mathcal{M}$, other relations of the same length can be obtained by multiplying by elements of $\mathcal{M}$, and then shorter relations can be obtained by subtraction. Eventually this leads to a contradiction, so the matrices in $\mathcal{M}$ are linearly independent.

The Fifty-Fourth William Lowell Putnam Mathematical Competition
December 4, 1993

A1. Let (b, c) be the rightmost intersection point. Interpret $\int_0^b ((2x - 3x^3) - c)\, dx$ in terms of areas.

A2. Let $a_n = (x_{n+1} + x_{n-1})/x_n$. Show that two consecutive instances of the given identity imply $a_{n+1} = a_n$.

A3. Each f is determined by its values on $S = \{1, 2, \ldots, n\}$ and $S_i = S - \{i\}$ for $i = 1, 2, \ldots, n$.

A4. Let $X_k = \sum_{i=1}^{k} x_i$ and $Y_\ell = \sum_{j=1}^{\ell} y_j$. Without loss of generality $X_{19} \geq Y_{93}$. Let $g(\ell)$ be the distance from Y_ℓ to the largest X_k (possibly $X_0 = 0$) satisfying $X_k \leq Y_\ell$. Apply the Pigeonhole Principle to the values of $g(\ell)$.

A5. Substitute $x = -1/(t - 1)$ and $x = 1 - 1/t$, respectively, into the second and third integral.

A6. Define a sequence $(a_n)_{n \geq 0}$ by $a_n = 2$ if $n = \lfloor (2 + \sqrt{3})m \rfloor$ for some integer $m \geq 0$, and $a_n = 3$ otherwise. Prove that (a_n) satisfies the self-generation property.

B1. To guess the answer, try the problem with 1993 replaced by smaller numbers, and look for a pattern. Which m forces n to be large?

B2. Player B wins by making each move so that A cannot possibly win on the next move.

B3. Sketch the set of (x, y) in the unit square for which the integer nearest x/y is even. Evaluate its area by comparing to Leibniz's formula

$$\frac{\pi}{4} = 1 - \frac{1}{3} + \frac{1}{5} - \frac{1}{7} + \cdots.$$

B4. Define the linear operator T by

$$(Th)(x) = \int_0^1 h(y)K(x, y)\, dy.$$

Let r be the minimum value of f/g on $[0, 1]$. If $f - rg$ is not identically zero, then $T^2(f - rg)$ is positive on $[0, 1]$, contradicting $T^2(f - rg) = f - rg$.

B5. Find a polynomial identity with integer coefficients satisfied by the six distances. Obtain a contradiction modulo a small power of 2.

B6. It suffices to show that (a, b, c) with $0 < a \leq b \leq c$ can be transformed into (b', r, c') where $b = qa + r$ and $0 \leq r < a$. This is accomplished by a sequence of transformations dictated by the binary expansion of q.

The Fifty-Fifth William Lowell Putnam Mathematical Competition
December 3, 1994

A1. Let $b_m = \sum_{i=2^{m-1}}^{2^m-1} a_i$.

A2. Transform the ellipse into a circle by a change of variables.

A3. Let the vertices be $(0,0)$, $(1,0)$, and $(0,1)$. Consider $(\sqrt{2}-1,0)$, $(0,\sqrt{2}-1)$, $(2-\sqrt{2},\sqrt{2}-1)$, and $(\sqrt{2}-1,2-\sqrt{2})$.

A4. The quadratic polynomial $\det(A+tB)$ takes the value ± 1 at $t = 0,1,2,3,4$.

A5. The set $C = \{r_n : n \geq 0\} \cup \{0\}$ is compact, so the image of C^{1994} under the "sum the coordinates" map $\mathbb{R}^{1994} \to \mathbb{R}$ is a countable compact set.

A6. Show that if more than 2^{n-1} of the functions $f_1^{e_1} \circ \cdots \circ f_n^{e_n}$ map A to itself, then f_n maps A to itself and more than 2^{n-2} of the functions $f_1^{e_1} \circ \cdots \circ f_{n-1}^{e_{n-1}}$ map A to itself.

B1. If there are 15 squares within 250 of a given positive integer N, the squares are $m^2, (m+1)^2, \ldots, (m+14)^2$ for some integer $m \geq 0$. For each m, find the possibilities for N.

B2. By replacing x by $x - 9/4$ and adding a linear polynomial, reduce to the analogous problem for $x^4 + ax^2$.

B3. The conditions become simpler when rephrased in terms of $g(x) = \ln f(x)$ or even $h(x) = \ln f(x) - x$.

B4. Write $A = CDC^{-1}$ where D is diagonal. Then $A^n = CD^nC^{-1}$ gives explicit formulas for the entries of A^n.

B5. Use $\alpha = 1 - 1/n^2$ or $\alpha = e^{-1/n^2}$.

B6. Separate into congruences modulo 100 and modulo 101. Show that $2^a \equiv 1 \pmod{101}$ if and only if a is divisible by 100.

The Fifty-Sixth William Lowell Putnam Mathematical Competition
December 2, 1995

A1. Use proof by contradiction.

A2. Estimate the integrand, for example using Taylor series.

A3. Let $D = d_1d_2\ldots d_9$ and $E = e_1e_2\ldots e_9$. Write the given condition on D and E as congruences modulo 7 and sum them.

A4. Let $z_1, z_2, \ldots$ be the labels in order, and set $S_j = z_1 + \cdots + z_j - j(n-1)/n$.

A5. The coordinates of an eigenvector of the matrix (a_{ij}) will be the coefficients in a linear relation.

A6. Compare the number of such matrices with $b = a + 1$ and $c = a + 2$ to the number of such $3 \times (n+1)$ matrices with $a = b = c$. If the claim were false, the number of such matrices with $a = b = c$ would grow too slowly with n.

B1. For a given π, the function $\pi(x)$ takes at most three different values.

B2. Rolling without slipping implies an equality of arc lengths.

B3. Use the antisymmetry and multilinearity of the determinant.

B4. Express L in terms of itself.

B5. Use parity.

B6. Consider the spacing between consecutive members of $S(\alpha)$, $S(\beta)$ or $S(\gamma)$.

The Fifty-Seventh William Lowell Putnam Mathematical Competition December 7, 1996

A1. Use a trigonometric substitution.

A2. Use vectors.

A3. In how many ways can a student choose 3 courses?

A4. If A were a subset of a circle, and S were the set of (a, b, c) such that $a, b, c \in A$ occur in that order going clockwise around the circle, how could an ordering on A be defined in terms of S?

A5. Show that

$$\sum_{n=1}^{k} \frac{1}{p}\binom{p}{n} \equiv \sum_{n=1}^{k} \frac{(-1)^{n-1}}{n} = \sum_{n=1}^{k} \frac{1}{n} - 2\sum_{n=1}^{\lfloor k/2 \rfloor} \frac{1}{2n} \pmod{p},$$

and substitute $n = p - m$ in one of the sums.

A6. Examine the behavior of the sequence $x,\ x^2 + c,\ (x^2 + c)^2 + c,\ \ldots$ for various values of x and c.

B1. Find a recursion for the number of minimal selfish subsets.

B2. Take the logarithm and estimate the resulting sum as an integral.

B3. Find a transformation on any nonoptimal arrangement that increases the sum.

B4. Recall that $\sin A$ and $\cos A$ are defined by power series; that definition can be used to prove that certain trigonometric identities still hold for matrices. Alternatively, use a bit of linear algebra to conjugate A into a simple form before computing $\sin A$ and $\cos A$.

B5. Find a recursion for the number of balanced strings.

B6. The given expression is the gradient of a certain function.

The Fifty-Eighth William Lowell Putnam Mathematical Competition
December 6, 1997

A1. Use well-chosen Cartesian coordinates, or recall the relationship between H, O and the centroid of ABC.

A2. Determine how the game progresses by induction.

A3. Use integration by parts.

A4. For ψ to be a homomorphism, $\psi(e)$ must equal e, so try $a = \phi(e)^{-1}$.

A5. Discard solutions in pairs until almost nothing is left.

A6. Use the generating function $p(t) = \sum_{i \geq 0} x_{i+1} t^i$.

B1. Split the sum up into intervals on which the summand can be computed explicitly.

B2. Multiply by $f'(x)$.

B3. Separate the terms of the sum according to whether 5 divides the denominator.

B4. Find a recursion for the sum.

B5. See the solution to 1985A4.

B6. Start with five points spaced as far apart as possible; in the optimal arrangement, many of the distances between them will be equal.

The Fifty-Ninth William Lowell Putnam Mathematical Competition
December 5, 1998

A1. Take a diagonal cross-section.

A2. The result can be obtained simply by manipulating areas, without evaluating any integrals.

A3. A function cannot be positive and strictly concave-down over the entire real line.

A4. Find the number of digits in A_n, and find a recursion for A_n modulo 11.

A5. Use a greedy algorithm.

A6. Show that there is a point C' such that A, B, C' are vertices of a square and $|CC'| < 1$. To do this, work in a new coordinate system in which $B = (0,0)$ and $A = (s,0)$ for some $s > 0$.

B1. The numerator is a difference of squares.

B2. Use reflections.

B3. Find the area of a spherical cap in terms of its height.

B4. Combine terms from opposite ends of the sum.

B5. Use a Taylor expansion to approximate $\sqrt{N}$.

B6. Work modulo a suitable integer, or show that for suitable n, $\sqrt{P(n)}$ eventually falls between two consecutive integers.

The Sixtieth William Lowell Putnam Mathematical Competition
December 4, 1999

A1. Guess the form of f, g, h.

A2. Factor $p(x)$ over the real numbers.

A3. Find an explicit formula for a_n. Alternatively, guess from small examples a formula for m in terms of n.

A4. Add the series to itself.

A5. A continuous function on a compact set achieves a minimum value.

A6. Let $b_n = a_n/a_{n-1}$.

B1. Compute angles.

B2. Without loss of generality, P has a multiple zero at $x = 0$. Compare the largest powers of x dividing the two sides.

B3. Sum the "missing" terms instead, or sort the terms of the sum by congruence conditions on m and n.

B4. Integrate inequalities to obtain more inequalities.

B5. Find the eigenvectors, or at least the eigenvalues, of A.

B6. Let n be the smallest positive integer such that $\gcd(s, n) > 1$ for all $s \in S$. There exists $s \in S$ dividing n.

The Sixty-First William Lowell Putnam Mathematical Competition
December 2, 2000

A1. Use geometric series.

A2. Make $n+1$ a perfect square.

A3. Break up the area into triangles.

A4. Use integration by parts to mollify the rapidly oscillating factor.

A5. Relate the sides, area, and circumradius of the triangle formed by the points.

A6. If f is a polynomial with integer coefficients, then $m-n$ divides $f(m)-f(n)$ for all integers m and n.

B1. Use the Pigeonhole Principle.

B2. Recall that $\gcd(m,n)$ can be written as $am+bn$ for some integers a and b.

B3. Use Rolle's Theorem to count zeros. To establish the limit, look at the dominant term of $\frac{d^k f}{dt^k}$ as $k \to \infty$.

B4. Use a trigonometric substitution.

B5. Use generating functions modulo 2.

B6. There must exist an equilateral triangle of side length $2\sqrt{2}$.

Solutions

The 12-tuple $(n_{10}, n_9, n_8, n_7, n_6, n_5, n_4, n_3, n_2, n_1, n_0, n_\emptyset)$ following the problem number gives the performance of the top 200 or so competitors on that problem: n_i is the number who scored i, and $n_\emptyset$ is the number of blank papers.

The Forty-Sixth William Lowell Putnam Mathematical Competition
December 7, 1985

A1. (125, 6, 0, 0, 0, 0, 0, 0, 0, 0, 61, 9)

Determine, with proof, the number of ordered triples (A_1, A_2, A_3) of sets which have the property that

(i) $A_1 \cup A_2 \cup A_3 = \{1, 2, 3, 4, 5, 6, 7, 8, 9, 10\}$, **and**

(ii) $A_1 \cap A_2 \cap A_3 = \emptyset$,

where $\emptyset$ denotes the empty set. Express the answer in the form $2^a 3^b 5^c 7^d$, where a, b, c, and d are nonnegative integers.

Answer. The number of such triples of sets is $2^{10}3^{10}$.

Solution. There is a bijection between triples of subsets of $\{1, \ldots, 10\}$ and 10×3 matrices with $0, 1$ entries, sending (A_1, A_2, A_3) to the matrix $B = (b_{ij})$ with $b_{ij} = 1$ if $i \in A_j$ and $b_{ij} = 0$ otherwise. Under this bijection the set S of triples satisfying

$$A_1 \cup A_2 \cup A_3 = \{1, \ldots, 10\} \quad \text{and} \quad A_1 \cap A_2 \cap A_3 = \emptyset$$

maps onto the set T of 10×3 matrices with $0, 1$ entries such that no row is (000) or (111). The number of possibilities for each row of such a matrix is $2^3 - 2 = 6$, so $\#T = 6^{10}$. Hence $\#S = \#T = 2^{10}3^{10}$. ∎

Reinterpretation. Equivalently, this problem asks for the number of ways of placing the numbers 1 through 10 in the Venn diagram of Figure 1, where no numbers are placed in the two regions marked with an "×".

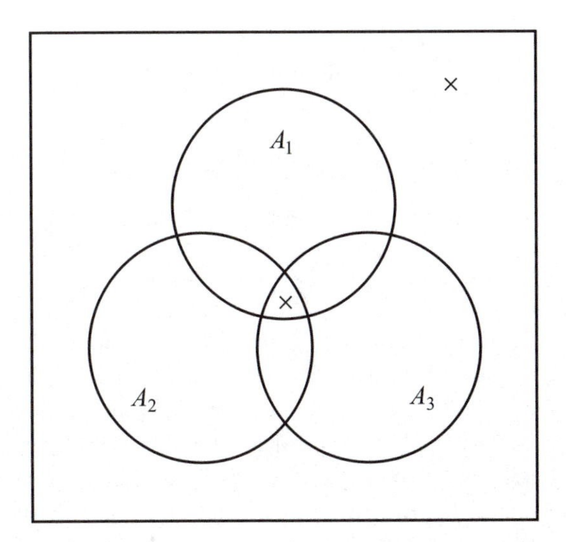

FIGURE 1.
Venn diagram interpretation of the solution to 1985A1.

A2. (29, 15, 31, 14, 38, 11, 2, 15, 6, 4, 21, 15)

Let T be an acute triangle. Inscribe a pair R, S of rectangles in T as shown:

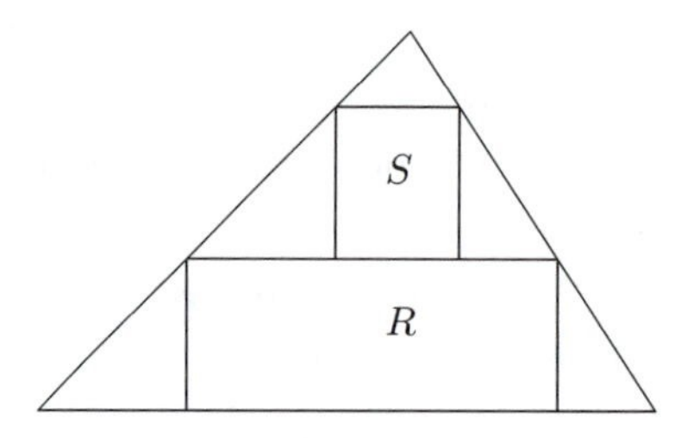

Let $A(X)$ denote the area of polygon X. Find the maximum value, or show that no maximum exists, of $\frac{A(R)+A(S)}{A(T)}$, where T ranges over all triangles and R, S over all rectangles as above.

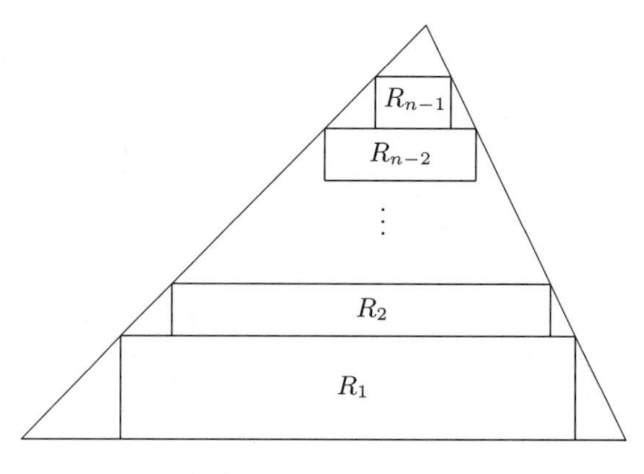

FIGURE 2.
A generalization of 1985A2.

Answer. The maximum value of $\frac{A(R)+A(S)}{A(T)}$ exists and equals $2/3$.

Solution. In fact, for any $n \geq 2$, we can find the maximum value of

$$\frac{A(R_1) + \cdots + A(R_{n-1})}{A(T)}$$

for any stack of rectangles inscribed in T as shown in Figure 2. The altitude of T divides T into right triangles U on the left and V on the right. For $i = 1, \ldots, n-1$, let U_i denote the small right triangle to the left of R_i, and let U_n denote the small right triangle above R_{n-1} and to the left of the altitude of T. Symmetrically define $V_1, \ldots,$ V_n to be the right triangles on the right. Each U_i is similar to U, so $A(U_i) = a_i^2 A(U)$, where a_i is the altitude of U_i, measured as a fraction of the altitude of T. Similarly, $A(V_i) = a_i^2 A(V)$. Hence

$$\begin{aligned} A(U_1) + \cdots + A(U_n) + A(V_1) + \cdots + A(V_n) &= (a_1^2 + \cdots + a_n^2)(A(U) + A(V)) \\ &= (a_1^2 + \cdots + a_n^2)A(T). \end{aligned}$$

Since T is the disjoint union of all the R_i, U_i, and V_i,

$$\frac{A(R_1)+\cdots+A(R_{n-1})}{A(T)} = 1 - \frac{A(U_1)+\cdots+A(U_n)+A(V_1)+\cdots+A(V_n)}{A(T)}$$
$$= 1-(a_1^2+\cdots+a_n^2).$$

The a_i must be positive numbers with sum 1, and conversely any such a_i give rise to a stack of rectangles in T.

It remains to *minimize* $a_1^2+\cdots+a_n^2$ subject to the constraints $a_i > 0$ for all i and $a_1+\cdots+a_n=1$. That the minimum is attained when $a_1=\cdots=a_n=1/n$ can be proved in many ways:

1. The identity

$$n(a_1^2+\cdots+a_n^2) = (a_1+\cdots+a_n)^2 + \sum_{i<j}(a_i-a_j)^2$$

   implies that

$$a_1^2+\cdots+a_n^2 \geq \frac{1}{n}(a_1+\cdots+a_n)^2 = \frac{1}{n},$$

   with equality if and only if $a_1=a_2=\cdots=a_n$.

2. Take $b_1=\cdots=b_n=1$ in the *Cauchy-Schwarz Inequality* [HLP, Theorem 7]

$$(a_1^2+\cdots+a_n^2)(b_1^2+\cdots+b_n^2) \geq (a_1b_1+\cdots+a_nb_n)^2,$$

   which holds for arbitrary $a_1,\ldots,a_n,b_1,\ldots,b_n\in\mathbb{R}$, with equality if and only if $(a_1,\ldots,a_n)$ and $(b_1,\ldots,b_n)$ are linearly dependent.

3. Take $b_i=a_i$ in *Chebychev's Inequality* [HLP, Theorem 43], which states that if $a_1\geq\cdots\geq a_n>0$ and $b_1\geq\cdots\geq b_n>0$, then

$$\left(\frac{\sum_{i=1}^n a_ib_i}{n}\right) \geq \left(\frac{\sum_{i=1}^n a_i}{n}\right)\left(\frac{\sum_{i=1}^n b_i}{n}\right),$$

   with equality if and only if all the a_i are equal or all the b_i are equal.

4. Take $r=2$ and $s=1$ in the *Power Mean Inequality* [HLP, Theorem 16], which states that for real numbers $a_1,\ldots,a_n>0$, if we define the rth *power mean* as

$$P_r = \left(\frac{a_1^r+\cdots+a_n^r}{n}\right)^{1/r},$$

   (and $P_0=\lim_{r\to 0}P_r=(a_1a_2\cdots a_n)^{1/n}$), then $P_r\geq P_s$ whenever $r>s$, with equality if and only if $a_1=\cdots=a_n$.

5. Take $f(x)=x^2$ in *Jensen's Inequality* [HLP, Theorem 90], which states that if $f(x)$ is a convex (concave-up) function on an interval I, then

$$\frac{f(a_1)+\cdots+f(a_n)}{n} \geq f\left(\frac{a_1+\cdots+a_n}{n}\right)$$

   for all $a_1,\ldots,a_n\in I$, with equality if and only if the a_i are all equal or f is linear on a closed interval containing all the a_i.

6. Let H denote the hyperplane $x_1+\cdots+x_n=1$ in $\mathbb{R}^n$. The line L through $\mathbf{0}=(0,\ldots,0)$ perpendicular to H is the one in the direction of $(1,\ldots,1)$, which

meets H at $P = (1/n, \dots, 1/n)$. The quantity $a_1^2 + \cdots + a_n^2$ can be viewed as the square of the distance from $\mathbf{0}$ to the point $(a_1, \dots, a_n)$ on H, and this is minimized when $(a_1, \dots, a_n) = P$.

In any case, we find that the minimum value of $a_1^2 + \cdots + a_n^2$ is $1/n$, so the maximum value of $\frac{A(R_1)+\cdots+A(R_{n-1})}{A(T)}$ is $1 - 1/n$. For the problem as stated, $n = 3$, so the maximum value is $2/3$. ■

Remark. The minimum is unchanged if instead of allowing T to vary, we fix a particular acute triangle T.

Remark. While we are on the subject of inequalities, we should also mention the very useful *Arithmetic-Mean–Geometric-Mean Inequality* (AM-GM), which states that for nonnegative real numbers $a_1, \dots, a_n$, we have

$$\frac{a_1 + a_2 + \cdots + a_n}{n} \geq (a_1 a_2 \cdots a_n)^{1/n},$$

with equality if and only if $a_1 = a_2 = \cdots = a_n$. This is the special case $P_1 \geq P_0$ of the Power Mean Inequality. It can also be deduced by taking $f(x) = \ln x$ in Jensen's Inequality.

A3. (100, 5, 18, 0, 0, 0, 0, 0, 2, 10, 19, 47)

Let d be a real number. For each integer $m \geq 0$, define a sequence $\{a_m(j)\}$, $j = 0, 1, 2, \dots$ by the condition

$$a_m(0) = d/2^m, \quad \textbf{and} \quad a_m(j+1) = (a_m(j))^2 + 2a_m(j), \quad j \geq 0.$$

Evaluate $\lim_{n\to\infty} a_n(n)$.

Answer. The value of $\lim_{n\to\infty} a_n(n)$ is $e^d - 1$.

Solution. We have $a_m(j+1) + 1 = (a_m(j) + 1)^2$, so by induction on j,

$$a_m(j) + 1 = (a_m(0) + 1)^{2^j}.$$

Hence

$$\lim_{n\to\infty} a_n(n) = \lim_{n\to\infty} \left(1 + \frac{d}{2^n}\right)^{2^n} - 1.$$

If $f(x) = \ln(1 + dx)$, then $f'(x) = d/(1 + dx)$, and in particular

$$\lim_{x\to 0} \frac{\ln(1 + dx)}{x} = f'(0) = d.$$

Applying the continuous function e^x yields

$$\lim_{x\to 0} (1 + dx)^{1/x} = e^d,$$

and taking the sequence $x_n = 1/2^n$ yields

$$\lim_{n\to\infty} \left(1 + \frac{d}{2^n}\right)^{2^n} = e^d,$$

so $\lim_{n\to\infty} a_n(n) = e^d - 1$. ■

A4. (72, 30, 6, 23, 0, 0, 0, 0, 1, 5, 33, 31)

Define a sequence $\{a_i\}$ by $a_1 = 3$ and $a_{i+1} = 3^{a_i}$ for $i \geq 1$. Which integers between 00 and 99 inclusive occur as the last two digits in the decimal expansion of infinitely many a_i?

Answer. Only 87 occurs infinitely often.

Solution. Let $\phi(n)$ denote the Euler ϕ-function, which equals the number of integers between 1 and n inclusive that are relatively prime to n, or more abstractly, the order of the multiplicative group $(\mathbb{Z}/n\mathbb{Z})^*$. If the prime factorization of n is $p_1^{e_1} \cdots p_k^{e_k}$, then $\phi(n)$ can be computed by the formula

$$\phi(n) = \prod_{i=1}^{k} \phi(p_i^{e_i}) = \prod_{i=1}^{k} p_i^{e_i-1}(p_i - 1).$$

Euler's Theorem [Lar1, p. 148], which is Lagrange's Theorem (the order of an element of a finite group divides the order of the group, [Lar1, p. 147]) applied to the group $(\mathbb{Z}/n\mathbb{Z})^*$, states that $a^{\phi(n)} \equiv 1 \pmod{n}$ whenever $\gcd(a, n) = 1$.

It shows that $a_i = 3^{a_{i-1}}$ modulo 100 is determined by a_{i-1} modulo $\phi(100) = 40$, for $i \geq 2$. Similarly a_{i-1} mod 40 is determined by a_{i-2} mod 16, which is determined by a_{i-3} mod 8, for $i \geq 4$. Finally a_{i-3} mod 8 is determined by a_{i-4} mod 2 for $i \geq 5$, since $3^2 \equiv 1 \pmod{8}$. For $i \geq 5$, a_{i-4} is odd, so

$$\begin{aligned}
a_{i-3} &= 3^{a_{i-4}} \equiv 3^1 \equiv 3 \pmod{8} \\
a_{i-2} &= 3^{a_{i-3}} \equiv 3^3 \equiv 11 \pmod{16} \\
a_{i-1} &= 3^{a_{i-2}} \equiv 3^{11} \equiv 27 \pmod{40} \\
a_i &= 3^{a_{i-1}} \equiv 3^{27} \equiv 87 \pmod{100}.
\end{aligned}$$

Thus the only integer that appears as the last two digits of infinitely many a_i is 87. ∎

Remark. Carmichael's lambda function. The exponent in Euler's Theorem is not always best possible. The function $\lambda(n)$ giving the best possible exponent for n, i.e., the least positive integer λ such that $a^\lambda \equiv 1 \pmod{n}$ whenever $\gcd(a, n) = 1$, is known as *Carmichael's lambda function* or as the *reduced totient function*. If $n = p_1^{e_1} \cdots p_k^{e_k}$, then

$$\lambda(n) = \operatorname{lcm}(\lambda(p_1^{e_1}), \ldots, \lambda(p_k^{e_k})),$$

where

$$\lambda(p^e) = \phi(p^e) = p^{e-1}(p - 1),$$

unless $p = 2$ and $e \geq 3$ in which case $\lambda(2^e) = 2^{e-2}$ instead of 2^{e-1}. For more information, see [Ros1, Section 9.6].

One can simplify the computations in the above solution by using $\lambda(n)$ in place of $\phi(n)$: iteration of λ maps 100 to 20 to 4 to 2, so starting from $a_{i-3} \equiv 1 \pmod{2}$ we obtain

$$\begin{aligned}
a_{i-2} &= 3^{a_{i-3}} \equiv 3^1 \equiv 3 \pmod{4} \\
a_{i-1} &= 3^{a_{i-2}} \equiv 3^3 \equiv 7 \pmod{20} \\
a_i &= 3^{a_{i-1}} \equiv 3^7 \equiv 87 \pmod{100}.
\end{aligned}$$

Stronger result. More generally, one can show that for any integer $c \geq 1$, the sequence defined by $a_1 = c$ and $a_{i+1} = c^{a_i}$ for $i \geq 1$ is eventually constant when reduced modulo a positive integer n. To prove this, one can use the Chinese Remainder Theorem to reduce to the case where n is a power of a prime p. If p divides c, then the sequence is eventually 0 modulo n; otherwise we can use Euler's Theorem and strong induction on n, since $\phi(n) < n$ for $n > 1$. See 1997B5 for a related problem, and further discussion.

A5. (44, 2, 7, 5, 0, 0, 0, 0, 10, 27, 46, 60)

Let $I_m = \int_0^{2\pi} \cos(x)\cos(2x)\cdots\cos(mx)\,dx$. For which integers m, $1 \leq m \leq 10$, is $I_m \neq 0$?

Answer. The values of m between 1 and 10 for which $I_m \neq 0$ are $3, 4, 7, 8$.

Solution. By de Moivre's Theorem ($\cos\theta + i\sin\theta = e^{i\theta}$, [Spv, Ch. 24]), we have

$$I_m = \int_0^{2\pi} \prod_{k=1}^{m} \left(\frac{e^{ikx} + e^{-ikx}}{2}\right) dx = 2^{-m} \sum_{\epsilon_k = \pm 1} \int_0^{2\pi} e^{i(\epsilon_1 + 2\epsilon_2 + \cdots + m\epsilon_m)x}\,dx,$$

where the sum ranges over the 2^m m-tuples $(\epsilon_1, \ldots, \epsilon_m)$ with $\epsilon_k = \pm 1$ for each k. For integers t,

$$\int_0^{2\pi} e^{itx}\,dx = \begin{cases} 2\pi, & \text{if } t = 0 \\ 0, & \text{otherwise.} \end{cases}$$

Thus $I_m \neq 0$ if and only 0 can be written as

$$\epsilon_1 + 2\epsilon_2 + \cdots + m\epsilon_m$$

for some $\epsilon_1, \ldots, \epsilon_m \in \{1, -1\}$. If such ϵ_k exist, then

$$0 = \epsilon_1 + 2\epsilon_2 + \cdots + m\epsilon_m \equiv 1 + 2 + \cdots + m = \frac{m(m+1)}{2} \pmod 2$$

so $m(m+1) \equiv 0 \pmod 4$, which forces $m \equiv 0$ or $3 \pmod 4$. Conversely, if $m \equiv 0 \pmod 4$, then

$$0 = (1 - 2 - 3 + 4) + (5 - 6 - 7 + 8) + \cdots + ((m-3) - (m-2) - (m-1) + m),$$

and if $m \equiv 3 \pmod 4$,

$$0 = (1 + 2 - 3) + (4 - 5 - 6 + 7) + (8 - 9 - 10 + 11) + ((m-3) - (m-2) - (m-1) + m).$$

Thus $I_m \neq 0$ if and only $m \equiv 0$ or $3 \pmod 4$. The integers m between 1 and 10 satisfying this condition are $3, 4, 7, 8$. ∎

Reinterpretation. This is a question in Fourier analysis. The function

$$\cos(x)\cos(2x)\cdots\cos(mx)$$

is continuous and periodic with period 2π, so it can be written as a Fourier series

$$\cos(x)\cos(2x)\cdots\cos(mx) = a_0 + \sum_{j=1}^{\infty} b_j \cos(jx) + \sum_{k=1}^{\infty} c_k \sin(kx).$$

The question asks: For which integers m between 1 and 10 is a_0 nonzero? By similar methods, one can show that:

(i) $c_k = 0$ for all k, and

(ii) $b_j = p/2^{m-1}$, where p is the number of ways to express j as $\epsilon_1 + 2\epsilon_2 + \cdots + m\epsilon_m$, where $\epsilon_1, \ldots, \epsilon_m \in \{1, -1\}$.

In particular, only finitely many b_j are nonzero, and $a_0 + \sum_j b_j = 1$.

A6. (8, 2, 0, 0, 0, 0, 0, 0, 3, 19, 22, 147)

If $p(x) = a_0 + a_1 x + \cdots + a_m x^m$ is a polynomial with real coefficients a_i, then set

$$\Gamma(p(x)) = a_0^2 + a_1^2 + \cdots + a_m^2.$$

Let $f(x) = 3x^2 + 7x + 2$. Find, with proof, a polynomial $g(x)$ with real coefficients such that

(i) $g(0) = 1$, and

(ii) $\Gamma(f(x)^n) = \Gamma(g(x)^n)$

for every integer $n \geq 1$.

Answer. One such $g(x)$ is $6x^2 + 5x + 1$.

Solution. For any polynomial $p(x)$, let $\gamma(p(x)) = p(x)p(x^{-1})$, which is a Laurent polynomial (an expression of the form $\sum_{j=m}^{n} a_j x^j$ where a_j are constants and m, n are integers, *not necessarily nonnegative*). Then $\Gamma(p(x))$ equals the coefficient of x^0 in $\gamma(p(x))$.

We have $f(x) = (3x+1)(x+2)$. Since

$$\gamma(x+2) = (x+2)(x^{-1}+2) = (1+2x^{-1})(1+2x) = \gamma(1+2x),$$

and $\gamma(p(x)q(x)) = \gamma(p(x))\gamma(q(x))$ for any polynomials $p(x)$ and $q(x)$, we find

$$\gamma(f(x)^n) = \gamma((3x+1)^n)\gamma((x+2)^n) = \gamma((3x+1)^n)\gamma((1+2x)^n) = \gamma(g(x)^n),$$

where $g(x) = (3x+1)(1+2x) = 6x^2 + 5x + 1$. Taking coefficients of x^0, we obtain $\Gamma(f(x)^n) = \Gamma(g(x)^n)$. Moreover $g(0) = 1$, so we are done. ∎

Remark. One could also show by brute force that $\Gamma\left((\sum a_i x^i)(\sum b_j x^j)\right)$ is unchanged by reversing the order of the b_j, without mentioning γ or Laurent polynomials. The coefficient of x^0 in a Laurent polynomial can also be expressed as a contour integral, via the residue theorem; hence one could begin a solution with the observation

$$\Gamma(p(x)) = \frac{1}{2\pi i} \oint_{|z|=1} p(z)p(z^{-1}) \frac{dz}{z}.$$

Remark. The solution is not unique: for any integer $k \geq 1$, the polynomials $g(x) = (3x^k + 1)(2x^k + 1)$ and $g(x) = (3x^k - 1)(2x^k - 1)$ also have the desired properties. We do not know if these are the only ones. Using the fact that the ring $\mathbb{R}[x, 1/x]$ of Laurent polynomials is a unique factorization domain (UFD), we could prove that these are the only ones if we could prove the following conjecture of Greg Kuperberg (communicated electronically):

Suppose $h_1, h_2 \in \mathbb{R}[x, 1/x]$ satisfy $h_1(x) = h_1(1/x)$ and $h_2(x) = h_2(1/x)$. Then the following are equivalent:

(a) The coefficient of x^0 in $h_1(x)^n$ equals the coefficient of x^0 of $h_2(x)^n$ for all positive integers n.

(b) There exists $j \in \mathbb{R}[x, 1/x]$ such that for $i = 1, 2$ we have $h_i(x) = j(\epsilon_i x^{k_i})$ for some $\epsilon_i \in \{1, -1\}$ and $k_i \geq 1$.

(It is easy to prove that (b) implies (a).)

B1. (112, 29, 0, 0, 0, 0, 0, 0, 0, 30, 13, 17)

Let k be the smallest positive integer with the following property:

There are distinct integers m_1, m_2, m_3, m_4, m_5 such that the polynomial

$$p(x) = (x - m_1)(x - m_2)(x - m_3)(x - m_4)(x - m_5)$$

has exactly k nonzero coefficients.

Find, with proof, a set of integers m_1, m_2, m_3, m_4, m_5 for which this minimum k is achieved.

Answer. The minimum is $k = 3$, and is attained for $\{m_1, m_2, m_3, m_4, m_5\} = \{-2, -1, 0, 1, 2\}$.

Solution. If $k = 1$, then $p(x)$ must be x^5, but this does not have five distinct integer zeros. If $k = 2$, then $p(x) = x^5 + ax^r$ for some nonzero $a \in \mathbb{Z}$ and $0 \leq r \leq 4$; this has $x = 0$ as double zero if $r \geq 2$ and has a nonreal zero if $r = 0$ or $r = 1$. Thus $k \geq 3$. The example

$$x(x-1)(x+1)(x-2)(x+2) = x(x^2-1)(x^2-4) = x^5 - 5x^3 + 4x$$

shows that in fact $k = 3$. ∎

Remark. More generally, we can prove that given $n \geq 1$, the smallest integer k for which there exist distinct integers $m_1, \ldots, m_n$ such that the polynomial

$$p(x) = (x - m_1) \cdots (x - m_n)$$

has exactly k nonzero coefficients is $k = \lceil (n+1)/2 \rceil = \lfloor n/2 \rfloor + 1$. The key is *Descartes' Rule of Signs*, which states that if $p(x) = a_1 x^{r_1} + a_2 x^{r_2} + \cdots + a_k x^{r_k}$ is a polynomial with $a_i \in \mathbb{R}^*$ and $r_1 > r_2 > \cdots > r_k$, then the number of positive real zeros of $p(x)$ counted with multiplicity is the number of sign changes in the sequence $a_1, a_2, \ldots, a_k$ minus a nonnegative even integer.

If $p(x)$ has k nonzero coefficients, $p(x)$ has at most $k-1$ positive real zeros. Applying Descartes' Rule of Signs to $p(-x)$ shows that $p(x)$ has at most $k-1$ negative real zeros. Hence the total number of distinct zeros of $p(x)$ is at most $(k-1)+(k-1)+1 = 2k-1$, where the $+1$ is for the possibility that 0 might be a root. If $p(x)$ has n distinct zeros all of which are integers, then $n \leq 2k - 1$, so $k \geq \lceil (n+1)/2 \rceil$.

On the other hand, given $n \geq 1$, we can exhibit a polynomial $p(x) = (x - m_1) \cdots (x - m_n)$ with distinct integer zeros $m_1, \ldots, m_n$ and with at most (hence exactly) $k = \lceil (n+1)/2 \rceil$ nonzero coefficients:

$$p(x) = \begin{cases} (x+1)(x-1)(x+2)(x-2)\cdots(x+(k-1))(x-(k-1)) & \text{if } n = 2k-2 \\ x(x+1)(x-1)(x+2)(x-2)\cdots(x+(k-1))(x-(k-1)) & \text{if } n = 2k-1. \end{cases}$$

B2. (3, 89, 3, 1, 0, 0, 0, 0, 2, 11, 37, 55)

Define polynomials $f_n(x)$ for $n \geq 0$ by $f_0(x) = 1$, $f_n(0) = 0$ for $n \geq 1$, and

$$\frac{d}{dx}(f_{n+1}(x)) = (n+1)f_n(x+1)$$

for $n \geq 0$. Find, with proof, the explicit factorization of $f_{100}(1)$ into powers of distinct primes.

Answer. The factorization of $f_{100}(1)$ is 101^{99}.

Solution. By induction, the given properties determine $f_n(x)$ uniquely. Computing and factoring $f_n(x)$ for the first few n suggests that $f_n(x) = x(x+n)^{n-1}$. We prove this by induction on n. The base case $n = 0$ is given: $f_0(x) = 1$. For $n \geq 0$, we indeed have $f_{n+1}(0) = 0$ and

$$\begin{aligned}\frac{d}{dx}f_{n+1}(x) &= (x+n+1)^n + nx(x+n+1)^{n-1} \\ &= (n+1)(x+1)(x+n+1)^{n-1} \\ &= (n+1)f_n(x+1),\end{aligned}$$

which completes the inductive step. Hence $f_{100}(1) = 101^{99}$. (Note that 101 is prime.) ∎

B3. (95, 15, 15, 11, 0, 0, 0, 0, 5, 3, 23, 34)

Let

$$\begin{matrix} a_{1,1} & a_{1,2} & a_{1,3} & \cdots \\ a_{2,1} & a_{2,2} & a_{2,3} & \cdots \\ a_{3,1} & a_{3,2} & a_{3,3} & \cdots \\ \vdots & \vdots & \vdots & \ddots \end{matrix}$$

be a doubly infinite array of positive integers, and suppose each positive integer appears exactly eight times in the array. Prove that $a_{m,n} > mn$ for some pair of positive integers (m, n).

Solution. Suppose not; i.e., suppose that $a_{m,n} \leq mn$ for all $m, n \geq 1$. Let

$$R(k) = \{\, (i,j) : a_{i,j} \leq k \,\}.$$

By hypothesis, $\#R(k) \leq 8k$. On the other hand, $R(k)$ contains all pairs (i, j) with $ij \leq k$, and there are

$$\left\lfloor \frac{k}{1} \right\rfloor + \left\lfloor \frac{k}{2} \right\rfloor + \cdots + \left\lfloor \frac{k}{k} \right\rfloor > \left(\frac{k}{1} - 1\right) + \left(\frac{k}{2} - 1\right) + \cdots + \left(\frac{k}{k} - 1\right) > k(\ln k - 1)$$

such pairs, since

$$\frac{1}{1} + \frac{1}{2} + \cdots + \frac{1}{k} > \int_1^k \frac{1}{x}\,dx = \ln k.$$

Hence $8k > k(\ln k - 1)$, which is a contradiction for $k > e^9 \approx 8103.08$. ∎

Remark. To solve the problem, the explicit lower bound on the rate of growth of $\frac{1}{1} + \frac{1}{2} + \cdots + \frac{1}{k}$ as $k \to \infty$ is not really needed: it suffices to know that this sum tends to ∞, i.e., that the harmonic series diverges.

B4. (115, 30, 9, 1, 2, 6, 4, 0, 9, 8, 11, 6)

Let C be the unit circle $x^2 + y^2 = 1$. A point p is chosen randomly on the circumference of C and another point q is chosen randomly from the interior of C (these points are chosen independently and uniformly over their domains). Let R be the rectangle with sides parallel to the x- and y-axes with diagonal pq. What is the probability that no point of R lies outside of C?

Answer. The probability is $4/\pi^2$.

Solution. Let $p = (\cos\theta, \sin\theta)$ and $q = (a, b)$. The other two vertices of R are $(\cos\theta, b)$ and $(a, \sin\theta)$. If $|a| \le |\cos\theta|$ and $|b| \le |\sin\theta|$, then each vertex (x, y) of R satisfies $x^2 + y^2 \le \cos^2\theta + \sin^2\theta = 1$, and no points of R can lie outside of C. Conversely, if no points of R lies outside of C, then applying this to the two vertices other than p and q, we find

$$\cos^2\theta + b^2 \le 1, \quad \text{and} \quad a^2 + \sin^2\theta \le 1,$$

or equivalently

$$|b| \le |\sin\theta|, \quad \text{and} \quad |a| \le |\cos\theta|. \tag{1}$$

These conditions imply that (a, b) lies inside or on C, so for any given θ, the probability that the random point $q = (a, b)$ satisfies (1) is

$$\frac{2|\cos\theta| \cdot 2|\sin\theta|}{\pi} = \frac{2}{\pi}|\sin(2\theta)|,$$

and the overall probability is

$$\frac{1}{2\pi}\int_0^{2\pi} \frac{2}{\pi}|\sin(2\theta)|\,d\theta = \frac{4}{\pi^2}\int_0^{\pi/2} \sin(2\theta)\,d\theta = \frac{4}{\pi^2}. \qquad \blacksquare$$

B5. (14, 8, 2, 2, 0, 0, 0, 0, 2, 2, 61, 110)

Evaluate $\int_0^\infty t^{-1/2}e^{-1985(t+t^{-1})}\,dt$. You may assume that $\int_{-\infty}^\infty e^{-x^2}\,dx = \sqrt{\pi}$.

Answer. The value of $\int_0^\infty t^{-1/2}e^{-1985(t+t^{-1})}\,dt$ is $\sqrt{\frac{\pi}{1985}}e^{-3970}$.

Solution (adapted from [Bernau]). For $a > 0$, let

$$I(a) = \int_0^\infty t^{-1/2}e^{-a(t+t^{-1})}\,dt.$$

The integral converges, since the integrand is bounded by $t^{-1/2}$ on $(0, 1]$ and by e^{-at} on $[1, \infty)$. Hence

$$I(a) = \lim_{B\to\infty}\left[\int_{1/B}^1 t^{-1/2}e^{-a(t+t^{-1})}\,dt + \int_1^B t^{-1/2}e^{-a(t+t^{-1})}\,dt\right].$$

Substitute $1/t$ for t in the first integral to conclude

$$I(a) = \lim_{B\to\infty}\int_1^B (t^{-1/2} + t^{-3/2})e^{-a(t+t^{-1})}\,dt.$$

Now use the substitution $u = a^{1/2}(t^{1/2} - t^{-1/2})$ to obtain

$$\begin{aligned} I(a) &= 2a^{-1/2} \lim_{B\to\infty} \int_0^{a^{1/2}(B^{1/2}-B^{-1/2})} e^{-u^2-2a}\,du \\ &= 2a^{-1/2}e^{-2a} \int_0^{\infty} e^{-u^2}\,du \\ &= \sqrt{\frac{\pi}{a}}e^{-2a}, \end{aligned}$$

so $I(1985) = \sqrt{\frac{\pi}{1985}}e^{-3970}$. ■

Remark. The *modified Bessel function of the second kind* (also known as *Macdonald's function*) has the integral representation

$$K_\nu(z) = \int_0^\infty e^{-z\cosh t}\cosh(\nu t)\,dt$$

for $\operatorname{Re}(z) > 0$ [O, p. 250]. When $\nu = 1/2$, the substitution $u = e^t$ relates this to expressions occurring in the solution above; to be precise, $I(a) = 2K_{1/2}(2a)$ for all $a > 0$. Thus $K_{1/2}(z) = \sqrt{\frac{\pi}{2z}}e^{-z}$ for $z > 0$. Similar formulas exist for $K_{n+1/2}(z)$ for each integer n. For arbitrary ν, the function $w = K_\nu(z)$ is a solution of the differential equation

$$z^2w'' + zw' - (z^2 + \nu^2)w = 0.$$

B6. (5, 0, 0, 0, 0, 4, 0, 0, 9, 7, 26, 150)

Let G be a finite set of real $n \times n$ matrices $\{M_i\}$, $1 \le i \le r$, which form a group under matrix multiplication. Suppose that $\sum_{i=1}^r \operatorname{tr}(M_i) = 0$, where $\operatorname{tr}(A)$ denotes the trace of the matrix A. Prove that $\sum_{i=1}^r M_i$ is the $n \times n$ zero matrix.

Solution 1. Let $S = \sum_{i=1}^r M_i$. For any j, the sequence $M_jM_1, M_jM_2, \ldots, M_jM_r$ is a permutation of the elements of G, and summing yields $M_jS = S$. Summing this from $j = 1$ to r yields $S^2 = rS$. Therefore the minimal polynomial of S divides $x^2 - rx$, and every eigenvalue of S is either 0 or r. But the eigenvalues counted with multiplicity sum to $\operatorname{tr}(S) = 0$, so they are all 0. At this point, we present three ways to finish the proof that $S = 0$:

1. Every eigenvalue of $S - rI$ is $-r \neq 0$, so $S - rI$ is invertible. Hence from $S(S - rI) = 0$ we obtain $S = 0$.
2. The minimal polynomial $p(x)$ of S must be x, $x - r$, or $x(x - r)$. Since every zero of the minimal polynomial is an eigenvalue, the minimal polynomial is x. By the Cayley-Hamilton Theorem [Ap2, Theorem 7.8], $p(S) = 0$; that is, $S = 0$.
3. The Jordan canonical form of S over the complex numbers has 0's (the eigenvalues) on the main diagonal, possible 1's just above the diagonal, and 0's elsewhere. The condition $S^2 = rS$ implies that there are no 1's, so the Jordan canonical form of S is 0. Thus $S = 0$. ■

Literature note. See [Ap2, Ch. 4] for a quick introduction to eigenvalues and eigenvectors.

Solution 2 (based on an idea of Dave Savitt).

Lemma. *Let G be a finite group of order r. Let $\rho : G \to \operatorname{Aut}(V)$ be a representation of G on some finite-dimensional complex vector space V. Then $\sum_{g \in G} \operatorname{tr} \rho(g)$ is a nonnegative integer divisible by r, and is zero if and only if $\sum_{g \in G} \rho(g) = 0$.*

Proof. Let $\eta_1, \dots, \eta_s$ be the irreducible characters of G. Theorem 3 on p. 15 of [Se2] implies that if $\chi = \sum_{i=1}^s a_i \eta_i$ and $\psi = \sum_{i=1}^s b_i \eta_i$ are arbitrary characters, then

$$\frac{1}{r} \sum_{g \in G} \chi(g)\overline{\psi(g)} = \sum_{i=1}^{s} a_i b_i. \tag{2}$$

Applying this to the character of ρ and the trivial character $\mathbf{1}$ shows that $\frac{1}{r} \sum_{g \in G} \operatorname{tr} \rho(g)$ equals the multiplicity of $\mathbf{1}$ in ρ, which is a nonnegative integer.

Now suppose that the matrix $S = \sum_{g \in G} \rho(g)$ is nonzero. Choose $v \in V$ with $Sv \neq 0$. The relation $\rho(h)S = S$ shows that Sv is fixed by $\rho(h)$ for all $h \in G$. In other words, Sv spans a trivial subrepresentation of ρ, so the nonnegative integer of the previous paragraph is positive. □

We now return to the problem at hand. Unfortunately the M_i do not necessarily define a representation of G, since the M_i need not be invertible. Instead we need to apply the lemma to the action of G on $\mathbb{C}^n/K$, for some subspace K. Given $1 \leq i, j \leq r$, there exists k such that $M_i = M_k M_j$, so $\ker M_j \subseteq \ker M_i$. This holds for all i and j, so the M_i have a common kernel, which we call K. Then the M_i and S also act on $\mathbb{C}^n/K$. If $v \in \mathbb{C}^n$ maps to an element of $\mathbb{C}^n/K$ in the kernel of M_i acting on $\mathbb{C}^n/K$, then $M_i M_i v \in M_i(K) = 0$, but $M_i M_i = M_j$ for some j, so $v \in \ker(M_j) = K$. Thus the M_i act invertibly on $\mathbb{C}^n/K$. We finish by applying the lemma to the representation $\mathbb{C}^n/K$ of G, using the observation that $\operatorname{tr} S$ is the sum of the traces of S acting on $\mathbb{C}^n/K$ and on K, with the trace on K being zero. ■

Remark. One can also give a elementary variant of this solution (which somewhat obscures the connection with representation theory). Namely, we prove by induction on n that $\operatorname{tr} S$ is a nonnegative integer divisible by r, which is nonzero if $S \neq 0$. The case $n = 1$ is straightforward; given the result for $n - 1$, choose as above a vector $v \in \mathbb{C}^n$ with $Sv \neq 0$, so that each of the matrices preserves v. Let V be the span of v; then the trace of S is equal to the sum of its trace on V and on the quotient $\mathbb{C}^n/V$. The former is r and the latter is a nonnegative integer divisible by r.

Literature note. See [Se2] for an introduction to representation theory. The relations given by (2) are known as the *orthogonality relations for characters.*

The Forty-Seventh William Lowell Putnam Mathematical Competition
December 6, 1986

A1. (152, 23, 10, 7, 0, 0, 0, 2, 2, 3, 1, 1)

Find, with explanation, the maximum value of $f(x) = x^3 - 3x$ on the set of all real numbers x satisfying $x^4 + 36 \leq 13x^2$.

Answer. The maximum value is 18.

Solution. The condition $x^4+36 \leq 13x^2$ is equivalent to $(x-3)(x-2)(x+2)(x+3) \leq 0$, which is satisfied if and only if $x \in [-3,-2] \cup [2,3]$. The function f is increasing on $[-3,-2]$ and on $[2,3]$, since $f'(x) = 3(x^2-1) > 0$ on these intervals. Hence the maximum value is $\max\{f(-2), f(3)\} = 18$. ■

A2. (155, 0, 0, 0, 0, 0, 0, 0, 0, 0, 33, 13)

What is the units (i.e., rightmost) digit of $\left\lfloor \frac{10^{20000}}{10^{100}+3} \right\rfloor$? Here $\lfloor x \rfloor$ is the greatest integer $\leq x$.

Answer. The units digit is 3.

Solution. Taking $x = 10^{100}$ and $y = -3$ in the factorization

$$x^{200} - y^{200} = (x-y)(x^{199} + x^{198}y + \cdots + xy^{198} + y^{199})$$

shows that the number

$$I = \frac{10^{20000} - 3^{200}}{10^{100}+3} = \left(10^{100}\right)^{199} - \left(10^{100}\right)^{198} 3 + \cdots + 10^{100}3^{198} - 3^{199} \tag{1}$$

is an integer. Moreover, $I = \left\lfloor \frac{10^{20000}}{10^{100}+3} \right\rfloor$, since

$$\frac{3^{200}}{10^{100}+3} = \frac{9^{100}}{10^{100}+3} < 1.$$

By (1),

$$I \equiv -3^{199} \equiv -3^3(81)^{49} \equiv -27 \equiv 3 \pmod{10},$$

so the units digit of I is 3. ■

A3. (53, 6, 15, 1, 0, 0, 0, 1, 12, 1, 26, 86)

Evaluate $\sum_{n=0}^{\infty} \operatorname{Arccot}(n^2+n+1)$, where $\operatorname{Arccot} t$ for $t \geq 0$ denotes the number θ in the interval $0 < \theta \leq \pi/2$ with $\cot\theta = t$.

Answer. The series converges to $\pi/2$.

Solution 1. If $\alpha = \operatorname{Arccot} x$ and $\beta = \operatorname{Arccot} y$ for some $x, y > 0$, then the addition formula

$$\cot(\alpha+\beta) = \frac{\cot\alpha\cot\beta - 1}{\cot\alpha + \cot\beta}$$

shows that

$$\operatorname{Arccot} x + \operatorname{Arccot} y = \operatorname{Arccot}\frac{xy-1}{x+y} \tag{1}$$

provided that $\operatorname{Arccot} x + \operatorname{Arccot} y \leq \pi/2$. The latter condition is equivalent to $\operatorname{Arccot} x \leq \operatorname{Arccot}(1/y)$, which is equivalent to $x \geq 1/y$, and hence equivalent to

$xy \geq 1$. Verifying the $xy \geq 1$ condition at each step, we use (1) to compute the first few partial sums

$$\begin{aligned} \text{Arccot}\, 1 &= \text{Arccot}\, 1 \\ \text{Arccot}\, 1 + \text{Arccot}\, 3 &= \text{Arccot}(1/2) \\ \text{Arccot}\, 1 + \text{Arccot}\, 3 + \text{Arccot}\, 7 &= \text{Arccot}(1/3), \end{aligned}$$

and guess that $\sum_{n=0}^{m-1} \text{Arccot}(n^2+n+1) = \text{Arccot}(1/m)$ for all $m \geq 1$. This is easily proved by induction on m: the base case is above, and the inductive step is

$$\begin{aligned} \sum_{n=0}^{m} \text{Arccot}(n^2+n+1) &= \text{Arccot}(m^2+m+1) + \sum_{n=0}^{m-1} \text{Arccot}(n^2+n+1) \\ &= \text{Arccot}(m^2+m+1) + \text{Arccot}\left(\frac{1}{m}\right) \quad \text{(inductive hypothesis)} \\ &= \text{Arccot}\left(\frac{(m^2+m+1)/m - 1}{m^2+m+1+1/m}\right) \quad \text{(by (1))} \\ &= \text{Arccot}\left(\frac{1}{m+1}\right). \end{aligned}$$

Hence

$$\sum_{n=0}^{\infty} \text{Arccot}(n^2+n+1) = \lim_{m\to\infty} \text{Arccot}\left(\frac{1}{m}\right) = \text{Arccot}(0) = \pi/2. \qquad \blacksquare$$

Solution 2. For real $a \geq 0$ and $b \neq 0$, $\text{Arccot}(a/b)$ is the argument (between $-\pi/2$ and $\pi/2$) of the complex number $a+bi$. Therefore, if any three complex numbers satisfy $(a+bi)(c+di) = (e+fi)$, where $a,c,e \geq 0$ and $b,d,f \neq 0$, then then $\text{Arccot}(a/b) + \text{Arccot}(c/d) = \text{Arccot}(e/f)$. Factoring the polynomial $n^2+n+1+i$ yields

$$(n^2+n+1+i) = (n+i)(n+1-i).$$

Taking arguments, we find that $\text{Arccot}(n^2+n+1) = \text{Arccot}\, n - \text{Arccot}(n+1)$. (This identity can also be proved from the difference formula for cot, but then one needs to guess the identity in advance.) The series $\sum_{n=0}^{\infty} \text{Arccot}(n^2+n+1)$ telescopes to $\lim_{n\to\infty} (\text{Arccot}(0) - \text{Arccot}(n+1)) = \pi/2$. $\blacksquare$

Related question. Evaluate the infinite series:

$$\sum_{n=1}^{\infty} \text{Arctan}\left(\frac{2}{n^2}\right), \qquad \sum_{n=1}^{\infty} \text{Arctan}\left(\frac{8n}{n^4-2n^2+5}\right).$$

The first is problem 26 of [WH], and both appeared in a problem due to J. Anglesio in the *Monthly* [Mon3, Mon5].

Literature note. The value of series such as these have been known for a long time. In particular, $\sum_{n=1}^{\infty} \text{Arctan}\left(\frac{2}{n^2}\right)$ and some similar series were evaluated at least as early as 1878 [Gl]. Ramanujan [Berndt, p. 37] independently evaluated this and other

Arctan series shortly after 1903. The generalizations

$$\sum_{n=1}^{\infty} \operatorname{Arctan}\left(\frac{2xy}{n^2 - x^2 + y^2}\right) \equiv \operatorname{Arctan}\frac{y}{x} - \operatorname{Arctan}\frac{\tanh \pi y}{\tan \pi x} \pmod{\pi},$$

$$\sum_{n=1}^{\infty} \operatorname{Arctan}\left(\frac{2xy}{(2n-1)^2 - x^2 + y^2}\right) \equiv \operatorname{Arctan}\left(\tan\frac{\pi x}{2}\tanh\frac{\pi y}{2}\right) \pmod{\pi}$$

appear in [GK]. For further history see Chapter 2 of [Berndt].

A4. (21, 3, 4, 4, 5, 7, 0, 6, 3, 7, 23, 118)

A *transversal* of an $n\times n$ matrix A consists of n entries of A, no two in the same row or column. Let $f(n)$ be the number of $n \times n$ matrices A satisfying the following two conditions:

(a) Each entry $\alpha_{i,j}$ of A is in the set $\{-1, 0, 1\}$.

(b) The sum of the n entries of a transversal is the same for all transversals of A.

An example of such a matrix A is

$$A = \begin{pmatrix} -1 & 0 & -1 \\ 0 & 1 & 0 \\ 0 & 1 & 0 \end{pmatrix}.$$

Determine with proof a formula for $f(n)$ of the form

$$f(n) = a_1 b_1^n + a_2 b_2^n + a_3 b_3^n + a_4,$$

where the a_i's and b_i's are rational numbers.

Answer. The value of $f(n)$ is $4^n + 2\cdot 3^n - 4\cdot 2^n + 1$.

Solution (Doug Jungreis).

Lemma. *Condition (b) is equivalent to the statement that any two rows of the matrix differ by a constant vector, i.e., a vector of the form $(c, c, \ldots, c)$.*

Proof. If two rows differ by a constant vector, then (b) holds. Conversely, if (b) holds, for any i, j, k, l in $\{1, \ldots, n\}$, take a transversal containing a_{ik} and a_{jl}, and then switch a_{il} and a_{jk} to get a new transversal. Since these two transversals have the same sum, $a_{ik} + a_{jl} = a_{il} + a_{jk}$, or equivalently, $a_{ik} - a_{jk} = a_{il} - a_{jl}$. Thus rows i and j differ by the constant vector with all components equal to $a_{i1} - a_{j1}$. Since i and j were arbitrary, each pair of rows differs by a constant vector. □

We compute $f(n)$ by considering four cases.

Case 1: the first row of the matrix is a constant vector. Then each row is constant by the Lemma, so each row is $(0, \ldots, 0)$, $(1, \ldots, 1)$, or $(-1, \ldots, -1)$. Thus there are 3^n such matrices.

Case 2: both 0 *and* 1 *appear in the first row, but not* -1. Then there are $2^n - 2$ possibilities for the first row. Each other row must differ from the first by either $(0, \ldots, 0)$ or $(-1, ..., -1)$, by the Lemma and condition (a), so there are 2^{n-1} possibilities for these rows. This gives a total of $2^{n-1}(2^n - 2)$ possibilities in this case.

Case 3: both 0 *and* -1 *appear in the first row, but not* 1. These are just the negatives of the matrices in case 2, so we again have $2^{n-1}(2^n - 2)$ possibilities.

Case 4: Both 1 *and* -1 *(and possibly also* 0*) appear in the first row.* This covers all other possibilities for the first row, i.e., the remaining $3^n - 2 \cdot 2^n + 1$ possibilities. Then every row must be equal to the first, by the Lemma and condition (a), so we have a total of $3^n - 2 \cdot 2^n + 1$ possibilities in this case.

Adding the four cases gives $f(n) = 4^n + 2 \cdot 3^n - 4 \cdot 2^n + 1$. ■

Related question.

> If an $n \times n$ matrix M with nonnegative integer entries satisfies condition (b), that the sum of the n entries of a transversal is the same number m for all transversals of A, show that M is the sum of m permutation matrices. (A *permutation matrix* is a matrix with one 1 in each row and each column, and all other entries 0.)

For a card trick related to this result, see [Kl].

This result is a discrete version of the following result.

Birkhoff-von Neumann Theorem. *The convex hull of the permutation matrices is precisely the set of* doubly stochastic matrices*: matrices with entries in* $[0, 1]$ *with each row and column summing to* 1.

A5. (13, 4, 0, 0, 0, 0, 0, 1, 0, 2, 39, 142)

Suppose $f_1(x)$, $f_2(x)$, $\dots$, $f_n(x)$ are functions of n real variables $x = (x_1, \dots, x_n)$ with continuous second-order partial derivatives everywhere on $\mathbb{R}^n$. Suppose further that there are constants c_{ij} such that

$$\frac{\partial f_i}{\partial x_j} - \frac{\partial f_j}{\partial x_i} = c_{ij}$$

for all i and j, $1 \le i \le n$, $1 \le j \le n$. Prove that there is a function $g(x)$ on $\mathbb{R}^n$ such that $f_i + \partial g/\partial x_i$ is linear for all i, $1 \le i \le n$. (A linear function is one of the form

$$a_0 + a_1 x_1 + a_2 x_2 + \cdots + a_n x_n.)$$

Solution. Note that $c_{ij} = -c_{ji}$ for all i and j. Let $h_i = \frac{1}{2}\sum_j c_{ij} x_j$, so $\partial h_i/\partial x_j = \frac{1}{2} c_{ij}$. Then

$$\frac{\partial h_i}{\partial x_j} - \frac{\partial h_j}{\partial x_i} = \frac{1}{2} c_{ij} - \frac{1}{2} c_{ji} = c_{ij} = \frac{\partial f_i}{\partial x_j} - \frac{\partial f_j}{\partial x_i},$$

so

$$\frac{\partial (h_i - f_i)}{\partial x_j} = \frac{\partial (h_j - f_j)}{\partial x_i}$$

for all i and j. Hence $(h_1 - f_1, \dots, h_n - f_n)$ is a gradient, i.e., there is a differentiable function g on $\mathbb{R}^n$ such that $\partial g/\partial x_i = h_i - f_i$ for each i. Then $f_i + \partial g/\partial x_i = h_i$ is linear for each i. ■

A6. (1, 4, 1, 1, 0, 1, 0, 0, 6, 4, 64, 119)

Let $a_1, a_2, \ldots, a_n$ be real numbers, and let $b_1, b_2, \ldots, b_n$ be distinct positive integers. Suppose there is a polynomial $f(x)$ satisfying the identity

$$(1-x)^n f(x) = 1 + \sum_{i=1}^{n} a_i x^{b_i}.$$

Find a simple expression (not involving any sums) for $f(1)$ in terms of $b_1, b_2, \ldots, b_n$ and n (but independent of $a_1, a_2, \ldots, a_n$).

Answer. The number $f(1)$ equals $b_1 b_2 \cdots b_n / n!$.

Solution 1. For $j \geq 1$, let $(b)_j$ denote $b(b-1)\cdots(b-j+1)$. For $0 \leq j \leq n$, differentiating the identity j times and putting $x = 1$ (or alternatively substituting $x = y + 1$ and equating coefficients) yields

$$\begin{aligned}
0 &= 1 + \sum a_i, \\
0 &= \sum a_i b_i, \\
0 &= \sum a_i (b_i)_2, \\
&\vdots \\
0 &= \sum a_i (b_i)_{n-1}, \\
(-1)^n n! f(1) &= \sum a_i (b_i)_n.
\end{aligned}$$

In other words, $A\mathbf{v} = 0$, where

$$A = \begin{pmatrix} -1 & 1 & 1 & \cdots & 1 \\ 0 & b_1 & b_2 & \cdots & b_n \\ 0 & (b_1)_2 & (b_2)_2 & \cdots & (b_n)_2 \\ \vdots & \vdots & \vdots & \ddots & \vdots \\ 0 & (b_1)_{n-1} & (b_2)_{n-1} & \cdots & (b_n)_{n-1} \\ (-1)^n n! f(1) & (b_1)_n & (b_2)_n & \cdots & (b_n)_n \end{pmatrix} \quad \text{and} \quad \mathbf{v} = \begin{pmatrix} -1 \\ a_1 \\ a_2 \\ \vdots \\ a_n \end{pmatrix}.$$

Since $\mathbf{v} \neq 0$, $\det A = 0$. Since $(b)_j$ is a monic polynomial of degree j in b with no constant term, we can add a linear combination of rows $2, 3, \ldots, k$ to row $k+1$, for $2 \leq k \leq n$, to obtain

$$A' = \begin{pmatrix} -1 & 1 & 1 & \cdots & 1 \\ 0 & b_1 & b_2 & \cdots & b_n \\ 0 & b_1^2 & b_2^2 & \cdots & b_n^2 \\ \vdots & \vdots & \vdots & \ddots & \vdots \\ 0 & b_1^{n-1} & b_2^{n-1} & \cdots & b_n^{n-1} \\ (-1)^n n! f(1) & b_1^n & b_2^n & \cdots & b_n^n \end{pmatrix}.$$

Expanding by minors along the first column yields $0 = \det A' = -\det(V') + n!f(1)\det(V)$ where

$$V = \begin{pmatrix} 1 & 1 & \cdots & 1 \\ b_1 & b_2 & \cdots & b_n \\ b_1^2 & b_2^2 & \cdots & b_n^2 \\ \vdots & \vdots & \ddots & \vdots \\ b_1^{n-1} & b_2^{n-1} & \cdots & b_n^{n-1} \end{pmatrix} \text{ and } V' = \begin{pmatrix} b_1 & b_2 & \cdots & b_n \\ b_1^2 & b_2^2 & \cdots & b_n^2 \\ \vdots & \vdots & \ddots & \vdots \\ b_1^{n-1} & b_2^{n-1} & \cdots & b_n^{n-1} \\ b_1^n & b_2^n & \cdots & b_n^n \end{pmatrix}.$$

Factoring b_i out of the ith column of V' shows that $\det(V') = b_1 b_2 \cdots b_n \det(V)$. Hence $-b_1 b_2 \cdots b_n \det(V) + n!f(1)\det(V) = 0$. Since the b_i are distinct, $\det(V) \neq 0$ (see below). Thus $f(1) = b_1 b_2 \cdots b_n / n!$. ■

Remark (The Vandermonde determinant). The matrix V is called the Vandermonde matrix. Its determinant D is a polynomial of total degree $\binom{n}{2}$ in the b_i, and D vanishes whenever $b_i = b_j$ for $i > j$, so D is divisible by $b_i - b_j$ whenever $i > j$. These $b_i - b_j$ have no common factor, so $\prod_{i>j}(b_i - b_j)$ divides D. But this product also has total degree $\binom{n}{2}$, and the coefficient of $b_2 b_3^2 \cdots b_n^{n-1}$ in D and in $\prod_{i>j}(b_i - b_j)$ both equal 1, so $D = \prod_{i>j}(b_i - b_j)$. In particular, if the b_i are distinct numbers, then $D \neq 0$. See Problem 1941/14(ii) [PutnamI, p. 17] for an extension, and 1999B5 for another application.

Solution 2. Subtract 1 from both sides, set $x = e^t$, and expand the left-hand side in a power series. Since $1 - e^t = t +$ (higher order terms), we get

$$-1 + (-1)^n f(1) t^n + \text{(higher order terms)} = \sum_{i=1}^{n} a_i e^{b_i t}. \tag{1}$$

The right-hand side $F(t)$ satisfies the linear differential equation

$$F^{(n)}(t) - (b_1 + \cdots + b_n) F^{(n-1)}(t) + \cdots + (-1)^n b_1 b_2 \cdots b_n F(t) = 0 \tag{2}$$

with characteristic polynomial $p(z) = (z - b_1)(z - b_2) \cdots (z - b_n)$. On the other hand, from the left-hand side of (1) we see that $F(0) = -1$, $F^{(i)} = 0$ for $i = 1, \ldots, n-1$, and $F^{(n)}(0) = (-1)^n f(1) n!$. Hence taking $t = 0$ in (2) yields

$$(-1)^n f(1) n! - 0 + 0 - 0 + \cdots + (-1)^n b_1 b_2 \cdots b_n (-1) = 0,$$

so $f(1) = b_1 b_2 \cdots b_n / n!$. ■

B1. (183, 3, 7, 0, 0, 0, 0, 0, 4, 2, 1, 1)

Inscribe a rectangle of base b and height h and an isosceles triangle of base b in a circle of radius one as shown. For what value of h do the rectangle and triangle have the same area?†

Answer. The only such value of h is $2/5$.

Solution. The radius OX (see Figure 3) has length equal to $h/2$ plus the altitude of the triangle, so the altitude of the triangle is $1 - h/2$. If the rectangle and triangle have the same area, then $bh = \frac{1}{2} b(1 - h/2)$. Cancel b and solve for h to get $h = 2/5$. ■

---

† The figure is omitted here, since it is almost identical to Figure 3 used in the solution.

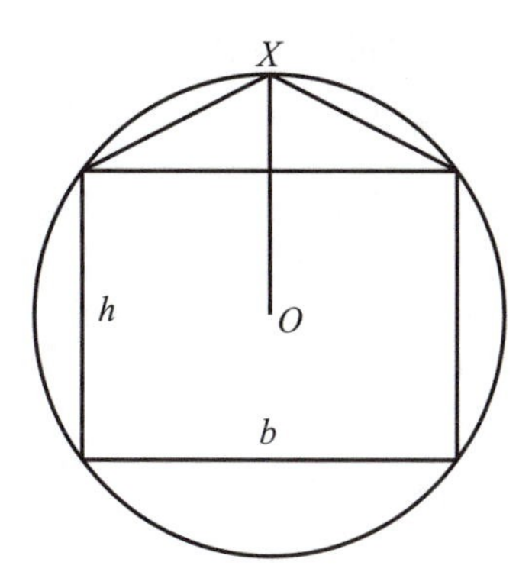

FIGURE 3.

B2. (123, 31, 16, 3, 0, 0, 0, 0, 16, 5, 2, 5)

Prove that there are only a finite number of possibilities for the ordered triple $T = (x-y, y-z, z-x)$, where x, y, and z are complex numbers satisfying the simultaneous equations

$$x(x-1) + 2yz = y(y-1) + 2zx = z(z-1) + 2xy,$$

and list all such triples T.

Answer. The possibilities for T are $(0,0,0)$, $(0,-1,1)$, $(1,0,-1)$, $(-1,1,0)$.

Solution. Subtracting $y(y-1)+2zx$ from $x(x-1)+2yz$, $z(z-1)+2xy$ from $y(y-1)+2zx$, and $x(x-1)+2yz$ from $z(z-1)+2xy$, we find that the given system is equivalent to

$$(x-y)(x+y-1-2z) = 0$$
$$(y-z)(y+z-1-2x) = 0$$
$$(z-x)(z+x-1-2y) = 0.$$

If no two of x, y, z are equal, then $x+y-1-2z = y+z-1-2x = z+x-1-2y = 0$, and adding gives $-3 = 0$. Hence at least two of x, y, z are equal. If $x = y$ and $y \neq z$, then $z = 2x+1-y = x+1$, so $T = (x-x, x-z, z-x) = (0,-1,1)$. By symmetry, the only possibilities for T are $(0,0,0)$, $(0,-1,1)$, $(1,0,-1)$, $(-1,1,0)$. Finally we give examples of (x,y,z) giving rise to each of the four possibilities, respectively: $(0,0,0)$, $(0,0,1)$, $(1,0,0)$, $(0,1,0)$. ∎

B3. (26, 5, 4, 1, 0, 1, 0, 4, 3, 5, 33, 119)

Let Γ consist of all polynomials in x with integer coefficients. For f and g in Γ and m a positive integer, let $f \equiv g \pmod{m}$ mean that every coefficient of $f-g$ is an integral multiple of m. Let n and p be positive integers with p prime. Given that f, g, h, r, and s are in Γ with $rf + sg \equiv 1 \pmod{p}$ and $fg \equiv h \pmod{p}$, prove that there exist F and G in Γ with $F \equiv f \pmod{p}$, $G \equiv g \pmod{p}$, and $FG \equiv h \pmod{p^n}$.

Solution. We prove by induction on k that there exist polynomials $F_k, G_k \in \Gamma$ such that $F_k \equiv f \pmod{p}$, $G_k \equiv g \pmod{p}$, and $F_k G_k \equiv h \pmod{p^k}$. For the base case $k = 1$, we take $F_1 = f$, $G_1 = g$.

For the inductive step, we assume the existence F_k, G_k as above, and try to construct F_{k+1}, G_{k+1}. By assumption, $h - F_k G_k = p^k t$, for some $t \in \Gamma$. We will try $F_{k+1} = F_k + p^k \Delta_1$ and $G_{k+1} = G_k + p^k \Delta_2$, where $\Delta_1, \Delta_2 \in \Gamma$ are yet to be chosen. Then $F_{k+1} \equiv F_k \equiv f \pmod{p}$, $G_{k+1} \equiv G_k \equiv g \pmod{p}$, and

$$\begin{aligned} F_{k+1}G_{k+1} &= F_k G_k + p^k(\Delta_2 F_k + \Delta_1 G_k) + p^{2k}\Delta_1\Delta_2 \\ &\equiv F_k G_k + p^k(\Delta_2 F_k + \Delta_1 G_k) \pmod{p^{k+1}}. \end{aligned}$$

If we choose $\Delta_2 = tr$ and $\Delta_1 = ts$, then

$$\Delta_2 F_k + \Delta_1 G_k \equiv trf + tsg = t(rf + sg) \equiv t \pmod{p},$$

so $p^k(\Delta_2 F_k + \Delta_1 G_k) \equiv p^k t \pmod{p^{k+1}}$, and $F_{k+1}G_{k+1} \equiv F_k G_k + p^k t = h \pmod{p^{k+1}}$, completing the inductive step. ∎

Remark. This problem is a special case of a version of Hensel's Lemma [Ei, p. 208], a fundamental result in number theory. Here is a different version [NZM, Theorem 2.23], which can be thought of as a p-adic analogue of Newton's method for solving polynomial equations via successive approximation:

Hensel's Lemma. *Suppose that $f(x)$ is a polynomial with integral coefficients. If $f(a) \equiv 0 \pmod{p^j}$ and $f'(a) \not\equiv 0 \pmod{p}$, then there is a unique $t \pmod{p}$ such that $f(a + tp^j) \equiv 0 \pmod{p^{j+1}}$.*

B4. (22, 8, 6, 6, 0, 0, 0, 0, 4, 7, 59, 89)

For a positive real number r, let $G(r)$ be the minimum value of $\left|r - \sqrt{m^2 + 2n^2}\right|$ for all integers m and n. Prove or disprove the assertion that $\lim_{r\to\infty} G(r)$ exists and equals 0.

Answer. The limit exists and equals 0.

Solution (Doug Jungreis). First,

$$0 \le |r - \sqrt{m^2 + 2n^2}| = \frac{|r^2 - m^2 - 2n^2|}{r + \sqrt{m^2 + 2n^2}} \le \frac{|r^2 - m^2 - 2n^2|}{r},$$

so it will suffice to bound the latter expression. Select the largest integer $m \ge 0$ such that $r^2 - m^2 \ge 0$. Then $m^2 \le r^2 < (m+1)^2$, so $m \le r$ and $r^2 - m^2 < 2m + 1$. Next select the largest integer $n \ge 0$ such that $r^2 - m^2 - 2n^2 \ge 0$. Then $2n^2 \le r^2 - m^2 < 2(n+1)^2$. This implies $n \le \sqrt{(r^2 - m^2)/2}$ and

$$\begin{aligned} |r^2 - m^2 - 2n^2| &= r^2 - m^2 - 2n^2 \\ &< 2(2n + 1) \\ &\le 2 + 4\sqrt{(r^2 - m^2)/2} \\ &\le 2 + \sqrt{2m + 1} \\ &\le 2 + \sqrt{2r + 1}. \end{aligned}$$

Hence $G(r) \le (2 + \sqrt{2r + 1})/r$ and $\lim_{r\to\infty} G(r) = 0$. ∎

B5. (10, 0, 0, 0, 0, 0, 0, 0, 0, 0, 58, 133)

Let $f(x,y,z) = x^2 + y^2 + z^2 + xyz$. Let $p(x,y,z), q(x,y,z), r(x,y,z)$ be polynomials with real coefficients satisfying

$$f(p(x,y,z), q(x,y,z), r(x,y,z)) = f(x,y,z).$$

Prove or disprove the assertion that the sequence p, q, r consists of some permutation of $\pm x, \pm y, \pm z$, where the number of minus signs is 0 or 2.

Solution. The assertion is false, since $(p,q,r) = (x, y, -xy-z)$ satisfies $f(p,q,r) = f(x,y,z)$. ∎

Motivation. Take $p = x$, $q = y$, and view

$$x^2 + y^2 + r^2 + xyr = x^2 + y^2 + z^2 + xyz,$$

as an equation to be solved for the polynomial r. It is equivalent to the quadratic equation

$$r^2 + (xy)r + (z^2 - xyz) = 0$$

and we already know one solution, namely $r = z$, so the quadratic is easy to factor:

$$(r - z)(r + (xy + z)) = 0.$$

Thus $r = -xy - z$ is the other solution.

Remark. We now describe the set of all solutions (p,q,r). First, there is (x,y,z). Second, choose a real number r with $|r| > 2$, factor $p^2 + rpq + q^2$ as $(p - \alpha q)(p - \beta q)$ for distinct real numbers α and β, choose $c \in \mathbb{R}^*$, and solve the system

$$\begin{aligned} p - \alpha q &= c \\ p - \beta q &= (x^2 + y^2 + z^2 + xyz - r^2)/c \end{aligned}$$

for p and q as polynomials in x, y, z to obtain a solution (p,q,r). Third, consider all triples obtainable from the two types above by iterating the following operations: permuting p, q, r, changing the signs of an even number of p, q, r, and replacing (p,q,r) by $(-qr - p, q, r)$. All such triples are solutions.

We claim that all solutions arise in this way. It suffices to show that given a solution (p,q,r), either it is of one of the first two types, or one of the operations above transforms it to another solution with $\deg p + \deg q + \deg r$ lower, where degree of a polynomial means its total degree in x, y, z, so $x^i y^j z^k$ has degree $i + j + k$.

Let (p,q,r) be a solution. We may assume $\deg p \geq \deg q \geq \deg r$. Also we may assume $\deg(p + qr) \geq \deg p$, since otherwise we perform the transformation replacing p with $-qr - p$. Since (p,q,r) is a solution,

$$p(p + qr) - (x^2 + y^2 + z^2 + xyz) = -(q^2 + r^2). \tag{1}$$

Suppose $\deg p \geq 2$. Then $\deg(p + qr) \geq \deg p \geq 2$ and $\deg p(p + qr) \geq 4$, so (1) implies the middle equality in

$$\deg(p^2) \leq \deg p(p + qr) = \deg(q^2 + r^2) \leq \deg(p^2).$$

The ends are equal, so equality holds everywhere. In particular, $\deg(p + qr) = \deg p$ and $\deg q = \deg p$. Then $\deg(qr) \leq \max\{\deg(p + qr), \deg p\} = \deg p = \deg q$, so r is a constant. The equation can be rewritten as

$$p^2 + rpq + q^2 = x^2 + y^2 + z^2 + xyz - r^2.$$

The quadratic form in p, q on the left must take on negative values, since the right-hand side is negative for x, y, and z chosen equal and sufficiently negative. Thus its discriminant $\Delta = r^2 - 4$ is positive, so $|r| > 2$. The right-hand side is irreducible in $\mathbb{R}[x, y, z]$, since as a polynomial in z it is a monic quadratic with discriminant $\Delta' = (xy)^2 - 4(x^2 + y^2 - r^2)$ which cannot be the square of any polynomial, since Δ' as a quadratic polynomial in x has nonzero x^2 and x^0 coefficients but zero x^1 coefficient. Therefore the factorization $(p - \alpha q)(p - \beta q)$ of the left-hand side matches a *trivial* factorization of the right-hand side, so (p, q, r) is a solution of the second type described above.

It remains to consider the case $\deg p < 2$. Then p, q, r are at most linear. Equating the homogeneous degree 3 parts in

$$p^2 + q^2 + r^2 + pqr = x^2 + y^2 + z^2 + xyz$$

shows that after permutation, $(p, q, r) = (a_1 x + b_1, a_2 y + b_2, a_3 z + b_3)$ where the a_i and b_i are real numbers with $a_1 a_2 a_3 = 1$. Equating coefficients of xy yields $b_3 = 0$. Similarly $b_1 = b_2 = 0$. Equating coefficients of x^2 yields $a_1 = \pm 1$. Similarly $a_2 = \pm 1$ and $a_3 = \pm 1$. Since $a_1 a_2 a_3 = 1$, (p, q, r) is obtained from (x, y, z) by an even number of sign changes.

Remark. The preceding analysis is similar to the proof that the positive integer solutions to the *Markov equation*

$$x^2 + y^2 + z^2 = 3xyz$$

are exactly those obtained from $(1, 1, 1)$ by iterations of $(x, y, z) \mapsto (x, y, 3xy - z)$ and permutations. For the connection of this equation to binary quadratic forms and continued fractions and much more, see [CF]. For an unsolved problem about the set of solutions, see [Guy, p. 166].

Related question. The key idea in this problem is also essential to the solution to Problem 6 of the 1988 International Mathematical Olympiad [IMO88, p. 38]:

> Let a and b be positive integers such that $ab + 1$ divides $a^2 + b^2$. Show that $\frac{a^2+b^2}{ab+1}$ is the square of an integer.

B6. (3, 0, 0, 0, 0, 0, 0, 0, 0, 0, 32, 166)

Suppose A, B, C, D are $n \times n$ matrices with entries in a field F, satisfying the conditions that AB^t and CD^t are symmetric and $AD^t - BC^t = I$. Here I is the $n \times n$ identity matrix, and if M is an $n \times n$ matrix, M^t is the transpose of M. Prove that $A^t D - C^t B = I$.

Solution. The conditions of the problem are

(1) $AB^t = (AB^t)^t = BA^t$,

(2) $CD^t = (CD^t)^t = DC^t$,

(3) $AD^t - BC^t = I$.

Taking the transpose of (3) gives $DA^t - CB^t = I$. These four equations are the entries in the block matrix identity

$$\begin{pmatrix} A & B \\ C & D \end{pmatrix} \begin{pmatrix} D^t & -B^t \\ -C^t & A^t \end{pmatrix} = \begin{pmatrix} I & 0 \\ 0 & I \end{pmatrix}.$$

(Here the matrices should be considered $(2n) \times (2n)$ matrices in the obvious way.) If X, Y are $m \times m$ matrices with $XY = I_m$, the $m \times m$ identity matrix, then $Y = X^{-1}$ and $YX = I_m$ too. Applying this to our product with $m = 2n$, we obtain

$$\begin{pmatrix} D^t & -B^t \\ -C^t & A^t \end{pmatrix} \begin{pmatrix} A & B \\ C & D \end{pmatrix} = \begin{pmatrix} I & 0 \\ 0 & I \end{pmatrix},$$

and equating the lower right blocks shows that $-C^tB + A^tD = I$, as desired. ■

The Forty-Eighth William Lowell Putnam Mathematical Competition
December 5, 1987

A1. (72, 24, 22, 0, 0, 0, 0, 0, 5, 0, 45, 36)
Curves A, B, C, and D are defined in the plane as follows:[†]

$$A = \left\{ (x,y) : x^2 - y^2 = \frac{x}{x^2+y^2} \right\},$$
$$B = \left\{ (x,y) : 2xy + \frac{y}{x^2+y^2} = 3 \right\},$$
$$C = \left\{ (x,y) : x^3 - 3xy^2 + 3y = 1 \right\},$$
$$D = \left\{ (x,y) : 3x^2y - 3x - y^3 = 0 \right\}.$$

Prove that $A \cap B = C \cap D$.

Solution 1. Let $z = x + iy$. The equations defining A and B are the real and imaginary parts of the equation $z^2 = z^{-1} + 3i$, and similarly the equations defining C and D are the real and imaginary parts of $z^3 - 3iz = 1$. Hence for all real x and y, we have

$$(x,y) \in A \cap B \iff z^2 = z^{-1} + 3i \iff z^3 - 3iz = 1 \iff (x,y) \in C \cap D.$$

Thus $A \cap B = C \cap D$. ∎

Solution 2. Let $F = x^2 - y^2 - \frac{x}{x^2+y^2}$, $G = 2xy + \frac{y}{x^2+y^2} - 3$, $H = x^3 - 3xy^2 + 3y - 1$, and $J = 3x^2y - 3x - y^3$ be the rational functions whose sets of zeros are A, B, C, and D, respectively. The identity

$$\begin{pmatrix} x & -y \\ y & x \end{pmatrix} \begin{pmatrix} F \\ G \end{pmatrix} = \begin{pmatrix} H \\ J \end{pmatrix} \tag{1}$$

shows immediately that $F = G = 0$ implies $H = J = 0$. Conversely if $H = J = 0$ at (x,y), then $(x,y) \neq (0,0)$, so $\det \begin{pmatrix} x & -y \\ y & x \end{pmatrix} = x^2 + y^2$ is nonzero, so (1) implies $F = G = 0$ at (x,y). ∎

Remark. Solution 1 multiplies $z^2 - z^{-1} - 3i$ by z to obtain $z^3 - 3iz - 1$. Solution 2 is simply doing this key step in terms of real and imaginary parts, so it is really the same solution, less elegantly written.

A2. (117, 28, 22, 2, 0, 0, 0, 3, 10, 5, 10, 7)
The sequence of digits

$$1234567891011121314151617181920 21\ldots$$

is obtained by writing the positive integers in order. If the 10^nth digit in this sequence occurs in the part of the sequence in which the m-digit numbers are placed, define $f(n)$ to be m. For example, $f(2) = 2$ because the

---

[†] The equations defining A and B are indeterminate at $(0,0)$. The point $(0,0)$ belongs to neither.

100th digit enters the sequence in the placement of the two-digit integer 55. Find, with proof, $f(1987)$.

Answer. The value of $f(1987)$ is 1984.

Solution. Let $g(m)$ denote the total number of digits in the integers with m or fewer digits. Then $f(n)$ equals the integer m such that $g(m-1) < 10^n \leq g(m)$.

There are $10^r - 10^{r-1}$ numbers with exactly r digits, so $g(m) = \sum_{r=1}^{m} r(10^r - 10^{r-1})$. We have

$$g(1983) \leq \sum_{r=1}^{1983} 1983(10^r - 10^{r-1}) \leq 1983 \cdot 10^{1983} < 10^{1987}$$

and

$$g(1984) \geq 1984(10^{1984} - 10^{1983}) = 1984 \cdot 9 \cdot 10^{1983} > 10^4 \cdot 10^{1983} = 10^{1987}$$

so $f(1987) = 1984$. ∎

Motivation. Based on the growth of geometric series one might guess that $g(m)$ has size roughly equal to its top term, which is $9 \cdot m \cdot 10^{m-1}$. Thus we seek m such that $9 \cdot m \cdot 10^{m-1} \approx 10^{1987}$. Using $9 \approx 10$, the condition becomes $m \approx 10^{1987-m}$. This leads to the guess $m = 1984$, since $1984 \approx 10^3$.

Remark. There is a closed form for $g(m)$:

$$\begin{aligned} g(m) &= \sum_{r=1}^{m} r10^r - \sum_{r=1}^{m} r10^{r-1} \\ &= \sum_{r=1}^{m} r10^r - \sum_{s=0}^{m-1} (s+1)10^s \\ &= \left(m10^m + \sum_{r=0}^{m-1} r10^r \right) - \sum_{s=0}^{m-1} (s+1)10^s \\ &= m10^m - \sum_{s=0}^{m-1} 10^s \\ &= m10^m - (10^m - 1)/9. \end{aligned}$$

More generally, there is a closed form for $\sum_{r=a}^{b} P(r)x^r$ for any integers $a \leq b$, fixed polynomial P, and number x. Rearrangement as above shows that

$$(1-x)\sum_{r=a}^{b} P(r)x^r = P(a)x^a - P(b)x^{b+1} + \sum_{r=a+1}^{b} (P(r) - P(r-1))x^r,$$

and the last sum is of the same type but with a polynomial $P(r) - P(r-1)$ of lower degree than P, or zero if P was constant to begin with. Hence one can evaluate the sum by induction on $\deg P$.

Alternatively, $\sum_{r=a}^{b} P(r)x^r$ can be evaluated by repeatedly differentiating the formula for the geometric series

$$\sum_{r=a}^{b} x^r = \frac{x^{b+1} - x^a}{x - 1}$$

and taking linear combinations of the resulting identities.

A3. (119, 32, 1, 0, 1, 4, 2, 0, 0, 5, 11, 29)

For all real x, the real-valued function $y = f(x)$ satisfies

$$y'' - 2y' + y = 2e^x.$$

(a) If $f(x) > 0$ for all real x, must $f'(x) > 0$ for all real x? Explain.

(b) If $f'(x) > 0$ for all real x, must $f(x) > 0$ for all real x? Explain.

Answers. (a) No. (b) Yes.

Solution. One solution to the differential equation is x^2e^x, and the general solution to the differential equation $y'' - 2y' + y = 0$ with characteristic equation $(r-1)^2$ is $(bx+c)e^x$, so the general solution to the original equation is $f(x) = (x^2+bx+c)e^x$. Then $f'(x) = \left(x^2 + (b+2)x + (b+c)\right)e^x$. Since the leading coefficient of $x^2 + bx + c$ is positive, we have

$$f(x) > 0 \text{ for all } x \iff x^2 + bx + c > 0 \text{ for all } x \iff b^2 - 4c < 0.$$

Similarly

$$f'(x) > 0 \text{ for all } x \iff (b+2)^2 - 4(b+c) < 0 \iff b^2 - 4c + 4 < 0.$$

Clearly $b^2 - 4c < 0$ does not imply $b^2 - 4c + 4 < 0$. (Take $b = 1$, $c = 1$ for instance.) But $b^2 - 4c + 4 < 0$ does imply $b^2 - 4c < 0$. ■

A4. (14, 8, 6, 4, 0, 0, 0, 1, 28, 9, 25, 109)

Let P be a polynomial, with real coefficients, in three variables and F be a function of two variables such that

$$P(ux, uy, uz) = u^2F(y-x, z-x) \quad \textbf{for all real } x, y, z, u,$$

and such that $P(1,0,0) = 4$, $P(0,1,0) = 5$, and $P(0,0,1) = 6$. Also let A, B, C be complex numbers with $P(A,B,C) = 0$ and $|B - A| = 10$. Find $|C - A|$.

Answer. The value of $|C - A|$ is $(5/3)\sqrt{30}$.

Solution. Letting $u = 1$ and $x = 0$, we have that $F(y,z) = P(0,y,z)$ is a polynomial. Also, $F(uy,uz) = P(0,uy,uz) = u^2P(0,y,z) = u^2F(y,z)$, so F is homogeneous of degree 2. Therefore

$$P(x,y,z) = F(y-x, z-x) = a(y-x)^2 + b(y-x)(z-x) + c(z-x)^2$$

for some real a, b, c. Then $4 = P(1,0,0) = a + b + c$, $5 = P(0,1,0) = a$, and $6 = P(0,0,1) = c$, so $b = -7$. Now

$$0 = P(A,B,C) = 5(B-A)^2 - 7(B-A)(C-A) + 6(C-A)^2,$$

so the number $m = (C-A)/(B-A)$ satisfies $5 - 7m + 6m^2 = 0$. The zeros of $6m^2 - 7m + 5$ are complex conjugate with product $5/6$, so $|m| = \sqrt{5/6}$. Thus $|C - A| = \sqrt{5/6}|B - A| = (5/3)\sqrt{30}$. ■

A5. (8, 0, 2, 0, 0, 0, 0, 0, 1, 1, 57, 135)

Let

$$\vec{G}(x,y) = \left(\frac{-y}{x^2+4y^2}, \frac{x}{x^2+4y^2}, 0\right).$$

Prove or disprove that there is a vector-valued function

$$\vec{F}(x,y,z) = (M(x,y,z), N(x,y,z), P(x,y,z))$$

with the following properties:

(i) M, N, P **have continuous partial derivatives for all** $(x,y,z) \neq (0,0,0)$;

(ii) Curl $\vec{F} = \vec{0}$ **for all** $(x,y,z) \neq (0,0,0)$;

(iii) $\vec{F}(x,y,0) = \vec{G}(x,y)$.

Answer. There is no such $\vec{F}$.

Solution. Let S be a surface not containing $(0,0,0)$ whose boundary ∂S is the ellipse $x^2+4y^2-4=z=0$ parameterized by $(2\cos\theta, \sin\theta, 0)$ for $0 \le \theta \le 2\pi$. (For instance, S could be the half of the ellipsoid $x^2+4y^2+z^2=4$ with $z \ge 0$.) If F exists, then

$$\begin{aligned}
0 &= \iint_S (\operatorname{Curl}\vec{F})\cdot\vec{n}\,dS \qquad (\text{since } \operatorname{Curl}\vec{F}=\vec{0} \text{ on } S)\\
&= \int_{\partial S} \vec{F}\cdot d\vec{r} \qquad (\text{by Stokes' Theorem, e.g. [Ap2, Theorem 12.3]})\\
&= \int_{\partial S} \vec{G}\cdot d\vec{r} \qquad (\text{since } \vec{F}=\vec{G} \text{ on } \partial S)\\
&= \int_0^{2\pi} \left(\frac{-\sin\theta}{4}, \frac{2\cos\theta}{4}, 0\right)\cdot(-2\sin\theta, \cos\theta, 0)\,d\theta\\
&= \int_0^{2\pi} \frac{1}{2}\,d\theta\\
&= \pi,
\end{aligned}$$

a contradiction. ∎

A6. (5, 10, 2, 0, 1, 0, 0, 3, 1, 4, 80, 98)

For each positive integer n**, let** $a(n)$ **be the number of zeros in the base 3 representation of** n**. For which positive real numbers** x **does the series**

$$\sum_{n=1}^{\infty} \frac{x^{a(n)}}{n^3}$$

converge?

Answer. For positive real x, the series converges if and only if $x < 25$.

Solution. The integer $n \ge 1$ has exactly $k+1$ digits in base 3 if and only if $3^k \le n < 3^{k+1}$. Define

$$S_k = \sum_{n=3^k}^{3^{k+1}-1} \frac{x^{a(n)}}{n^3}, \qquad \text{and} \qquad T_k = \sum_{n=3^k}^{3^{k+1}-1} x^{a(n)}.$$

The given series $\sum_{n=1}^{\infty} x^{a(n)}/n^3$ has all terms positive, so it will converge if and only if $\sum_{k=0}^{\infty} S_k$ converges. For $3^k \le n < 3^{k+1}$, we have $3^{3k} \le n^3 < 3^{3k+3}$, so $T_k/3^{3k+3} \le S_k \le T_k/3^{3k}$. Therefore $\sum_{k=0}^{\infty} S_k$ converges if and only if $\sum_{k=0}^{\infty} T_k/3^{3k}$ converges. The number of n with $k+1$ digits base 3 and satisfying $a(n) = i$ is $\binom{k}{i}2^{k+1-i}$, because there are $\binom{k}{i}$ possibilities for the set of positions of the i zero digits (since the leading digit cannot be zero), and then 2^{k+1-i} ways to select 1 or 2 as each of the remaining digits. Therefore

$$T_k = \sum_{i=0}^{k} \binom{k}{i} 2^{k+1-i} x^i = (x+2)^k.$$

Hence

$$\sum_{k=0}^{\infty} T_k/3^{3k} = \sum_{k=0}^{\infty} \left(\frac{x+2}{27}\right)^k,$$

which converges if and only if $|(x+2)/27| < 1$. For positive x, this condition is equivalent to $0 < x < 25$. ■

Remark. More generally, let $\alpha_k(n)$ be the number of zeros in the base k expansion of n, and let $A_k(x) = \sum_{n=1}^{\infty} x^{\alpha_k(n)}/n^k$. Then $A_k(x)$ converges at a positive real number x if and only if $x < k^k - k + 1$.

Literature note. For more other convergent sums involving digits in base b representations, see [BB]. This article contains exact formulas for certain sums, as well as approximations by "nice" numbers that agree to a remarkable number of decimal places.

B1. (148, 5, 4, 1, 0, 0, 0, 0, 1, 0, 15, 30)

Evaluate

$$\int_2^4 \frac{\sqrt{\ln(9-x)}\,dx}{\sqrt{\ln(9-x)} + \sqrt{\ln(x+3)}}.$$

Answer. The value of the integral is 1.

Solution. The integrand is continuous on $[2,4]$. Let I be the value of the integral. As x goes from 2 to 4, $9-x$ and $x+3$ go from 7 to 5, and from 5 to 7, respectively. This symmetry suggests the substitution $x = 6 - y$ reversing the interval $[2,4]$. After interchanging the limits of integration, this yields

$$I = \int_2^4 \frac{\sqrt{\ln(y+3)}\,dy}{\sqrt{\ln(y+3)} + \sqrt{\ln(9-y)}}.$$

Thus

$$2I = \int_2^4 \frac{\sqrt{\ln(x+3)} + \sqrt{\ln(9-x)}}{\sqrt{\ln(x+3)} + \sqrt{\ln(9-x)}}\,dx = \int_2^4 dx = 2,$$

and $I = 1$. ■

Remark. The same argument applies if $\sqrt{\ln x}$ is replaced by any continuous function such that $f(x+3) + f(9-x) \neq 0$ for $2 \le x \le 4$.

B2. (47, 3, 4, 0, 0, 0, 0, 0, 1, 3, 62, 84)

Let r, s, and t be integers with $0 \le r$, $0 \le s$, and $r+s \le t$. Prove that

$$\frac{\binom{s}{0}}{\binom{t}{r}} + \frac{\binom{s}{1}}{\binom{t}{r+1}} + \cdots + \frac{\binom{s}{s}}{\binom{t}{r+s}} = \frac{t+1}{(t+1-s)\binom{t-s}{r}}.$$

(Note: $\binom{n}{k}$ denotes the binomial coefficient $\frac{n(n-1)\cdots(n+1-k)}{k(k-1)\cdots 3\cdot 2\cdot 1}$.)

Let $F(r,s,t)$ be the left-hand side.

Solution 1. We prove

$$F(r,s,t) = \frac{t+1}{(t+1-s)\binom{t-s}{r}}$$

by induction on s. The base case $s=0$ is trivial since $\binom{0}{0}=1$.

For $s \ge 1$, $r \ge 0$ and $r+s \le t$,

$$\begin{aligned} F(r,s,t) &= \frac{\binom{s-1}{0}}{\binom{t}{r}} + \frac{\binom{s-1}{0}+\binom{s-1}{1}}{\binom{t}{r+1}} + \cdots + \frac{\binom{s-1}{s-2}+\binom{s-1}{s-1}}{\binom{t}{r+s-1}} + \frac{\binom{s-1}{s-1}}{\binom{t}{r+s}} \\ &= F(r,s-1,t) + F(r+1,s-1,t). \end{aligned}$$

Applying the inductive hypothesis to the two terms on the right gives

$$F(r,s,t) = \frac{t+1}{(t+2-s)\binom{t+1-s}{r}} + \frac{t+1}{(t+2-s)\binom{t+1-s}{r+1}}.$$

The definition of binomial coefficients in terms of factorials lets us express $\binom{t+1-s}{r}$ and $\binom{t+1-s}{r+1}$ in terms of $\binom{t-s}{r}$; this leads to

$$\begin{aligned} F(r,s,t) &= \frac{t+1}{(t+2-s)\left(\frac{t+1-s}{t+1-s-r}\right)\binom{t-s}{r}} + \frac{t+1}{(t+2-s)\left(\frac{t+1-s}{r+1}\right)\binom{t-s}{r}} \\ &= \frac{t+1}{(t+2-s)\binom{t-s}{r}}\left(\frac{t+1-s-r}{t+1-s} + \frac{r+1}{t+1-s}\right) \\ &= \frac{t+1}{(t+1-s)\binom{t-s}{r}}, \end{aligned}$$

completing the inductive step. ∎

Solution 2. Writing the binomial coefficients in terms of factorials and regrouping, we find

$$F(r,s,t) = \frac{s!\,r!\,(t-r-s)!}{t!} \sum_{i=0}^{s} \binom{r+i}{r}\binom{t-r-i}{t-r-s}. \tag{1}$$

If we could prove

$$\sum_{i=0}^{s} \binom{r+i}{r}\binom{t-r-i}{t-r-s} = \binom{t+1}{t-s+1}, \tag{2}$$

then substituting into (1) would yield

$$F(r,s,t) = \frac{s!\,r!\,(t-r-s)!}{t!} \cdot \frac{(t+1)!}{(t-s+1)!\,s!} = \frac{t+1}{(t+1-s)\binom{t-s}{r}}.$$

We now provide three proofs of (2):

Proof 1 (Vandermonde's identity). We have

$$\begin{aligned}\binom{r+i}{r} &= \binom{r+i}{i} \\ &= \frac{(r+i)(r+i-1)\cdots(r+1)}{i!} \\ &= (-1)^i \frac{(-r-1)(-r-2)\cdots(-r-i)}{i!} \\ &= (-1)^i \binom{-r-1}{i}\end{aligned}$$

and similarly

$$\binom{t-r-i}{t-r-s} = \binom{(t-r-s)+(s-i)}{s-i} = \cdots = (-1)^{s-i}\binom{-t+r+s-1}{s-i}$$

and

$$\binom{t+1}{t-s+1} = \binom{(t-s+1)+s}{s} = \cdots = (-1)^s \binom{-t+s-2}{s}.$$

Thus (2) can be rewritten as

$$\sum_{i=0}^{s} \binom{-r-1}{i}\binom{-t+r+s-1}{s-i} = \binom{-t+s-2}{s},$$

by multiplying both sides by $(-1)^s$. This is a special case of *Vandermonde's identity*, which in general states that for integers m, n, s with $s \geq 0$,

$$\sum_{i=0}^{s} \binom{m}{i}\binom{n}{s-i} = \binom{m+n}{s}.$$

Proof 2 (generating functions). The binomial expansion gives

$$(1-x)^{-(n+1)} = \sum_{i=0}^{\infty} \binom{-n-1}{i}(-x)^i = \sum_{i=0}^{\infty}\binom{n+i}{n}x^i,$$

so taking coefficients of x^s in the identity $(1-x)^{-(r+1)}(1-x)^{-(t-r-s+1)} = (1-x)^{-(t-s+2)}$ yields (2).

Proof 3 (bijective proof). We will show that the two sides of (2) count something in two different ways. First set $j = r+i$ to rewrite (2) as

$$\sum_{j=r}^{r+s} \binom{j}{r}\binom{t-j}{t-r-s} = \binom{t+1}{s}. \tag{3}$$

We will show that both sides of (3) count the number of sequences of s zeros and $t-s$ ones, punctuated by a comma such that the number of ones occurring before the comma is r. On one hand, the number of such sequences of $t+1$ symbols (including the comma) equals the right-hand side $\binom{t+1}{s}$ of (3), because they can be constructed by choosing the s positions for the zeros: of the *remaining* positions, the $(r+1)^{\text{st}}$ must contain the comma and the others must contain ones. On the other hand, we can count the sequences according to the position of the comma: given that there are exactly j digits before the comma, there are $\binom{j}{r}$ possibilities for the digits before the comma (since r of them are to be ones and the rest are to be zeros), and $\binom{t-j}{t-r-s}$ possibilities for the digits after the comma (since one needs $t-r-s$ more ones to bring the total number of ones to $t-s$). Summing over j shows that the total number of sequences is the left-hand side of (3). ■

Motivation. To find the generating function solution, first observe that the sum in (1) looks like the coefficient of x^s in a product of two series, and then figure out what the two series must be.

Remark. Vandermonde's identity can also be used in 1991B4.

Literature note. Generating functions are a powerful method for proving combinatorial identities. A comprehensive introduction to this method is [Wi].

Related question. Problem 20 of [WH] is similar:

Evaluate the sum

$$S = \sum_{k=0}^{n} \frac{\binom{n}{k}}{\binom{2n-1}{k}}$$

for all positive integers n.

(Hint: the answer does not depend on n.)

B3. (27, 31, 5, 0, 0, 0, 0, 0, 3, 11, 57, 70)

Let F be a field in which $1+1 \neq 0$. Show that the set of solutions to the equation $x^2+y^2=1$ with x and y in F is given by $(x,y)=(1,0)$ and

$$(x,y) = \left(\frac{r^2-1}{r^2+1}, \frac{2r}{r^2+1}\right),$$

where r runs through the elements of F such that $r^2 \neq -1$.

Solution 1. For $r^2 \neq -1$, let $(x_r, y_r) = \left(\frac{r^2-1}{r^2+1}, \frac{2r}{r^2+1}\right)$. Clearly $(1,0)$ and (x_r,y_r) for $r^2 \neq -1$ are solutions.

Conversely suppose $x^2+y^2=1$. If $x=1$, then $y=0$ and we have $(1,0)$. Otherwise define $r = y/(1-x)$. (The reason for this is that $y_r/(1-x_r) = r$.) Then $1-x^2 = y^2 = r^2(1-x)^2$, but $x \neq 1$, so we may divide by $1-x$ to obtain $1+x = r^2(1-x)$, and $(r^2+1)x = (r^2-1)$. If $r^2=-1$, then this says $0=-2$, contradicting $1+1 \neq 0$. Thus $r^2 \neq -1$, $x = \frac{r^2-1}{r^2+1} = x_r$, and $y = r(1-x) = \frac{2r}{r^2+1} = y_r$. Hence every solution to $x^2+y^2=1$ not equal to $(1,0)$ is of the form (x_r,y_r) for some $r \in F$ with $r^2 \neq -1$. ■

Remark. If instead F is a field in which $1+1=0$ (i.e., the characteristic of F is 2), then $x^2+y^2=1$ is equivalent to $(x+y+1)^2=0$, and the set of solutions is $\{\,(t,t+1) : t \in F\,\}$.

Solution 2. Essentially the same solution can be motivated by geometry. The only solution to $x^2+y^2=1$ with $x=1$ is $(1,0)$. For each $(x,y) \in F^2$ satisfying $x^2+y^2=1$ with $x \neq 1$, the line through (x,y) and $(1,0)$ has slope $y/(x-1)$ in F. Hence we can find all solutions to $x^2+y^2=1$ with $x \neq 1$ by intersecting each nonvertical line through $(1,0)$ with the circle. (See Figure 4.)

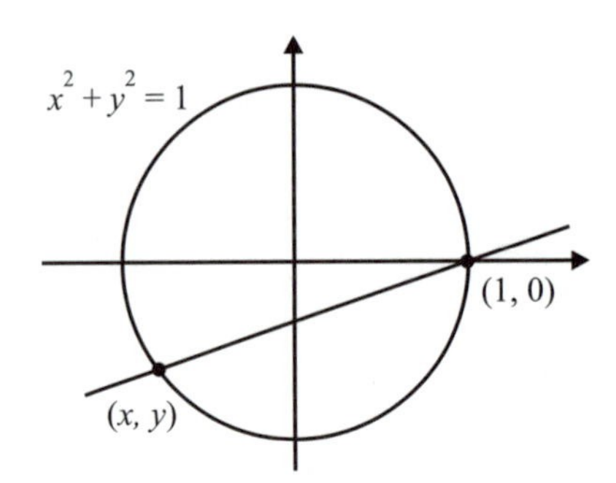

FIGURE 4.
Points with coordinates in F correspond to lines through $(1,0)$ with slopes in F.

If L_s is the line through $(1,0)$ with slope $s \in F$, its intersection with $x^2+y^2=1$ can be computed by substituting $y = s(x-1)$ into $x^2+y^2=1$, and solving the resulting equation

$$x^2 + s^2(x-1)^2 = 1.$$

This equation is guaranteed to have the solution $x = 1$, because $(1,0)$ is in the intersection. Therefore we obtain a factorization,

$$(x-1)\left((s^2+1)x + (1-s^2)\right) = 0,$$

which yields the solutions $x = 1$ and, if $s^2 \neq -1$, also $x = (s^2-1)/(s^2+1)$. (If $s^2 = -1$, then $1 - s^2 = -2 \neq 0$, and the second factor gives no solution.) Using $y = s(x-1)$, we find that these give $(1,0)$ and $\left(\frac{s^2-1}{s^2+1}, \frac{-2s}{s^2+1}\right)$, which, as we verify, do satisfy $x^2+y^2=1$ and $y = s(x-1)$. We finish by substituting $s = -r$. ∎

Remark. Let C be any nondegenerate conic over F with an F-rational point P: this means that C is given by a polynomial $f(x,y) \in F[x,y]$ of total degree 2 that does not factor into linear polynomials over any field extension of F, and $P = (a,b) \in F^2$ is a point such that $f(a,b) = 0$. The same method (of drawing all lines through P with slope in F, and seeing where they intersect $f(x,y) = 0$ other than at P) lets one parameterize the set of F-rational points of C in terms of a single parameter s. In the language of algebraic geometry, one says that any conic over k with a k-rational point is *birationally equivalent* to the line [Shaf, p. 11]. The same method works on certain equations of higher degree [NZM, Section 5.6].

The parameterization in the preceding paragraph is the reason why indefinite integrals of rational functions in t and $\sqrt{p(t)}$ for a single quadratic polynomial $p(t)$ can be expressed in terms of elementary functions [Shaf, p. 7]. It also explains why $\sin\theta$ and $\cos\theta$ can be expressed as rational functions of a single function:

$$(\cos\theta, \sin\theta) = \left(\frac{1-t^2}{1+t^2}, \frac{2t}{1+t^2}\right) \quad \text{where } t = \tan(\theta/2).$$

This, in turn, explains why rational functions in $\sin\theta$ and $\cos\theta$, such as

$$\frac{\sin^3\theta - 7\sin\theta\cos\theta}{1+\cos^3\theta},$$

have elementary antiderivatives.

Remark. The parameterization of solutions to $x^2+y^2=1$ over $\mathbb{Q}$ is closely linked to the parameterization of primitive Pythagorean triples, i.e., positive integer solutions to $a^2+b^2=c^2$ with $\gcd(a,b,c)=1$. In any primitive Pythagorean triple, exactly one of a and b is even; the set of primitive Pythagorean triples (a,b,c) with b even equals the set of triples $(m^2-n^2, 2mn, m^2+n^2)$ with m,n ranging over positive integers of opposite parity satisfying $m>n$ and $\gcd(m,n)=1$. A more number-theoretic (less geometric) approach to this classification is given in [NZM, Section 5.3].

B4. (13, 14, 15, 1, 0, 0, 0, 0, 11, 23, 79, 48)

Let $(x_1,y_1)=(0.8,0.6)$ and let $x_{n+1}=x_n\cos y_n - y_n \sin y_n$ and $y_{n+1}=x_n \sin y_n + y_n \cos y_n$ for $n=1,2,3,\ldots$. For each of $\lim_{n\to\infty} x_n$ and $\lim_{n\to\infty} y_n$, prove that the limit exists and find it or prove that the limit does not exist.

Answer. Both limits exist: $\lim_{n\to\infty} x_n=-1$ and $\lim_{n\to\infty} y_n=0$.

Solution. Since $(0.8)^2+(0.6)^2=1$, we have $(x_1,y_1)=(\cos\theta_1,\sin\theta_1)$ where $\theta_1=\cos^{-1}(0.8)$. If $(x_n,y_n)=(\cos\theta_n,\sin\theta_n)$ for some $n\geq 1$ and number θ_n, then by the trigonometric addition formulas, $(x_{n+1},y_{n+1})=(\cos(\theta_n+y_n),\sin(\theta_n+y_n))$. Hence by induction, $(x_n,y_n)=(\cos\theta_n,\sin\theta_n)$ for all $n\geq 1$, where θ_2, θ_3, ... are defined recursively by $\theta_{n+1}=\theta_n+y_n$ for $n\geq 1$. Thus $\theta_{n+1}=\theta_n+\sin\theta_n$.

For $0<\theta<\pi$, $\sin\theta>0$ and $\sin\theta=\sin(\pi-\theta)<\pi-\theta$ (see remark below for explanation), so $0<\theta+\sin\theta<\pi$. By induction, $0<\theta_n<\pi$ for all $n\geq 1$. Also $\theta_{n+1}=\theta_n+\sin\theta_n>\theta_n$, so the bounded sequence θ_1, θ_2, ... is also increasing, and hence has a limit $L\in[0,\pi]$. Since $\sin t$ is a continuous function, taking the limit as $n\to\infty$ in $\theta_{n+1}=\theta_n+\sin\theta_n$ shows that $L=L+\sin L$, so $\sin L=0$. But $L\in[0,\pi]$ and $L\geq\theta_1>0$, so $L=\pi$. By continuity of $\cos t$ and $\sin t$, $\lim_{n\to\infty} x_n=\cos L=\cos\pi=-1$ and $\lim_{n\to\infty} y_n=\sin L=\sin\pi=0$. ∎

Remark. To show that $\sin x<x$ for $x>0$, integrate $\cos t\leq 1$ from $t=0$ to $t=x$, and note that $\cos t<1$ for $t\in(0,2\pi)$.

Reinterpretation. This problem is about the limiting behavior of a dynamical system. For more examples of dynamical systems, see 1992B3, 1995B4, and 1996A6.

B5. (10, 6, 3, 2, 2, 0, 0, 0, 8, 3, 16, 154)

Let O_n be the n-dimensional vector $(0,0,\ldots,0)$. Let M be a $2n\times n$ matrix of complex numbers such that whenever $(z_1,z_2,\ldots,z_{2n})M=O_n$, with complex z_i, not all zero, then at least one of the z_i is not real. Prove that for arbitrary real numbers $r_1,r_2,\ldots,r_{2n}$, there are complex numbers $w_1,w_2,\ldots,w_n$ such that

$$\operatorname{Re}\left[M\begin{pmatrix} w_1\\ \vdots \\ w_n\end{pmatrix}\right]=\begin{pmatrix} r_1\\ \vdots \\ r_{2n}\end{pmatrix}.$$

(Note: if C is a matrix of complex numbers, $\operatorname{Re}(C)$ is the matrix whose entries are the real parts of the entries of C.)

Solution. Write $M = A + iB$ where A and B are real $2n \times n$ matrices. If $z = \begin{pmatrix} z_1 & z_2 & \cdots & z_{2n} \end{pmatrix}$ is a row vector with real entries such that $z\begin{pmatrix} A & B \end{pmatrix} = 0$, then $zM = zA + izB = 0$, so $z = 0$ by hypothesis. Hence $\begin{pmatrix} A & B \end{pmatrix}$ is an *invertible* real $2n \times 2n$ matrix.

Let r be the real column vector $(r_1, \ldots, r_{2n})$. If w is a complex column vector of length n, and we write $w = u + iv$ where u and v are the real and imaginary parts of w, then the condition $\operatorname{Re}[Mw] = r$ is equivalent to $Au - Bv = r$, and to $\begin{pmatrix} A & B \end{pmatrix}\begin{pmatrix} u \\ -v \end{pmatrix} = r$. Since $\begin{pmatrix} A & B \end{pmatrix}$ is invertible, we can find a real column vector $\begin{pmatrix} u \\ -v \end{pmatrix}$ satisfying this. ■

B6. (8, 1, 1, 1, 0, 0, 1, 2, 1, 3, 62, 124)

Let F be the field of p^2 elements where p is an odd prime. Suppose S is a set of $(p^2-1)/2$ distinct nonzero elements of F with the property that for each $a \neq 0$ in F, exactly one of a and $-a$ is in S. Let N be the number of elements in the intersection $S \cap \{\, 2a : a \in S \,\}$. Prove that N is even.

We write $2S$ for $\{\, 2a : a \in S \,\}$.

Solution 1. For $a \in S$, there is a unique way to write $2a = \epsilon_a s_a$ where $\epsilon_a = \pm 1$ and $s_a \in S$. Then $S \cap 2S = \{\, a \in S : \epsilon_a = 1 \,\}$, so $\prod_{a \in S} \epsilon_a = (-1)^{\#S - N} = (-1)^N$, since $\#S = (p-1) \cdot (p+1)/2$ is even. In F, we have

$$2^{(p^2-1)/2} \prod_{a \in S} a = \prod_{a \in S} \epsilon_a s_a = (-1)^N \prod_{a \in S} a$$

so $(-1)^N = 2^{(p^2-1)/2} = (2^{p-1})^{(p+1)/2} = 1^{(p+1)/2} = 1$, by Fermat's Little Theorem [Lar1, p. 148]. Hence N is even. ■

Remark. Gauss based one of his proofs of quadratic reciprocity on the following lemma[†] [NZM, Theorem 3.2], whose proof uses the same method as Solution 1:

Lemma. *Let p be an odd prime, and suppose that a is an integer prime to p. Consider the least positive residues modulo p of a, $2a$, $3a$, ... , $((p-1)/2)a$. If n is the number of these that exceed $p/2$, then the Legendre symbol $\left(\frac{a}{p}\right)$ equals $(-1)^n$.*

The case $a = 2$, which is especially close to Solution 1, easily implies the formula $\left(\frac{2}{p}\right) = (-1)^{(p^2-1)/8}$, see [NZM, Theorem 3.3].

Solution 2. Let $\{1, x\}$ be a basis for F over the field $\mathbb{F}_p$ of p elements. Let $H = \{1, 2, \ldots, (p-1)/2\} \subset \mathbb{F}_p$, and let

$$S_0 = \{\, a + bx : a \in H, b \in \mathbb{F}_p \text{ or } a = 0, b \in H \,\}.$$

For each nonzero $a \in F$, exactly one of a and $-a$ is in S_0. Also,

$$S_0 \cap 2S_0 = \{\, a + bx : a \in Q, b \in \mathbb{F}_p \text{ or } a = 0, b \in Q \,\}$$

[†] This lemma rarely appears outside the context of the proof of quadratic reciprocity. It should not be confused with the more important result called Gauss's Lemma that is stated after Solution 3 to 1998B6.

where $Q = H \cap 2H$, so $\#(S_0 \cap 2S_0) = (\#Q)p + (\#Q)$, which is divisible by $p + 1$, hence even.

Every other possible S can obtained by repeatedly replacing some $\alpha \in S$ by $-\alpha$, so it suffices to show that the parity of $N = \#(S \cap 2S)$ is unchanged by such an operation on S. Suppose S is as in the problem, and S' is the same as S except with α replaced by $-\alpha$. Define N' analogously. We will show that $N' - N$ is even.

Note that

(i) If $\beta \in S \cap 2S$, then $\beta \in S' \cap 2S'$ unless $\beta = \alpha$ or 2α.

(ii) If $\beta \in S' \cap 2S'$, then $\beta \in S \cap 2S$ unless $\beta = -\alpha$ or -2α.

In other words, N' can be computed from N by subtracting 1 for each of α and 2α that belongs to $S \cap 2S$, and adding 1 for each of $-\alpha$ and -2α that belongs to $S \cap 2S$. These adjustments are determined according to the four cases in the table below.

Case	In $S \cap 2S$?				$N' - N$
	α	2α	$-\alpha$	-2α	
(1) $\alpha/2 \in S$, $2\alpha \in S$	yes	yes	no	no	-2
(2) $\alpha/2 \in S$, $-2\alpha \in S$	yes	no	no	yes	0
(3) $-\alpha/2 \in S$, $2\alpha \in S$	no	yes	yes	no	0
(4) $-\alpha/2 \in S$, $-2\alpha \in S$	no	no	yes	yes	2

In each row of the table, $N' - N$ equals the number of times "yes" appears under the $-\alpha$ and -2α headers *minus* the number of times "yes" appears under the α and 2α headers. Hence $N' - N$ is even in each case, as desired. ■

Related question. The following problem, proposed by Iceland for the 1985 International Mathematical Olympiad, is susceptible to the approach of the second solution.

Suppose $x_1, \ldots, x_n \in \{-1, 1\}$, and

$$x_1x_2x_3x_4 + x_2x_3x_4x_5 + \cdots + x_{n-3}x_{n-2}x_{n-1}x_n$$
$$+ x_{n-2}x_{n-1}x_nx_1 + x_{n-1}x_nx_1x_2 + x_nx_1x_2x_3 = 0.$$

Show that n is divisible by 4.

The Forty-Ninth William Lowell Putnam Mathematical Competition December 3, 1988

A1. (202, 0, 0, 0, 0, 0, 0, 0, 4, 2, 0, 0)

Let R be the region consisting of the points (x, y) of the cartesian plane satisfying both $|x| - |y| \le 1$ and $|y| \le 1$. Sketch the region R and find its area.

Remark. For this problem, the average of the scores of the top 200 or so participants was approximately 9.76, higher than for any other problem in this volume. The second easiest problem by this measure was 1988B1, from the same year, with an average of 9.56. At the other extreme, the average for each of 1999B4 and 1999B5 was under 0.01.

Answer. The area of R is 6.

Solution. The part of R in the first quadrant is defined by the inequalities $x \ge 0$, $0 \le y \le 1$, and $x - y \le 1$. This is the trapezoid T with vertices $(0,0)$, $(1,0)$, $(2,1)$, $(0,1)$. It is the union of the unit square with vertices $(0,0)$, $(1,0)$, $(1,1)$, $(0,1)$ and the half-square (triangle) with vertices $(1,0)$, $(2,1)$, $(1,1)$, so the area of T is $3/2$. The parts of R in the other quadrant are obtained by symmetry, reflecting T in both axes (see Figure 5), so the area of R is $4(3/2) = 6$. ■

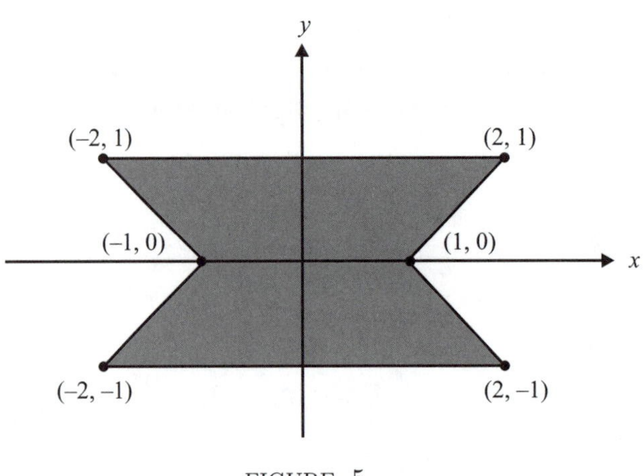

FIGURE 5.
The region R.

A2. (174, 1, 7, 1, 0, 0, 0, 3, 3, 0, 15, 4)

A not uncommon calculus mistake is to believe that the product rule for derivatives says that $(fg)' = f'g'$. If $f(x) = e^{x^2}$, determine, with proof, whether there exists an open interval (a, b) and a nonzero function g defined on (a, b) such that this wrong product rule is true for x in (a, b).

Answer. Yes, (a, b) and g exist.

Solution. We are asked for a solution g to $fg' + gf' = f'g'$, which is equivalent to $g' + \left(\frac{f'}{f-f'}\right) g = 0$ if we avoid the zeros of $f - f' = (1-2x)e^{x^2}$, i.e., avoid $x = 1/2$. By the existence and uniqueness theorem for first order linear ordinary differential equations [BD, Theorem 2.4.1], if $x_0 \neq 1/2$, and y_0 is any real number, then there exists a unique solution $g(x)$, defined in some neighborhood (a, b) of x_0, with $g(x_0) = y_0$. By taking y_0 nonzero, we obtain a nonzero solution g. ∎

Remark. One can solve the differential equation by separation of variables, to find all possible g. If g is nonzero, the differential equation is equivalent to

$$\frac{g'}{g} = \frac{f'}{f'-f} = \frac{2xe^{x^2}}{(2x-1)e^{x^2}} = 1 + \frac{1}{2x-1},$$
$$\ln|g(x)| = x + \frac{1}{2}\ln|2x-1| + c,$$

from which one finds that the nonzero solutions are of the form $g(x) = Ce^x|2x-1|^{1/2}$ for any nonzero number C, on any interval not containing $1/2$.

Related question. What other pairs of functions f and g satisfy $(fg)' = f'g'$? For some discussion, see [Hal, Problem 2G].

A3. (42, 17, 10, 0, 0, 0, 0, 6, 5, 3, 34, 91)

Determine, with proof, the set of real numbers x for which

$$\sum_{n=1}^{\infty} \left(\frac{1}{n}\csc\frac{1}{n} - 1\right)^x$$

converges.

Answer. The series converges if and only if $x > 1/2$.

Solution. Define

$$a_n = \frac{1}{n}\csc\frac{1}{n} - 1 = \frac{1}{n\sin\frac{1}{n}} - 1.$$

Taking $t = 1/n$ in the inequality $0 < \sin t < t$ for $t \in (0, \pi)$, we obtain $a_n > 0$, so each term a_n^x of the series is defined for any real x. Using $\sin t = t - t^3/3! + O(t^5)$ as $t \to 0$, we have, as $n \to \infty$,

$$\begin{aligned} a_n &= \frac{1}{n\left(\frac{1}{n} - \frac{1}{6n^3} + O\left(\frac{1}{n^5}\right)\right)} - 1 \\ &= \frac{1}{1 - \frac{1}{6n^2} + O\left(\frac{1}{n^4}\right)} - 1 \\ &= \frac{1}{6n^2} + O\left(\frac{1}{n^4}\right). \end{aligned}$$

In particular, if $b_n = 1/n^2$, then a_n^x/b_n^x has a finite limit as $n \to \infty$, so by the Limit Comparison Test [Spv, Ch. 22, Theorem 2], $\sum_{n=1}^{\infty} a_n^x$ converges if and only if $\sum_{n=1}^{\infty} b_n^x = \sum_{n=1}^{\infty} n^{-2x}$ converges, which by the Integral Comparison Test [Spv, Ch. 22, Theorem 4] holds if and only if $2x > 1$, i.e., $x > 1/2$. ∎

Remark (big-O and little-o notation). Recall that $O(g(n))$ is a stand-in for a function $f(n)$ for which there exists a constant C such that $|f(n)| \leq C|g(n)|$ for all sufficiently large n. (This does not necessarily imply that $\lim_{n\to\infty} f(n)/g(n)$ exists.)

Similarly "$f(t) = O(g(t))$ as $t \to 0$" means that there exists a constant C such that $|f(t)| \le C|g(t)|$ for sufficiently small nonzero t.

On the other hand, $o(g(n))$ is a stand-in for a function $f(n)$ such that

$$\lim_{n\to\infty} \frac{f(n)}{g(n)} = 0.$$

One can similarly define "$f(t) = o(g(t))$ as $t \to 0$".

A4. (45, 29, 2, 2, 1, 25, 11, 0, 3, 3, 25, 62)

(a) If every point of the plane is painted one of three colors, do there necessarily exist two points of the same color exactly one inch apart?

(b) What if "three" is replaced by "nine"?

Justify your answers.

Answers. (a) Yes. (b) No.

Solution.

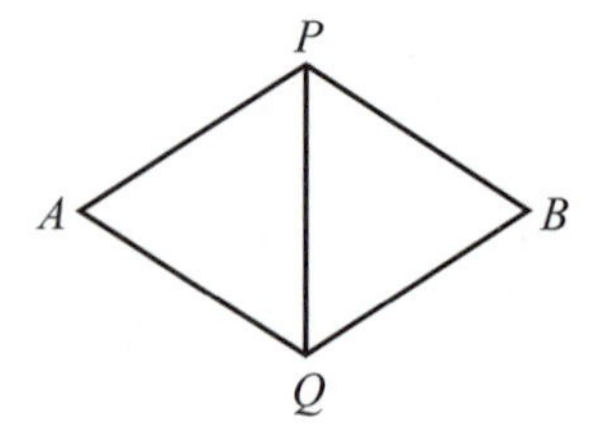

FIGURE 6.

(a) Suppose, for the sake of obtaining a contradiction, that we have a coloring of the plane with three colors such that any two points at distance 1 have different colors. If A and B are two points at distance $\sqrt{3}$, then the circles of radius 1 centered at A and B meet in two points P, Q forming equilateral triangles APQ and BPQ. (See Figure 6.) The three points of each equilateral triangle have different colors, and this forces A and B to have equal colors. Now consider a triangle CDE with $CD = CE = \sqrt{3}$ and $DE = 1$. (See Figure 7.) We know that C, D have the same color and C, E have the same color, so D, E have the same color, contradicting our hypothesis about points at distance 1.

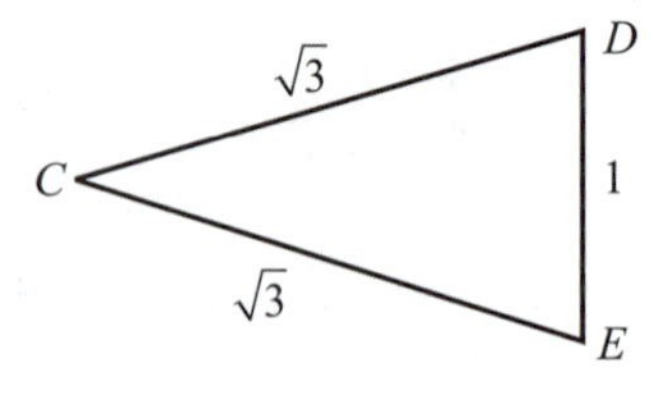

FIGURE 7.

(b) For $P = (x, y) \in \mathbb{R}^2$, define

$$f(P) = (\lfloor (3/2)x \rfloor \bmod 3, \lfloor (3/2)y \rfloor \bmod 3) \in \{0, 1, 2\}^2.$$

Color the nine elements of $\{0,1,2\}^2$ with the nine different colors, and give P the color of $f(P)$. (Thus the plane is covered by blocks of 3×3 squares, where the little squares have side length $2/3$. See Figure 8.) If P and Q are points with the same color, either they belong to the same little square, in which case $PQ \leq (2/3)\sqrt{2} < 1$, or to different little squares. In the latter case, either their x-coordinates differ by at least $4/3$, or their y-coordinates differ by at least $4/3$, so $PQ \geq 4/3 > 1$. ■

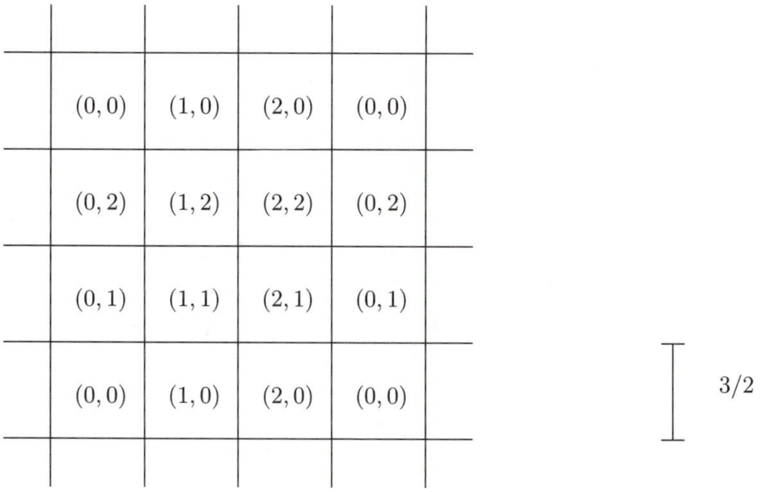

$(0,0)$	$(1,0)$	$(2,0)$	$(0,0)$
$(0,2)$	$(1,2)$	$(2,2)$	$(0,2)$
$(0,1)$	$(1,1)$	$(2,1)$	$(0,1)$
$(0,0)$	$(1,0)$	$(2,0)$	$(0,0)$

3/2

FIGURE 8.

Remark. Part (a) appears in [New, p. 7], as does the following variation:

> Assume that the points of the plane are each colored red or blue. Prove that one of these colors contains pairs of points at *every* distance.

Remark. Figure 9 shows a well-known coloring of the plane with *seven* colors such that points at distance 1 always have different colors. Such a coloring can be constructed as follows. Let $\omega = \frac{-1+\sqrt{-3}}{2}$, and view the ring of *Eisenstein integers* $\mathbb{Z}[\omega] = \{\, a + b\omega : a, b \in \mathbb{Z}\,\}$ as a lattice in the complex plane. Then $\mathfrak{p} = (2-\omega)\mathbb{Z}[\omega]$ is a sublattice of index $|2-\omega|^2 = 7$. (In fact, $\mathfrak{p}$ is one of the two prime ideals of $\mathbb{Z}[\omega]$ that divide (7).) Give each coset of $\mathfrak{p}$ in $\mathbb{Z}[\omega]$ its own color. Next color each point in $\mathbb{C}$ according to the color of a nearest point in $\mathbb{Z}[\omega]$. This gives a coloring by hexagons. The diameter of each hexagon equals $\sqrt{4/3}$, and if two distinct hexagons have the same color, the smallest distance between points of their closures is $\sqrt{7/3}$. Hence if $\sqrt{4/3} < d < \sqrt{7/3}$, no two points in the plane at distance d have the same color. Scaling the whole picture by $1/d$, we find a coloring in which no two points in the plane at distance 1 have the same color.

Remark. Let $\chi(n)$ denote the minimum number of colors needed to color the points in $\mathbb{R}^n$ so that each pair of points separated by distance 1 have different colors. Parts (a) and (b) show $\chi(2) > 3$ and $\chi(2) \leq 9$, respectively. Because of the hexagonal

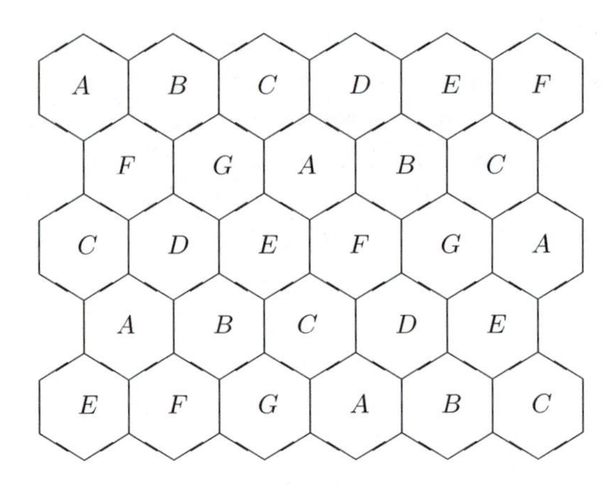

FIGURE 9.

coloring in the previous remark, we know $4 \le \chi(2) \le 7$. Ronald L. Graham is offering \$1000 for an improvement of either bound.

The study of $\chi(n)$ goes back at least to [Had] in 1944. As $n \to \infty$, one has

$$(1+o(1))(1.2)^n \le \chi(n) \le (3+o(1))^n,$$

the lower and upper bounds being due to [FW] and [LR], respectively. (See the remark after the solution to 1988A3 for the meaning of the little-o notation.) For more on such questions, consult [Gr2] and the references listed there, and browse issues of the journal *Geombinatorics*.

A5. (32, 8, 4, 0, 0, 0, 0, 1, 7, 96, 9, 51)

Prove that there exists a *unique* function f from the set $\mathbb{R}^+$ of positive real numbers to $\mathbb{R}^+$ such that

$$f(f(x)) = 6x - f(x) \qquad \textbf{and} \qquad f(x) > 0 \qquad \textbf{for all } x > 0.$$

Solution. We will show that $f(x) = 2x$ for $x > 0$. Fix $x > 0$, and consider the sequence (a_n) of positive numbers defined by $a_0 = x$ and $a_{n+1} = f(a_n)$. The given functional equation implies that $a_{n+2} + a_{n+1} - 6a_n = 0$. The zeros of the characteristic polynomial $t^2 + t - 6$ of this linear recursion are -3 and 2, so there exist real constants c_1, c_2 such that $a_n = c_1 2^n + c_2(-3)^n$ for all $n \ge 0$. If $c_2 \ne 0$, then for large n, a_n has the same sign as $c_2(-3)^n$, which alternates with n; this contradicts $a_n > 0$. Therefore $c_2 = 0$, and $a_n = c_1 2^n$. In particular $f(x) = a_1 = 2a_0 = 2x$. This holds for any x. Finally, the function $f(x) = 2x$ does satisfy the conditions of the problem. ■

Related question. The following, which was Problem 5 of the 1989 Asian-Pacific Mathematical Olympiad [APMO], can be solved using a similar method:

Determine all functions f from the reals to the reals for which

(1) $f(x)$ is strictly increasing,

(2) $f(x) + g(x) = 2x$ for all real x,

where $g(x)$ is the composition inverse function to $f(x)$. (Note: f and g are said to be composition inverses if $f(g(x)) = x$ and $g(f(x)) = x$ for all real x.)

Remark (linear recursive sequences with constant coefficients). We can describe the general sequence $x_0, x_1, x_2, \ldots$ satisfying the *linear recursion*

$$x_{n+k} + b_{k-1}x_{n+k-1} + \cdots + b_1 x_{n+1} + b_0 x_n = 0$$

for all $n \geq 0$, where b_0, b_1, ..., b_{k-1} are constants, by considering the *characteristic polynomial*

$$f(t) = t^k + b_{k-1}t^{k-1} + \cdots + b_1 t + b_0$$

over a field large enough to contain all its zeros. If the zeros r_1, ..., r_k of $f(t)$ are distinct, then the general solution is $x_n = \sum_{i=1}^k c_i r_i^n$ where the c_i are arbitrary constants not depending on n. More generally, if $f(t) = \prod_{i=1}^s (t - r_i)^{m_i}$, then, provided that we are working over a field of characteristic zero, the general solution is $x_n = \sum_{i=1}^s c_i(n) r_i^n$ where $c_i(n)$ is a polynomial of degree less than m_i. For any characteristic, the latter statement remains true if we replace the polynomial $c_i(n)$ by a general linear combination of the binomial coefficients $\binom{n}{0}$, $\binom{n}{1}$, ..., $\binom{n}{m_i - 1}$. (These combinations are the same as the polynomials in n of degree less than m_i if the characteristic is zero, or if the characteristic is at least as large as m_i.) All of these statements can be proved by showing that the generating function $\sum_{i=0}^\infty x_i t^i$ is a rational function of t, and then decomposing it as a sum of partial fractions and expanding each partial fraction in a power series to get a formula for x_i. For more discussion, see [NZM, Appendix A.4].

For example, these results can be used to show that the Fibonacci sequence, defined by $F_0 = 0$, $F_1 = 1$, and $F_{n+2} = F_{n+1} + F_n$ for $n \geq 0$, satisfies

$$F_n = \frac{\left(\frac{1+\sqrt{5}}{2}\right)^n - \left(\frac{1-\sqrt{5}}{2}\right)^n}{\sqrt{5}}.$$

The formula for the general solution to linear recursions with constant coefficients can be thought of as the discrete analogue of the general solution to homogeneous linear ordinary differential equations with constant coefficients.

A6. (59, 14, 10, 7, 0, 0, 0, 0, 5, 0, 21, 92)

If a linear transformation A on an n-dimensional vector space has $n + 1$ eigenvectors such that any n of them are linearly independent, does it follow that A is a scalar multiple of the identity? Prove your answer.

Answer. Yes, A must be a scalar multiple of the identity.

Solution 1. Let x_1, x_2, ..., x_{n+1} be the given eigenvectors, and let λ_1, λ_2, ..., λ_{n+1} be their eigenvalues. The set $B_i = \{x_1, \ldots, x_{i-1}, x_{i+1}, \ldots, x_{n+1}\}$ is a linearly independent set of n vectors in an n-dimensional vector space, so B_i is a basis, with respect to which A is represented by the diagonal matrix $\operatorname{diag}(\lambda_1, \ldots, \lambda_{i-1}, \lambda_{i+1}, \ldots, \lambda_{n+1})$. Thus the trace of A equals $S - \lambda_i$ where $S = \sum_{i=1}^{n+1} \lambda_i$. But the trace is independent of the basis chosen, so $S - \lambda_i = S - \lambda_j$ for all i, j. Hence all the λ_i are equal. With respect to the basis B_1, A is represented by a diagonal matrix with equal entries on the diagonal, so A is a scalar multiple of the identity. ■

Remark. One could have worked with the multiset of roots of the characteristic polynomial, instead of their sum (the trace).

Solution 2 (Lenny Ng). Let $x_1, \ldots, x_{n+1}$ be the eigenvectors of A, with eigenvalues $\lambda_1, \ldots, \lambda_{n+1}$. Since $x_1, \ldots, x_n$ are linearly independent, they span the vector space; hence

$$x_{n+1} = \sum_{i=1}^{n} \alpha_i x_i$$

for some $\alpha_1, \ldots, \alpha_n$. Multiply by λ_{n+1}, or apply A to both sides, and compare:

$$\sum_{i=1}^{n} \alpha_i \lambda_{n+1} x_i = \lambda_{n+1} x_{n+1} = \sum_{i=1}^{n} \alpha_i \lambda_i x_i.$$

Thus $\alpha_i \lambda_{n+1} = \alpha_i \lambda_i$ for all i between 1 and n. If $\alpha_i = 0$ for some i, then x_{n+1} can be expressed as a linear combination of $x_1, \ldots, x_{i-1}, x_{i+1}, \ldots, x_n$, contradicting the linear independence hypothesis. Hence $\alpha_i \neq 0$ for all i, so $\lambda_{n+1} = \lambda_i$ for all i. This implies $A = \lambda_{n+1} I$. ■

B1. (176, 25, 0, 0, 0, 0, 0, 0, 1, 1, 1, 4)

A *composite* (positive integer) is a product ab with a and b not necessarily distinct integers in $\{2, 3, 4, \ldots\}$. Show that every composite is expressible as $xy + xz + yz + 1$, with x, y, and z positive integers.

Solution. Substituting $z = 1$ yields $(x+1)(y+1)$, so to represent the composite number $n = ab$ with $a, b \geq 2$, let $(x, y, z) = (a-1, b-1, 1)$. ■

Remark. Although the problem asks only about representing composite numbers, all but finitely many prime numbers are representable too. Theorem 1.1 of [BC] proves that the only positive integers not of the form $xy + xz + yz + 1$ for integers $x, y, z > 0$ are the 19 integers 1, 2, 3, 5, 7, 11, 19, 23, 31, 43, 59, 71, 79, 103, 131, 191, 211, 331, and 463, and possibly a 20th integer greater than 10^{11}. Moreover, if the Generalized Riemann Hypothesis (GRH) is true, then the 20th integer does not exist. (See [Le] for earlier work on this problem.)

The situation is analogous to that of the *class number* 1 *problem*: for many years it was known that the squarefree integers $d > 0$ such that $\mathbb{Q}(\sqrt{-d})$ has class number 1 were

$$d = 1, 2, 3, 7, 11, 19, 43, 67, 163$$

and possibly one more; the existence of this tenth imaginary quadratic field of class number 1 was eventually ruled out: see the appendix to [Se3] for the history and the connection of this problem to integer points on modular curves.

In fact, researchers in the 19th century connected the problem of determining the positive integers representable by $xy + xz + yz + 1$ to problems about class numbers of quadratic imaginary fields, or equivalently class numbers of binary quadratic forms: [Mord1] mentions that the connection is present in comments by Liouville, in *Jour. de maths.*, series 2, tome 7, 1862, page 44, on a paper by Hermite. See also [Bel],

[Wh], and [Mord2, p. 291]. The GRH implies the nonexistence of a Siegel zero for the Dirichlet L-functions associated to these fields, and this is what is used in the proof of Theorem 1.1 of [BC].

B2. (88, 38, 30, 0, 0, 0, 0, 0, 26, 4, 18, 4)

Prove or disprove: if x and y are real numbers with $y \geq 0$ and $y(y+1) \leq (x+1)^2$, then $y(y-1) \leq x^2$.

Solution 1. If $0 \leq y \leq 1$, then $y(y-1) \leq 0 \leq x^2$ as desired, so assume $y > 1$. If $x \leq y - 1/2$ then

$$y(y-1) = y(y+1) - 2y \leq (x+1)^2 - 2y = x^2 + 2x + 1 - 2y \leq x^2.$$

If $x \geq y - 1/2$ then

$$x^2 \geq y^2 - y + 1/4 > y(y-1).$$

■

Solution 2. As in Solution 1, we may assume $y > 1$. We are given $y(y+1) \leq (x+1)^2$, so $|x+1| \geq \sqrt{y(y+1)}$ and $|x| \geq \sqrt{y(y+1)} - 1 \geq \sqrt{y(y-1)}$. (The last inequality follows from taking $a = y-1$ and $b = y$ in the inequality $\sqrt{(a+1)(b+1)} \geq \sqrt{ab} + 1$ for $a, b > 0$, which is equivalent (via squaring) to $a + b \geq 2\sqrt{ab}$, the AM-GM Inequality mentioned at the end of 1985A2.) Squaring gives the result. ■

Solution 3. As in Solution 1, we may assume $y > 1$. Let $f(y) = y^2 - y$ and $g(x) = x^2$. For $y > 1$, we are asked to prove that $f(y+1) \leq g(x+1)$ implies $f(y) \leq g(x)$, or equivalently that $f(y) > g(x)$ implies $f(y+1) > g(x+1)$.

In this paragraph we show that for any x and y with $y > 1$, the inequality $f(y) \geq g(x)$ implies $f'(y) > g'(x)$. If $y^2 - y \geq x^2$ and $y > 1$, then $(2y-1)^2 > 4y^2 - 4y \geq 4x^2 = (2x)^2$ and $2y - 1 > 0$, so $2y - 1 > |2x| \geq 2x$, i.e., $f'(y) > g'(x)$.

Now fix x and y. Let $h(t) = f(y+t) - g(x+t)$. Given $h(0) > 0$, and that $h(t) > 0$ implies $h'(t) > 0$, we must show that $h(1) > 0$. If $h(1) \leq 0$, then by compactness there exists a smallest $u \in [0,1]$ such that $h(u) \leq 0$. For $t \in (0, u)$, $h(t) > 0$, so $h'(t) > 0$. But $h(0) > 0 \geq h(u)$, so h cannot be increasing on $[0, u]$. This contradiction shows $h(1) > 0$, i.e., $f(y+1) > g(x+1)$. ■

Remark. The problem is asking us to decide the truth of the *first order sentence*

$$\forall x \forall y((0 \leq y) \wedge (y(y+1) \leq (x+1)(x+1))) \implies (y(y-1) \leq x \cdot x)$$

in the language $(\mathbb{R}, 0, 1, +, -, \cdot, \leq)$. Roughly, a first order sentence in this language is an expression such as the above, involving logical operations $\wedge$ ("and"), $\vee$ ("or"), $\neg$ ("not"); binary operations $+, -, \cdot$; the relations $=, \leq$; variables x, y, ... bound by quantifiers $\exists$ ("there exists") and $\forall$ ("for all"); and parentheses. For precise definitions, see Chapter II of [EFT].

Tarski [Tar] proved that the first order theory of the field $\mathbb{R}$ is decidable; this means that there exists a deterministic algorithm (i.e., Turing machine, computer program) that takes as input any first order sentence and outputs YES or NO according to whether it is true over the real numbers or not. On the other hand, the first order theory of $\mathbb{Z}$ is undecidable by the work of Gödel [Göd], and J. Robinson [Robi] combined Gödel's result and Hasse's work on quadratic forms to prove that the first order theory of $\mathbb{Q}$ also is undecidable. Moreover, it is known that there is no algorithm for deciding the truth of first order sentences not involving $\forall$ or $\neg$: this was Matiyasevich's negative solution of Hilbert's Tenth Problem [Mat]. The analogous question for $\mathbb{Q}$ is still open. See [PZ] for more.

B3. (20, 16, 17, 2, 0, 0, 0, 0, 52, 33, 27, 41)

For every n in the set $\mathbb{Z}^+ = \{1, 2, \dots\}$ of positive integers, let r_n be the minimum value of $|c - d\sqrt{3}|$ for all nonnegative integers c and d with $c + d = n$. Find, with proof, the smallest positive real number g with $r_n \le g$ for all $n \in \mathbb{Z}^+$.

Answer. The smallest such g is $(1 + \sqrt{3})/2$.

Solution. Let $g = (1 + \sqrt{3})/2$. For each fixed n, the sequence n, $(n-1) - \sqrt{3}$, $(n-2) - 2\sqrt{3}$, ... , $-n\sqrt{3}$ is an arithmetic sequence with common difference $-2g$ and with terms on both sides of 0, so there exists a unique term x_n in it with $-g \le x_n < g$. Then $r_n = |x_n| \le g$.

For $x \in \mathbb{R}$, let "$x \bmod 1$" denote $x - \lfloor x \rfloor \in [0, 1)$. Since $\sqrt{3}$ is irrational, $\{(-d\sqrt{3}) \bmod 1 : d \in \mathbb{Z}^+\}$ is dense in $[0, 1)$. (See the remark below.) Hence for any $\epsilon > 0$, we can find a positive integer d such that $((-d\sqrt{3}) \bmod 1) \in (g - 1 - \epsilon, g - 1)$. Then $c - d\sqrt{3} \in (g - \epsilon, g)$ for some integer $c \ge 0$. Let $n = c + d$. Then $r_n = x_n = c - d\sqrt{3} > g - \epsilon$ by the uniqueness of x_n above. Thus g cannot be lowered. ■

Remark. Let α be irrational, and for $n \ge 1$, let $a_n = (n\alpha \bmod 1) \in [0, 1)$. Let us explain why $\{a_n : n = 1, 2, \dots\}$ is dense in $[0, 1]$ when α is irrational. Given a large integer $N > 0$, the Pigeonhole Principle [Lar1, Ch. 2.6] produces two integers $p, q \in \{1, 2, \dots, N+1\}$ such that a_p and a_q fall into the same subinterval $[i/N, (i+1)/N)$ for some $0 \le i \le N - 1$. Assume $q > p$. Then $(q - p)\alpha$ is congruent modulo 1 to a real number r with $|r| < 1/N$. Since α is irrational, $r \ne 0$. The multiples of $(q - p)\alpha$, taken modulo 1, will then pass within $1/N$ of any number in $[0, 1]$. This argument applies for any N, so any nonempty open subset of $[0, 1]$ will contain some a_n.

In fact, one can prove more, namely that the sequence $a_1, a_2, \dots$ is *equidistributed* in $[0, 1]$: this means that for each subinterval $[a, b] \subseteq [0, 1]$,

$$\lim_{M \to \infty} \frac{\#\{n : 1 \le n \le M \text{ and } a_n \in [a, b]\}}{M} = b - a.$$

One way to show this is to observe that the range $\{1, \dots, M\}$ can, up to an error of $o(M)$ terms if M is much larger than $N(q-p)$, be partitioned into N-term arithmetic sequences of the shape $c, c+(q-p), \dots, c+(N-1)(q-p)$ (notation as in the previous paragraph), and the a_n for n in this sequence will be approximately evenly spaced over $[0, 1]$ when N is large.

Equidistribution can also be deduced from Weyl's Equidistribution Theorem [Kör, Theorem 3.1′], which states that a sequence a_1, a_2, ... of real numbers is equidistributed modulo 1 if and only if

$$\lim_{n \to \infty} \frac{1}{n} \sum_{k=1}^{n} \chi_m(a_k) = 0 \tag{1}$$

for all nonzero integers m, where $\chi_m(x) = e^{2\pi imx}$. In our application, if we set $\omega = e^{2\pi i\sqrt{3}}$, then the limit in (1) is

$$\lim_{n\to\infty} \frac{1}{n}\sum_{k=1}^{n} \omega^{km} = \lim_{n\to\infty} \frac{1}{n}\left(\frac{1-\omega^{(n+1)m}}{1-\omega^m}\right) = 0,$$

since $|1-\omega^{(n+1)m}| \leq 2$, while $1-\omega^m \neq 0$, since $\sqrt{3}$ is irrational. (See Solution 4 to 1990A2 for another application of the density result, and see Solution 2 to 1995B6 for an application of the equidistribution of ($n\alpha$ mod 1) for irrational α. See 1995B6 also for a multidimensional generalization of the equidistribution result.)

B4. (17, 1, 0, 0, 0, 0, 0, 0, 3, 2, 73, 112)

Prove that if $\sum_{n=1}^{\infty} a_n$ is a convergent series of positive real numbers, then so is $\sum_{n=1}^{\infty}(a_n)^{n/(n+1)}$.

Solution. If $a_n \geq 1/2^{n+1}$, then

$$a_n^{n/(n+1)} = \frac{a_n}{a_n^{1/(n+1)}} \leq 2a_n.$$

If $a_n \leq 1/2^{n+1}$, then $a_n^{n/(n+1)} \leq 1/2^n$. Hence

$$a_n^{n/(n+1)} \leq 2a_n + \frac{1}{2^n}.$$

But $\sum_{n=1}^{\infty}(2a_n + 1/2^n)$ converges, so $\sum_{n=1}^{\infty} a_n^{n/(n+1)}$ converges by the Comparison Test [Spv, Ch. 22, Theorem 1]. ∎

B5. (9, 0, 10, 1, 0, 0, 0, 0, 1, 2, 45, 140)

For positive integers n, let $\mathbf{M}_n$ be the $2n+1$ by $2n+1$ skew-symmetric matrix for which each entry in the first n subdiagonals below the main diagonal is 1 and each of the remaining entries below the main diagonal is -1. Find, with proof, the rank of $\mathbf{M}_n$. (According to one definition, the rank of a matrix is the largest k such that there is a $k \times k$ submatrix with nonzero determinant.)

One may note that

$$\mathbf{M}_1 = \begin{pmatrix} 0 & -1 & 1 \\ 1 & 0 & -1 \\ -1 & 1 & 0 \end{pmatrix} \quad \textbf{and} \quad \mathbf{M}_2 = \begin{pmatrix} 0 & -1 & -1 & 1 & 1 \\ 1 & 0 & -1 & -1 & 1 \\ 1 & 1 & 0 & -1 & -1 \\ -1 & 1 & 1 & 0 & -1 \\ -1 & -1 & 1 & 1 & 0 \end{pmatrix}.$$

Answer. The rank of $\mathbf{M}_n$ equals $2n$.

Solution 1. We use induction on n to prove that $\operatorname{rank}(\mathbf{M}_n) = 2n$. We check the $n = 1$ case by Gaussian elimination.

Suppose $n \geq 2$, and that $\operatorname{rank}(\mathbf{M}_{n-1}) = 2(n-1)$ is known. Adding multiples of the first two rows of $\mathbf{M}_n$ to the other rows transforms $\mathbf{M}_n$ to a matrix of the form

$$\begin{pmatrix} \begin{matrix} 0 & -1 \\ 1 & 0 \end{matrix} & * \\ \mathbf{0} & -\mathbf{M}_{n-1} \end{pmatrix}$$

in which $\mathbf{0}$ and $*$ represent blocks of size $(2n-1) \times 2$ and $2 \times (2n-1)$, respectively. Thus $\operatorname{rank}(\mathbf{M}_n) = 2 + \operatorname{rank}(\mathbf{M}_{n-1}) = 2 + 2(n-1) = 2n$. ■

Solution 2. Let $e_1, \dots, e_{2n+1}$ be the standard basis of $\mathbb{R}^{2n+1}$, and let $v_1, \dots, v_{2n+1}$ be the rows of $\mathbf{M}_n$. Let $V = \{(a_1, \dots, a_{2n+1}) \in \mathbb{R}^{2n+1} : \sum a_i = 0\}$. We will show that the row space $\operatorname{RS}(\mathbf{M}_n)$ equals V.

We first verify the following:

(a) For all m, $v_m \in V$.

(b) The set $\{e_m - e_{m-1} : 2 \leq m \leq 2n+1\}$ is a basis of V.

(c) For $2 \leq m \leq 2n+1$, the vector $e_m - e_{m-1}$ is a linear combination of the v_i.

Proof of (a): $v_m = \sum_{i=1}^{n}(e_{m-i} - e_{m+i})$.
(All subscripts are to be considered modulo $2n+1$.)

Proof of (b): The $(2n) \times (2n+1)$ matrix with the $e_m - e_{m-1}$ as rows is in row echelon form (with nonzero rows).

Proof of (c): By the formula in the proof of (a),

$$\begin{aligned} v_m + v_{m+n} &= \sum_{i=1}^{n}(e_{m-i} - e_{m+i}) + \sum_{i=1}^{n}(e_{m+n-i} - e_{m+n+i}) \\ &= \sum_{i=1}^{n}(e_{m-i} + e_{m+n-i}) - \sum_{i=1}^{n}(e_{m+i} + e_{m+n+i}) \\ &= (E - e_{m+n}) - (E - e_m) \qquad (\text{where } E = \textstyle\sum e_m) \\ &= e_m - e_{m+n} \end{aligned}$$

so

$$\begin{aligned} e_m - e_{m-1} &= (e_m - e_{m+n}) + (e_{m+n} - e_{m+2n}) \\ &= (v_m + v_{m+n}) + (v_{m+n} + v_{m+2n}). \end{aligned}$$

Now, (a) implies $\operatorname{RS}(\mathbf{M}_n) \subseteq V$, and (b) and (c) imply $V \subseteq \operatorname{RS}(\mathbf{M}_n)$. Thus $\operatorname{RS}(\mathbf{M}_n) = V$. Hence $\operatorname{rank}(\mathbf{M}_n) = \dim V = 2n$. ■

Solution 3. The matrix is *circulant*, i.e., the entry m_{ij} depends only on $j-i$ modulo $2n+1$. Write $a_{j-i} = m_{ij}$, where all subscripts are considered modulo $2n+1$. (Thus a_i equals $0, -1, 1$ according as $i = 0$, $1 \leq i \leq n$, or $n+1 \leq i \leq 2n$.) For each of the $2n+1$ complex numbers ζ satisfying $\zeta^{2n+1} = 1$, let $v_\zeta = (1, \zeta, \zeta^2, \dots, \zeta^{2n})$. The v_ζ form a basis for $\mathbb{C}^{2n+1}$, since they are the columns of a Vandermonde matrix with nonzero determinant: see 1986A6. Since $\mathbf{M}_n$ is circulant, $\mathbf{M}_n v_\zeta = \lambda_\zeta v_\zeta$ where $\lambda_\zeta = a_0 + a_1\zeta + \cdots + a_{2n}\zeta^{2n}$. Thus $\{\lambda_\zeta : \zeta^{2n+1} = 1\}$ are all the eigenvalues of $\mathbf{M}_n$

with multiplicity. For our $\mathbf{M}_n$, $\lambda_1 = 0$ and for $\zeta \neq 1$,

$$\lambda_\zeta = \zeta + \cdots + \zeta^n - \zeta^{n+1} - \cdots - \zeta^{2n} = \frac{\zeta(1-\zeta^n)^2}{1-\zeta} \neq 0,$$

since $\gcd(n, 2n+1) = 1$. It follows that the image of $\mathbf{M}_n$ (as an endomorphism of $\mathbb{C}^{2n+1}$) is the span of the v_ζ with $\zeta \neq 1$, which is $2n$-dimensional, so $\operatorname{rank}(\mathbf{M}_n) = 2n$. ■

Remark. This method lets one compute the eigenvalues and eigenvectors (and hence also the determinant) of any circulant matrix. For an application of similar ideas, see 1999B5. For introductory material on circulant matrices, see [Bar, Sect. 13.2] and [Da]. Circulant matrices (and more general objects known as group determinants) played an early role in the development of representation theory for finite groups: see [Co] for a historical overview.

Solution 4. The sum of the rows of $\mathbf{M}_n$ is 0, so $\mathbf{M}_n$ is singular. (Alternatively, this follows since $\mathbf{M}_n$ is skew-symmetric of odd dimension.) Hence the rank can be at most $2n$.

To show that the rank is $2n$, we will prove that the submatrix $A = (a_{ij})$ obtained by deleting row $2n+1$ and column $2n+1$ of $\mathbf{M}_n$ has nonzero determinant. By definition, $\det A = \sum_{\pi \in S_{2n}} \operatorname{sgn}(\pi) a_{1\pi(1)} \cdots a_{(2n)\pi(2n)}$, where S_{2n} is the group of permutations of $\{1, \ldots, 2n\}$, and $\operatorname{sgn}(\pi) = \pm 1$ is the sign of the permutation π. We will prove that $\det A$ is nonzero by proving that it is an odd integer. Since a_{ij} is odd unless $i = j$, the term in the sum corresponding to π is 0 if $\pi(i) = i$ for some i, and odd otherwise. Thus $\det A \equiv f(2n) \pmod 2$, where for any integer $m \geq 1$, $f(m)$ denotes the number of permutations of $\{1, \ldots, m\}$ having no fixed points. We can compute $f(m)$ using the Inclusion-Exclusion Principle [Ros2, §5.5]: of the $m!$ permutations, $(m-1)!$ fix 1, $(m-1)!$ fix 2, and so on, but if we subtract all these, then we must add back the $(m-2)!$ permutations fixing 1 and 2 (since these have been subtracted twice), and so on for all other pairs, and then subtract $(m-3)!$ for each triple, and so on; this finally yields

$$\begin{aligned} f(m) &= m! - \binom{m}{1}(m-1)! + \binom{m}{2}(m-2)! - \cdots \\ &\quad + (-1)^{m-1}\binom{m}{m-1}1! + (-1)^m \binom{m}{m} 0! \\ &\equiv (-1)^{m-1} m + (-1)^m \pmod 2 \end{aligned}$$

so $f(2n)$ is odd, as desired. ■

Remark. Permutations without fixed points are called *derangements*. The formula for $f(m)$ can also be written as

$$f(m) = m!\left(1 - \frac{1}{1!} + \frac{1}{2!} - \frac{1}{3!} + \cdots + \frac{(-1)^m}{m!}\right),$$

which is the integer nearest to $m!/e$.

Solution 5. As in Solution 4, $\mathbf{M}_n$ is singular, and hence $\operatorname{rank}(\mathbf{M}_n) \leq 2n$. Let A be as in Solution 4. Then A^2 is equivalent modulo 2 to the $2n \times 2n$ identity matrix, so $\det A \neq 0$. Thus $\operatorname{rank}(\mathbf{M}_n) = 2n$. ■

B6. (38, 9, 4, 4, 24, 8, 0, 5, 5, 1, 17, 93)

Prove that there exist an infinite number of ordered pairs (a, b) of integers such that for every positive integer t the number $at+b$ is a triangular number if and only if t is a triangular number. (The triangular numbers are the $t_n = n(n+1)/2$ with n in $\{0, 1, 2, \dots\}$.)

Solution 1. It is easy to check that $t_{3n+1} = 9t_n + 1$, while $t_{3n} \equiv t_{3n+2} \equiv 0 \pmod{3}$ for any integer n. Hence for every positive integer t, t is a triangular number if and only if $9t + 1$ is triangular. If the nth iterate of the linear map $t \mapsto 9t + 1$ is $t \mapsto at + b$, then a chain of equivalences will show that t is a triangular number if and only if $at + b$ is triangular. We obtain infinitely many pairs of integers (a, b) in this way. ∎

Solution 2. If $t = n(n+1)/2$, then $8t+1 = (2n+1)^2$. Conversely if t is an integer such that $8t + 1$ is a square, then $8t + 1$ is the square of some odd integer $2n + 1$, and hence $t = n(n+1)/2$. Thus t is a triangular number if and only if $8t + 1$ is a square. If k is an odd integer, then $k^2 \equiv 1 \pmod{8}$, and

$$\begin{aligned} t \text{ is a triangular number} &\iff 8t+1 \text{ is a square} \\ &\iff k^2(8t+1) \text{ is a square} \\ &\iff 8\left(k^2t + \frac{k^2-1}{8}\right) + 1 \text{ is a square} \\ &\iff \left(k^2t + \frac{k^2-1}{8}\right) \text{ is a triangular number.} \end{aligned}$$

Hence we may take $(a, b) = (k^2, (k^2 - 1)/8)$ for any odd integer k. ∎

Remark. Let us call (a, b) a triangular pair if a and b are integers with the property that for positive integers t, t is a triangular number if and only if $at + b$ is a triangular number. Solution 2 showed that if k is an odd integer, then $(k^2, (k^2 - 1)/8)$ is a triangular pair. In other words, the triangular pairs are of the form $((2m + 1)^2, t_m)$, where m is any integer. We now show, conversely, that every triangular pair has this form.

Suppose that (a, b) is a triangular pair. For any integer $n \geq 0$, $at_n + b$ is triangular, so $8(at_n + b) + 1 = 4an^2 + 4an + (8b + 1)$ is a square. A polynomial $f(x) \in \mathbb{Z}[x]$ taking square integer values at all nonnegative integers must be the square of a polynomial in $\mathbb{Z}[x]$: see 1998B6 for a proof. Hence

$$4an^2 + 4an + (8b + 1) = \ell(n)^2 \tag{1}$$

for some linear polynomial $\ell(x)$. Completing the square shows that $\ell(x) = 2\sqrt{a}x + \sqrt{a}$. Since $\ell(n)^2$ is the square of an integer for any integer n, $a = \ell(0)^2 = k^2$ for some integer k, and equating constant coefficients in (1) shows that $a = 8b + 1$, so $b = (k^2 - 1)/8$. Finally, k must be odd, in order for b to be an integer.

The Fiftieth William Lowell Putnam Mathematical Competition
December 2, 1989

A1. (92, 2, 6, 7, 0, 0, 0, 9, 2, 8, 41, 32)
How many primes among the positive integers, written as usual in base 10, are such that their digits are alternating 1's and 0's, beginning and ending with 1?

Answer. There is only one such prime: 101.

Solution. Suppose that $N = 101\cdots0101$ with k ones, for some $k \geq 2$. Then

$$99N = 9999\cdots9999 = 10^{2k} - 1 = (10^k + 1)(10^k - 1).$$

If moreover N is prime, then N divides either $10^k + 1$ or $10^k - 1$, and hence one of $\frac{99}{10^k-1} = \frac{10^k+1}{N}$ and $\frac{99}{10^k+1} = \frac{10^k-1}{N}$ is an integer. For $k > 2$, $10^k - 1$ and $10^k + 1$ are both greater than 99, so we get a contradiction. Therefore $k = 2$ and $N = 101$ (which is prime). ∎

Remark. Essentially the same problem appeared on the 1979 British Mathematical Olympiad, reprinted in [Lar1, p. 123] as Problem 4.1.4:

Prove that there are no prime numbers in the infinite sequence of integers

$$10001, 100010001, 1000100010001, \ldots.$$

A2. (141, 6, 29, 0, 0, 0, 0, 0, 5, 7, 4, 7)
Evaluate $\int_0^a \int_0^b e^{\max\{b^2x^2, a^2y^2\}}\, dy\, dx$, **where** a **and** b **are positive.**

Answer. The value of the integral is $(e^{a^2b^2} - 1)/(ab)$.

Solution. Divide the rectangle into two parts by the diagonal line $ay = bx$ to obtain

$$\begin{aligned}
\int_0^a \int_0^b e^{\max\{b^2x^2, a^2y^2\}}\, dy\, dx &= \int_0^a \int_0^{bx/a} e^{b^2x^2}\, dy\, dx + \int_0^b \int_0^{ay/b} e^{a^2y^2}\, dx\, dy \\
&= \int_0^a \frac{bx}{a} e^{b^2x^2}\, dx + \int_0^b \frac{ay}{b} e^{a^2y^2}\, dy \\
&= \int_0^{a^2b^2} \frac{1}{2ab} e^u\, du + \int_0^{a^2b^2} \frac{1}{2ab} e^v\, dv \\
&= \frac{e^{a^2b^2} - 1}{ab}.
\end{aligned}$$
∎

A3. (13, 4, 4, 2, 0, 0, 0, 0, 6, 8, 56, 106)
Prove that if

$$11z^{10} + 10iz^9 + 10iz - 11 = 0,$$

then $|z| = 1$. **(Here** z **is a complex number and** $i^2 = -1$.**)**

Solution 1. Let $g(z) = 11z^{10} + 10iz^9 + 10iz - 11$. Let $p(w) = -w^{-5}g(iw) = 11w^5 + 10w^4 + 10w^{-4} + 11w^{-5}$. As θ increases from 0 to 2π, the real-valued function $p(e^{i\theta}) = 22\cos(5\theta) + 20\cos(4\theta)$ changes sign at least 10 times, since at $\theta = 2\pi k/10$

for $k = 0, 1, \dots, 9$ its value is $22(-1)^k + 20\cos(4\theta)$, which has the sign of $(-1)^k$. By the Intermediate Value Theorem [Spv, Ch. 7, Theorem 5], $p(e^{i\theta})$ has at least one zero between $\theta = 2\pi k/10$ and $\theta = 2\pi i(k+1)/10$ for $k = 0, \dots, 9$; this makes at least 10 zeros. Thus $g(z)$ also has at least 10 zeros on the circle $|z| = 1$, and these are all the zeros since $g(z)$ is of degree 10. ■

Solution 2. The equation can be rewritten as $z^9 = \frac{11-10iz}{11z+10i}$. If $z = a + bi$, then

$$|z|^9 = \left|\frac{11 - 10iz}{11z + 10i}\right| = \frac{\sqrt{11^2 + 220b + 10^2(a^2+b^2)}}{\sqrt{11^2(a^2+b^2) + 220b + 10^2}}.$$

Let $f(a, b)$ and $g(a, b)$ denote the numerator and denominator of the right-hand side. If $|z| > 1$, then $a^2 + b^2 > 1$, so $g(a, b) > f(a, b)$, making $|z^9| < 1$, a contradiction. If $|z| < 1$, then $a^2 + b^2 < 1$, so $g(a, b) < f(a, b)$, making $|z^9| > 1$, again a contradiction. Hence $|z| = 1$. ■

Solution 3. Rouché's Theorem [Ah, p. 153] states that if f and g are analytic functions on an open set containing a closed disc, and if $|g(z) - f(z)| < |f(z)|$ everywhere on the boundary of the disc, then f and g have the same number of zeros inside the disc. Let $f(z) = 10iz - 11$ and $g(z) = 11z^{10} + 10iz^9 + 10iz - 11$, and consider the discs $|z| \le \alpha$ with $\alpha \in (0, 1)$. Then

$$|g(z) - f(z)| = |11z^{10} + 10iz^9| = |z|^9|11z + 10i| < |10iz - 11| = |f(z)|$$

if $|z| < 1$, by the same calculation as in Solution 2. But f has its only zero at $11/(10i)$, outside $|z| = 1$, so g has no zeros with $|z| < \alpha$ for any $\alpha \in (0, 1)$, and hence no zeros with $|z| < 1$. Finally, $g(-1/z) = -z^{-10}g(z)$, so the nonvanishing of g for $|z| < 1$ implies the nonvanishing of g for $|z| > 1$. Therefore, if $g(z) = 0$, then $|z| = 1$. ■

A4. (54, 22, 16, 4, 0, 0, 0, 0, 4, 2, 53, 44)

If α is an irrational number, $0 < \alpha < 1$, is there a finite game with an honest coin such that the probability of one player winning the game is α? (An honest coin is one for which the probability of heads and the probability of tails are both $1/2$. A game is finite if with probability 1 it must end in a finite number of moves.)

Answer. Yes, such a game exists.

Solution. Let d_n be 0 or 1, depending on whether the nth toss yields heads or tails. Let $X = \sum_{n=1}^{\infty} d_n/2^n$. Then the distribution of X is uniform on $[0, 1]$, since for any rational number $c/2^m$ (i.e., any dyadic rational) in $[0, 1]$, the probability that $X \in [0, c/2^m]$ is exactly $c/2^m$.

Say that player 1 wins the game after N tosses, if it is guaranteed at that time that the eventual value of X will be less than α; this means that

$$\sum_{n=1}^{N} \frac{d_n}{2^n} + \sum_{n=N+1}^{\infty} \frac{1}{2^n} < \alpha.$$

Similarly, say that player 2 wins after N tosses, if it is guaranteed then that X will be greater than α.

The game will terminate if $X \neq \alpha$, which happens with probability 1; in fact it will terminate at the Nth toss or earlier if $|X - \alpha| > 1/2^N$. The probability that player 1 wins is the probability that $X \in [0, \alpha)$, which is α. ∎

Remark. The solution shows that the answer is yes for all real $\alpha \in [0, 1]$: there is no need to assume that α is irrational.

Remark. Essentially the same idea appears in [New, Problem 8]:

> Devise an experiment which uses only tosses of a fair coin, but which has success probability 1/3. Do the same for any success probability p, $0 \leq p \leq 1$.

Related question. Show that if $\alpha \in [0, 1]$ is not a dyadic rational (i.e., not a rational number with denominator equal to a power of 2), the expected number of tosses in the game in the solution equals 2.

Related question. For $\alpha \in [0, 1]$, let $f(\alpha)$ be the minimum over all games satisfying the conditions of the problem (such that player 1 wins with probability α) of the expected number of tosses in the game. (For some games, the expected number may be infinite; ignore those.) Prove that if $\alpha \in [0, 1]$ is a rational number whose denominator in lowest terms is 2^m for some $m \geq 0$, then $f(\alpha) = 2 - 1/2^{m-1}$. Prove that if α is any other real number in $[0, 1]$, then $f(\alpha) = 2$. (This is essentially [New, Problem 103].)

A5. (6, 1, 3, 0, 0, 0, 0, 1, 0, 2, 49, 137)

Let m be a positive integer and let $\mathcal{G}$ be a regular $(2m + 1)$-gon inscribed in the unit circle. Show that there is a positive constant A, independent of m, with the following property. For any point p inside $\mathcal{G}$ there are two distinct vertices v_1 and v_2 of $\mathcal{G}$ such that

$$\big|\, |p - v_1| - |p - v_2| \,\big| < \frac{1}{m} - \frac{A}{m^3}.$$

Here $|s - t|$ denotes the distance between the points s and t.

Solution 1. The greatest distance between two vertices of $\mathcal{G}$ is $w = 2\cos\left(\frac{\pi}{4m+2}\right)$, since these vertices with the center form an isosceles triangle with equal sides of length 1, with vertex angle $2\pi m/(2m + 1)$ and base angles $\pi/(4m + 2)$. (See Figure 10.) Hence for any vertices v_1 and v_2 of $\mathcal{G}$, the triangle inequality gives $||p - v_1| - |p - v_2|| \leq |v_1 - v_2| \leq w$. Thus the $2m + 1$ distances from p to the vertices lie in an interval of length at most w. Let the distances be $d_1 \leq d_2 \leq \cdots \leq d_{2m+1}$. Then $\sum_{i=1}^{2m}(d_{i+1} - d_i) = d_{2m+1} - d_1 \leq w$, so $d_{i+1} - d_i \leq w/(2m)$ for some i. It remains to show that there exists $A > 0$ independent of m such that $w/(2m) < 1/m - A/m^3$. In fact, the Taylor expansion of $\cos x$ gives

$$\frac{w}{2m} = \frac{1}{m}\left(1 - \frac{\pi^2}{2(4m+2)^2} + o(m^{-2})\right) = \frac{1}{m} - \frac{\pi^2}{32m^3} + o(m^{-3})$$

as $m \to \infty$, so any positive $A < \pi^2/32$ will work for all but finitely many m. We can shrink A to make $w/(2m) < 1/m - A/m^3$ for those finitely many m too, since $w/(2m) < 1/m$ for all m. ∎

Solution 2. We will prove an asymptotically stronger result, namely that for a regular n-gon $\mathcal{G}$ inscribed in a unit circle and for any p in the closed unit disc,

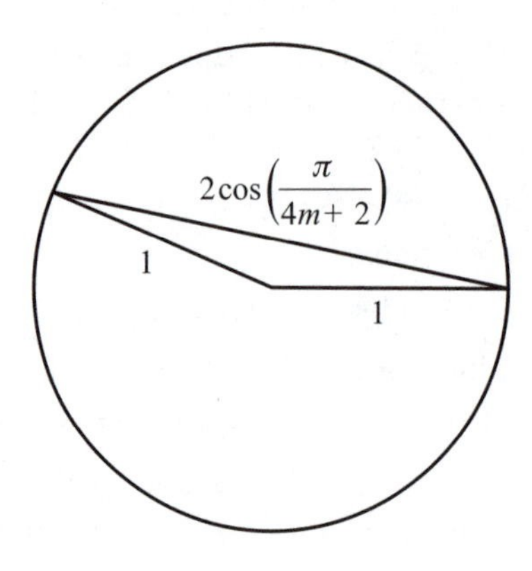

FIGURE 10.

there exist vertices v_1, v_2 of $\mathcal{G}$ such that $||p - v_1| - |p - v_2|| < \pi^2/n^2$. Center the polygon at $(0,0)$ and rotate to assume that $p = (-r, 0)$ with $0 \le r \le 1$. Let the two vertices of $\mathcal{G}$ closest to $(1,0)$ be $v_i = (\cos\theta_i, \sin\theta_i)$ for $i = 1, 2$ where $\theta_1 \le 0 \le \theta_2$ and $\theta_2 = \theta_1 + 2\pi/n$. Then

$$||p - v_1| - |p - v_2|| = |f_r(\theta_1) - f_r(\theta_2)|,$$

where

$$f_r(\theta) = |(-r, 0) - (\cos\theta, \sin\theta)| = \sqrt{r^2 + 2r\cos\theta + 1}.$$

Reflecting if necessary, we may assume $f_r(\theta_1) \ge f_r(\theta_2)$. A short calculation (for example using differentiation) shows that $f_r(\theta)$ is decreasing on $[0, \pi]$ and increasing on $[-\pi, 0]$. Thus for fixed r, $f_r(\theta_1) - f_r(\theta_2)$ is maximized when $\theta_1 = 0$ and $\theta_2 = 2\pi/n$. Next we claim that $f_r(0) - f_r(2\pi/n)$ is increasing with r, hence maximized at $r = 1$: this is because if v_2' is the point on line segment $\overline{pv_1}$ with $|p - v_2| = |p - v_2'|$, then as r increases, angle v_2pv_2' of the isosceles triangle shrinks, making angle $v_2v_2'p$ grow, putting v_2' farther from v_1, and $f_r(0) - f_r(2\pi/n) = |v_2' - v_1|$. See Figure 11. Hence

$$||p - v_1| - |p - v_2|| \le f_1(0) - f_1\left(\frac{2\pi}{n}\right) = 2 - 2\cos\left(\frac{\pi}{n}\right) < \frac{\pi^2}{n^2},$$

since $f_1(\theta) = 2\cos(\theta/2)$ for $-\pi \le \theta \le \pi$, and since the inequality $\cos x > 1 - x^2/2$ for $x \in (0, \pi/3]$ follows from the Taylor series of $\cos x$.

In order to solve the problem posed, we must deal with the case $n = 3$, i.e., $m = 1$, since the bound

$$||p - v_1| - |p - v_2|| \le 2 - 2\cos\left(\frac{\pi}{3}\right) = 1$$

is not quite good enough in that case. The proof shows, however, that equality holds for the chosen v_1 and v_2 only when p is on the circle and diametrically opposite v_1, and in this case p is equidistant from the other two vertices. Hence the minimum of $||p-v_1|-|p-v_2||$ over all choices of v_1 and v_2 is always less than 1, and by compactness there exists $A > 0$ such that it is less than $1 - A$ for all p in the disc, as desired. ■

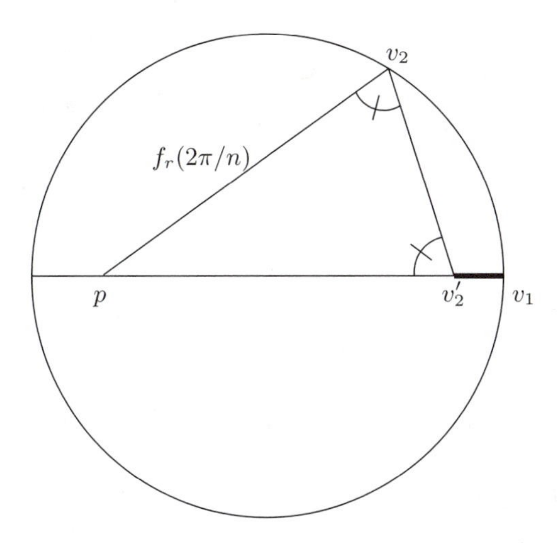

FIGURE 11.
Geometric interpretation of $f_r(0) - f_r(2\pi/n)$.

Remark. The π^2/n^2 improvement was first discovered by L. Crone and R. Holzsager, [Mon4, Solution to 10269]. (We guess that their solution was similar to ours.)

Stronger result. By considering the *three* vertices farthest from p instead of just v_1 and v_2, one can improve this to $(2/3)\pi^2/n^2$. Moreover, $(2/3)\pi^2$ cannot be replaced by any smaller constant, even if one insists that n be odd and that p be in $\mathcal{G}$.

First let us prove the improvement. As in Solution 2, assume that $p = (-r, 0)$. The vertex of $\mathcal{G}$ closest to $(1,0)$ is $q_\theta = (\cos\theta, \sin\theta)$ for some $\theta \in [-\pi/n, \pi/n]$. Reflecting if necessary, we may assume $\theta \in [0, \pi/n]$. Then $f_r(\theta - 2\pi/n)$, $f_r(\theta)$, and $f_r(\theta + 2\pi/n)$ are among the distances from p to the vertices of $\mathcal{G}$. We use a lemma that states that for fixed $\theta', \theta'' \in [-\pi, \pi]$, the function $|f_r(\theta') - f_r(\theta'')|$ of $r \in [0,1]$ is increasing (or zero if $|\theta'| = |\theta''|$): to prove this, we may assume that $0 \le \theta' < \theta'' \le \pi$ and observe that for fixed $r \in (0,1)$, the derivative

$$\frac{df_r(\theta')}{dr} = \frac{r + \cos\theta'}{\sqrt{(r+\cos\theta')^2 + \sin^2\theta'}},$$

equals the cosine of the angle $q_0 p q_{\theta'}$, whose measure increases with θ', so

$$\frac{df_r(\theta')}{dr} - \frac{df_r(\theta'')}{dr} > 0.$$

If $0 \leq \theta \leq \pi/(3n)$, then

$$\begin{aligned}\left|f_r\left(\theta - \frac{2\pi}{n}\right) - f_r\left(\theta + \frac{2\pi}{n}\right)\right| &\leq \left|f_1\left(\theta - \frac{2\pi}{n}\right) - f_1\left(\theta + \frac{2\pi}{n}\right)\right| \qquad \text{(by the lemma)}\\ &= \left|2\cos\left(\frac{\theta}{2} - \frac{\pi}{n}\right) - 2\cos\left(\frac{\theta}{2} + \frac{\pi}{n}\right)\right|\\ &= \left|4\sin\left(\frac{\theta}{2}\right)\sin\left(\frac{\pi}{n}\right)\right|\\ &< \frac{2\pi^2}{3n^2},\end{aligned}$$

since $0 < \sin x < x$ for $0 < x < \pi/2$. If instead $\pi/(3n) \leq \theta \leq \pi/n$, then

$$\begin{aligned}\left|f_r(\theta) - f_r\left(\theta - \frac{2\pi}{n}\right)\right| &\leq \left|f_1(\theta) - f_1\left(\theta - \frac{2\pi}{n}\right)\right| \qquad \text{(by the lemma)}\\ &= \left|4\sin\left(\frac{\pi}{2n} - \frac{\theta}{2}\right)\sin\left(\frac{\pi}{2n}\right)\right|\\ &< \frac{2\pi^2}{3n^2}.\end{aligned}$$

Thus there is always a pair of vertices of $\mathcal{G}$ whose distances to p differ by less than $(2/3)\pi^2/n^2$.

Now let us show that the constant $(2/3)\pi^2$ cannot be replaced by anything smaller, even if p is required to be inside $\mathcal{G}$. Without loss of generality, we may assume that the regular n-gon $\mathcal{G}$ is inscribed in the circle $x^2+y^2=1$ and has one vertex at $(\cos\theta, \sin\theta)$ where $\theta = \pi/(3n)$. Take $p = (-r, 0)$ where $r = \cos(\pi/n)$. Then r is the radius of the circle *inscribed* in $\mathcal{G}$, but since θ and $-\pi$ do not differ by an odd multiple of π/n, the point p lies strictly inside $\mathcal{G}$.

Finally we show that for this choice of p, if v_1 and v_2 are distinct vertices of $\mathcal{G}$, then

$$||p - v_1| - |p - v_2|| \geq \frac{2\pi^2}{3n^2} - O\left(\frac{1}{n^4}\right).$$

Since $f_r(\theta')$ is a decreasing function of $|\theta'|$, the distances from p to the vertices, in decreasing order, are

$$f_r(\theta),\ f_r(\theta - 2\pi/n),\ f_r(\theta + 2\pi/n),\ f_r(\theta - 4\pi/n),\ f_r(\theta + 4\pi/n),\ \ldots.$$

Since p is within $1 - \cos(\pi/n) \leq \pi^2/(2n^2)$ of $(-1, 0)$, this sequence is approximated by

$$f_1(\theta),\ f_1(\theta - 2\pi/n),\ f_1(\theta + 2\pi/n),\ f_1(\theta - 4\pi/n),\ f_1(\theta + 4\pi/n),\ \ldots,$$

with an error of at most $\pi^2/(2n^2)$ for each term. The latter sequence is

$$2\cos\left(\frac{\pi}{6n}\right),\ 2\cos\left(\frac{5\pi}{6n}\right),\ 2\cos\left(\frac{7\pi}{6n}\right),\ 2\cos\left(\frac{11\pi}{6n}\right),\ 2\cos\left(\frac{13\pi}{6n}\right),\ \ldots.$$

Hence the consecutive differences in the original sequence are within π^2/n^2 of the differences for the approximating sequence, which are of the form

$$\begin{aligned}2\cos\left(\frac{(6k-5)\pi}{6n}\right)-2\cos\left(\frac{(6k-1)\pi}{6n}\right)&=4\sin\left(\frac{(6k-3)\pi}{6n}\right)\sin\left(\frac{2\pi}{6n}\right)\\&=4\sin\left(\frac{(6k-3)\pi}{6n}\right)\left(\frac{2\pi}{6n}\right)-O\left(\frac{1}{n^3}\right)\end{aligned}$$

or of the form

$$\begin{aligned}2\cos\left(\frac{(6k-1)\pi}{6n}\right)-2\cos\left(\frac{(6k+1)\pi}{6n}\right)&=4\sin\left(\frac{6k\pi}{6n}\right)\sin\left(\frac{\pi}{6n}\right)\\&=4\sin\left(\frac{k\pi}{n}\right)\left(\frac{\pi}{6n}\right)-O\left(\frac{1}{n^3}\right),\end{aligned}$$

where $1\le k\le n/2+O(1)$. If $k\ge 3$, then bounding the sines from below by their values at $k=3$ and then applying $\sin x=x-O(x^3)$ as $x\to 0$ shows that these differences in the approximating sequence exceed $\frac{2\pi^2}{3n^2}+\frac{\pi^2}{n^2}$ so the corresponding differences in the original sequence exceed $\frac{2\pi^2}{3n^2}$. For $k<3$ (the first four differences), we forgo use of the approximating sequence and instead observe that for $r=\cos(\pi/n)$ and fixed integer j, Taylor series calculations give

$$f_r\left(\frac{j\pi}{3n}\right)=2-\frac{(18+j^2)\pi^2}{36n^2}+O\left(\frac{1}{n^4}\right),$$

which implies that the first four differences in the original sequence (between the terms corresponding to $j=1$, 5, 7, 11, 13) are at least

$$\frac{2\pi^2}{3n^2}-O\left(\frac{1}{n^4}\right).$$

Literature note. For an introduction to Taylor series, see [Spv, Ch. 23].

A6. (3, 1, 0, 0, 0, 0, 0, 1, 0, 4, 49, 141)

Let $\alpha=1+a_1x+a_2x^2+\cdots$ be a formal power series with coefficients in the field of two elements. Let

$$a_n=\begin{cases}1 & \text{if every block of zeros in the binary expansion of } n\\ & \quad\text{has an even number of zeros in the block,}\\ 0 & \text{otherwise.}\end{cases}$$

(For example, $a_{36}=1$ because $36=100100_2$, and $a_{20}=0$ because $20=10100_2$.) Prove that $\alpha^3+x\alpha+1=0$.

Solution. It suffices to prove that $\alpha^4+x\alpha^2+\alpha=0$, since the ring of formal power series over any field is an integral domain (a commutative ring with no nonzero zerodivisors), and $\alpha\ne 0$. Since $a_i^2=a_i$ for all i, and since cross terms drop out when we square in characteristic 2, we have

$$\alpha^2 = \sum_{n=0}^{\infty} a_n x^{2n}$$
$$\alpha^4 = \sum_{n=0}^{\infty} a_n x^{4n}, \text{ and}$$
$$x\alpha^2 = \sum_{n=0}^{\infty} a_n x^{2n+1}.$$

Let b_n be the coefficient of x^n in $\alpha^4 + x\alpha^2 + \alpha$. If n is odd, then the binary expansion of n is obtained from that of $(n-1)/2$ by appending a 1, $a_n = a_{(n-1)/2}$, and $b_n = a_{(n-1)/2} + a_n = 0$. If n is divisible by 2 but not 4, then $b_n = a_n = 0$, since n ends with a block of one zero. If n is divisible by 4, then the binary expansion of n is obtained from that of $n/4$ by appending two zeros, which does not change the evenness of the block lengths, so $a_{n/4} = a_n$, and $b_n = a_{n/4} + a_n = 0$. Thus $b_n = 0$ for all $n \geq 0$, as desired. ■

Remark. The fact that the coefficients of the algebraic power series α have a simple description is a special case of a much more general result of Christol. Let $\mathbb{F}_q[[x]]$ denote the ring of formal power series over the finite field $\mathbb{F}_q$ with q elements. Christol gives an automata-theoretic condition on a series $\beta \in \mathbb{F}_q[[x]]$ that holds if and only if β satisfies an algebraic equation with coefficients in $\mathbb{F}_q[x]$: see [C] and [CKMR]. For another application of Christol's result, and a problem similar to this one, see the remark following 1992A5. See Problem 1990A5 for a problem related to automata.

B1. (144, 10, 0, 0, 0, 0, 0, 0, 0, 0, 42, 3)

A dart, thrown at random, hits a square target. Assuming that any two parts of the target of equal area are equally likely to be hit, find the probability that the point hit is nearer to the center than to any edge. Express your answer in the form $(a\sqrt{b} + c)/d$, where a, b, c, d are positive integers.

Answer. The probability is $(4\sqrt{2} - 5)/3$

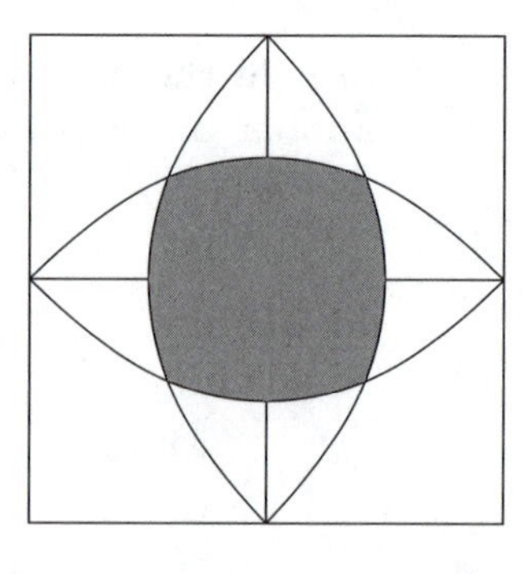

FIGURE 12.

Solution. We may assume that the dartboard has corners at $(\pm 1, \pm 1)$. A point (x, y) in the square is closer to the center than to the top edge if and only if $\sqrt{x^2+y^2} \leq 1-y$, which is equivalent to $x^2+y^2 \leq (1-y)^2$, and to $y \leq (1-x^2)/2$. This describes a region below a parabola. The region consisting of points in the board closer to the center than to any edge is the intersection of the four symmetrical parabolic regions inside the board: it is union of eight symmetric copies of the region A bounded by $x \geq 0$, $y \geq x$, $y \leq (1-x^2)/2$. (See Figure 12.) A short calculation shows that the bounding curves $y = x$ and $y = (1-x^2)/2$ intersect at $(x, y) = (\sqrt{2}-1, \sqrt{2}-1)$. Thus the desired probability is

$$\frac{8\,\text{Area}(A)}{\text{Area(board)}} = 2\,\text{Area}(A) = 2\int_0^{\sqrt{2}-1}\left(\frac{1-x^2}{2} - x\right)dx = \frac{4\sqrt{2}-5}{3}. \qquad \blacksquare$$

Related question. If a billiard table had the same shape as the region of points of the square closer to the center than to any edge, and a ball at the center were pushed in some direction not towards the corners, what would its path be?

B2. (49, 44, 17, 10, 0, 0, 0, 0, 3, 4, 41, 31)

Let S be a nonempty set with an associative operation that is left and right cancellative ($xy = xz$ implies $y = z$, and $yx = zx$ implies $y = z$). Assume that for every a in S the set $\{a^n : n = 1, 2, 3, \ldots\}$ is finite. Must S be a group?

Answer. Yes, S must be a group.

Solution. Choose $a \in S$. The finiteness hypothesis implies that some term in the sequence a, a^2, a^3, ... is repeated infinitely often, so we have $a^m = a^n$ for some integers $m, n \geq 1$ with $m - n \geq 2$. Let $e = a^{m-n}$. For any $x \in S$, $a^n ex = a^m x$, and cancelling $a^n = a^m$ shows that $ex = x$. Similarly $xe = x$, so e is an identity. Now $aa^{m-n-1} = a^{m-n} = e$ and $a^{m-n-1}a = a^{m-n} = e$, so a^{m-n-1} is an inverse of a. Since S is associative and has an identity, and since any $a \in S$ has an inverse, S is a group. $\blacksquare$

B3. (36, 4, 18, 48, 4, 10, 3, 1, 4, 1, 9, 61)

Let f be a function on $[0, \infty)$, differentiable and satisfying

$$f'(x) = -3f(x) + 6f(2x)$$

for $x > 0$. Assume that $|f(x)| \leq e^{-\sqrt{x}}$ for $x \geq 0$ (so that $f(x)$ tends rapidly to 0 as x increases). For n a nonnegative integer, define

$$\mu_n = \int_0^\infty x^n f(x)\,dx$$

(sometimes called the nth moment of f).

a. Express μ_n in terms of μ_0.

b. Prove that the sequence $\{\mu_n 3^n/n!\}$ always converges, and that the limit is 0 only if $\mu_0 = 0$.

Answer to part a. For each $n \geq 0$, $\mu_n = \frac{n!}{3^n}\left(\prod_{m=1}^n (1-2^{-m})\right)^{-1}\mu_0$.

Solution. a. As $x \to \infty$, $f(x)$ tends to 0 faster than any negative power of x, so the integral defining μ_n converges. Multiply the functional equation by x^n for some $n \geq 1$ and integrate from 0 to some finite $B > 0$:

$$\int_0^B x^n f'(x)\,dx = -3\int_0^B x^n f(x)\,dx + 6\int_0^B x^n f(2x)\,dx.$$

Using integration by parts [Spv, Ch. 18, Theorem 1] on the left, and the substitution $u = 2x$ on the last term converts this into

$$x^n f(x)\Big|_0^B - n\int_0^B f(x)x^{n-1}\,dx = -3\int_0^B x^n f(x)\,dx + \frac{6}{2^{n+1}}\int_0^{B/2} u^n f(u)\,du.$$

Taking limits as $B \to \infty$ yields

$$-n\mu_{n-1} = -3\mu_n + \frac{6}{2^{n+1}}\mu_n,$$

so

$$\mu_n = \frac{n}{3}(1-2^{-n})^{-1}\mu_{n-1}$$

for all $n \geq 1$. By induction, we obtain

$$\mu_n = \frac{n!}{3^n}\left(\prod_{m=1}^{n}(1-2^{-m})\right)^{-1}\mu_0.$$

b. Since $\sum_{m=1}^{\infty} 2^{-m}$ converges, $\prod_{m=1}^{\infty}(1-2^{-m})$ converges to some nonzero limit L, and

$$\mu_n\frac{3^n}{n!} = \left(\prod_{m=1}^{n}(1-2^{-m})\right)^{-1}\mu_0 \to L^{-1}\mu_0$$

as $n \to \infty$. This limit $L^{-1}\mu_0$ is finite, and equals zero if and only if $\mu_0 = 0$. ■

Remark. We show that nonzero functions $f(x)$ satisfying the conditions of the problem do exist. Given that $f(x)$ tends to zero rapidly as $x \to \infty$, one expects $f(2x)$ to be negligible compared to $f(x)$ for large x, and hence one can guess that $f(x)$ can be approximated by a solution to $y' = -3y$, for example e^{-3x}. But then the error in the differential equation $f'(x) = -3f(x) + 6f(2x)$ is of order e^{-6x} as $x \to \infty$, leading one to guess that $e^{-3x} + a_1e^{-6x}$ will be a better approximation, and so on, finally leading one to guess

$$f(x) = e^{-3x} + a_1e^{-6x} + a_2e^{-12x} + a_3e^{-24x} + \cdots, \tag{1}$$

where the coefficients a_i are to be solved for. Let $a_0 = 1$. If we differentiate (1) term by term, temporarily disregarding convergence issues, then for $k \geq 1$, the coefficient of $e^{-3\cdot 2^k x}$ in the expression $f'(x) + 3f(x) - 6f(2x)$ is $-3\cdot 2^k a_k + 3a_k - 6a_{k-1}$. If this is to be zero, then $a_k = \frac{-2}{2^k-1}a_{k-1}$.

All this so far has been motivation. Now we define the sequence a_0, a_1, ... by $a_0 = 1$ and $a_k = \frac{-2}{2^k-1}a_{k-1}$ for $k \geq 1$, so

$$a_k = \frac{(-2)^n}{(2-1)(2^2-1)\cdots(2^k-1)},$$

and *define* $f(x)$ by (1). The series $\sum_{k=0}^{\infty} |a_k|$ converges by the Ratio Test [Spv, Ch. 22, Theorem 3], so (1) converges to a continuous function on $[0, \infty)$. Moreover, $|a_k e^{-3\cdot 2^k x}| \le |a_k|$ for all complex x with $\mathrm{Re}(x) > 0$, so by the Weierstrass M-test and Weierstrass's Theorem [Ah, pp. 37, 176], (1) converges to a holomorphic function in this region. In particular, $f(x)$ is differentiable on $(0, \infty)$, and can be differentiated term by term, so $f'(x) = -3f(x) + 6f(2x)$ holds.

Finally, we will show that if $\epsilon > 0$ is sufficiently small, then $\epsilon f(x)$, which still satisfies the differential equation, now also satisfies $|\epsilon f(x)| \le e^{-\sqrt{x}}$. Since

$$f(x) = O(e^{-3x} + e^{-6x} + e^{-12x} + \cdots) = O(e^{-3x}) = o(e^{-\sqrt{x}})$$

as $x \to \infty$, we have $|f(x)| \le e^{-\sqrt{x}}$ for all x greater than some x_0. Let $M = \sup_{x \in [0, x_0]} \frac{|f(x)|}{e^{-\sqrt{x}}}$. If $\epsilon \in (0, 1)$ is chosen so that $\epsilon M \le 1$, then $|\epsilon f(x)| \le e^{-\sqrt{x}}$ both for $x \in [0, x_0]$ and for $x \in (x_0, \infty)$, as desired.

B4. (52, 7, 0, 0, 0, 0, 0, 0, 0, 1, 62, 77)

Can a countably infinite set have an uncountable collection of nonempty subsets such that the intersection of any two of them is finite?

Answer. Yes.

Solution 1. The set $\mathbb{Q}$ of rational numbers is countably infinite. For each real number α, choose a sequence of distinct rational numbers tending to α, and let S_α be the set of terms. If α, β are distinct real numbers, then $S_\alpha \cap S_\beta$ is finite, since otherwise a sequence obtained by listing its elements would converge to both α and β. In particular, $S_\alpha \ne S_\beta$. Thus $\{ S_\alpha : \alpha \in \mathbb{R} \}$ is an uncountable collection of nonempty subsets of $\mathbb{Q}$ with the desired property. ■

Remark. A minor variant on this solution would be to take the (countable) set of real numbers with terminating decimal expansions, and for each of the (uncountably many) irrational numbers a, let S_a be the set of decimal approximations to a obtained by truncating the decimal expansion at some point.

Solution 2. Let A denote the countably infinite set of finite strings of 0's and 1's. For each *infinite string* $a = a_1 a_2 a_3 \ldots$ of 0's and 1's, let $S_a = \{ a_1 a_2 \ldots a_n : n \ge 1 \}$ denote the set of finite initial substrings. There are uncountably many a, and if $a = a_1 a_2 \ldots$ and $b = b_1 b_2 \ldots$ are distinct infinite strings, say with $a_m \ne b_m$, then all strings in $S_a \cap S_b$ have length less than m, so $S_a \cap S_b$ is finite. ■

Solution 3. The set $\mathbb{Z}^2$ of lattice points in the plane is countably infinite. For each real α, let S_α denote the set of points in $\mathbb{Z}^2$ whose distance to the line $y = \alpha x$ is at most 1. If α, β are distinct real numbers, then $S_\alpha \cap S_\beta$ is a set of lattice points in a bounded region, so it is finite. ■

Solution 4. Let A be a disjoint union of a countably infinite number of countably infinite sets, so A is countably infinite. Call a collection of subsets $\mathcal{C}$ of A *good* if it consists of an infinite number of countably infinite subsets of A, and $S \cap T$ is finite for any distinct $S, T \in \mathcal{C}$. By construction of A, there exists a good collection (of disjoint subsets). Order the good collections by inclusion. For any chain of good collections, the union is also a good collection. Hence by Zorn's Lemma, there exists a maximal good collection $\mathcal{C}_{\max}$.

Suppose $\mathcal{C}_{\max}$ were countable, say $\mathcal{C}_{\max} = \{S_1, S_2, \dots\}$. Because S_n is infinite while $S_i \cap S_n$ is finite for $i \neq n$, there exists $b_n \in S_n - \bigcup_{i=1}^{n-1} S_i$ for each $n \geq 1$. For $m < n$, $b_m \in S_m$ but $b_n \notin S_m$, so $b_m \neq b_n$, and the set $B = \{b_1, b_2, \dots\}$ is countably infinite. Moreover, $b_N \notin S_n$ for $N > n$, so $B \cap S_n$ is finite. Hence $\{B, S_1, S_2, \dots\}$ is a good collection, contradicting the maximality of $\mathcal{C}_{\max}$. Thus $\mathcal{C}_{\max}$ is uncountable, as desired. ∎

Remark. This question appears as [New, Problem 49], where the countably infinite set is taken to be $\mathbb{Z}$.

Remark (Zorn's Lemma). A chain in a partially ordered set $(S, \leq)$ is a subset in which every two elements are comparable. Zorn's Lemma states that if S is a nonempty partially ordered set such that every chain in S has an upper bound in S, then S contains a maximal element, i.e., an element m such that the only element $s \in S$ satisfying $s \geq m$ is m itself. Zorn's Lemma is equivalent to the Axiom of Choice, which states that the product of a family of nonempty sets indexed by a nonempty set is nonempty. It is also equivalent to the Well Ordering Principle, which states that every set admits a well ordering. (A *well ordering* of a set S is a total ordering such that every nonempty subset $A \subseteq S$ has a least element.) See pages 151 and 196 of [En].

Related question. Show that the following similar question, a restatement of [Hal, Problem 11C], has a *negative* answer:

> Can a countably infinite set have an uncountable collection of nonempty subsets such that the intersection of any two of them has at most 1989 elements?

B5. (17, 17, 1, 1, 0, 0, 0, 1, 17, 29, 37, 79)

Label the vertices of a trapezoid T (quadrilateral with two parallel sides) inscribed in the unit circle as A, B, C, D so that AB is parallel to CD and A, B, C, D are in counterclockwise order. Let s_1, s_2, and d denote the lengths of the line segments AB, CD, and OE, where E is the point of intersection of the diagonals of T, and O is the center of the circle. Determine the least upper bound of $(s_1 - s_2)/d$ over all such T for which $d \neq 0$, and describe all cases, if any, in which it is attained.

Answer. The least upper bound of $(s_1 - s_2)/d$ equals 2. We have $(s_1 - s_2)/d = 2$ if and only if the diagonals BD and AC are perpendicular and $s_1 > s_2$.

Solution 1. (See Figure 13.) We may assume that AB and CD are horizontal, with AB below CD. Then by symmetry, $E = (0, e)$ for some e, and $d = |e|$. The diagonal AC has the equation $y = mx + e$ for some slope $m > 0$. Substituting $y = mx + e$ into $x^2 + y^2 = 1$ results in the quadratic polynomial

$$q(x) = x^2 + (mx + e)^2 - 1 = (m^2 + 1)x^2 + (2me)x + (e^2 - 1)$$

whose zeros are the x-coordinates of A and C, which also equal $-s_1/2$ and $s_2/2$, respectively. Hence $s_1 - s_2$ is -2 times the sum of the zeros of $q(x)$, so

$$s_1 - s_2 = -2\left(\frac{-2me}{m^2 + 1}\right). \tag{1}$$

If $d \neq 0$, then

$$\frac{s_1 - s_2}{d} = 2\left(\frac{2m}{m^2+1}\right)\operatorname{sgn}(e).$$

Now $(m-1)^2 \geq 0$ with equality if and only if $m = 1$, so $m^2 + 1 \geq 2m > 0$. Thus

$$\frac{s_1 - s_2}{d} \leq 2,$$

with equality if and only if $m = 1$ and $e > 0$, i.e., if the diagonals BD and AC are perpendicular *and* $s_1 > s_2$. (The latter is equivalent to $e > 0$ by (1).) ∎

Solution 2. (See Figure 13.) Again assume that AB and CD are horizontal. The x-coordinates of B and D are $s_1/2$ and $-s_2/2$, so their midpoint M has x-coordinate $(s_1 - s_2)/4$. Since line OM is the perpendicular bisector of $\overline{BD}$, $\angle OME$ is a right angle, and M lies on the circle with diameter $\overline{OE}$. For fixed $d = OE$, the x-coordinate $(s_1 - s_2)/4$ is maximized when M is at the rightmost point of this circle: then $(s_1 - s_2)/4 = d/2$ and $(s_1 - s_2)/d = 2$. This happens if and only if BD has slope -1 and $s_1 > s_2$, or equivalently if and only if the diagonals BD and AC are perpendicular and $s_1 > s_2$. ∎

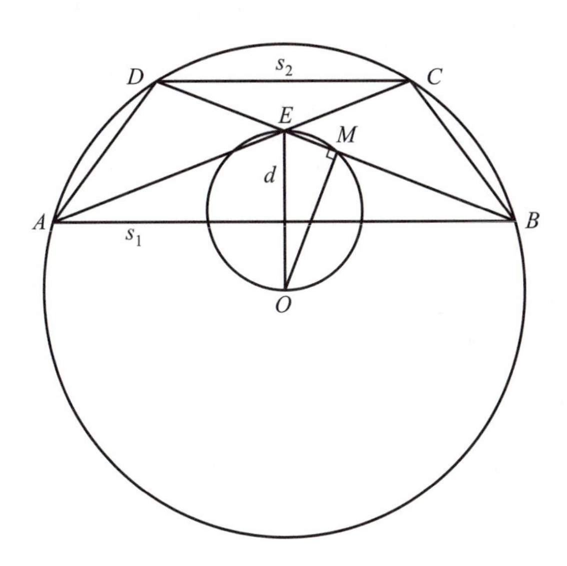

FIGURE 13.

Literature note. For further discussion of this problem, including more solutions, see [Lar2, pp. 33–38].

B6. (0, 1, 1, 1, 0, 0, 0, 2, 0, 11, 35, 148)

Let $(x_1, x_2, \ldots, x_n)$ be a point chosen at random from the n-dimensional region defined by $0 < x_1 < x_2 < \cdots < x_n < 1$. Let f be a continuous function on $[0,1]$ with $f(1) = 0$. Set $x_0 = 0$ and $x_{n+1} = 1$. Show that the expected

value of the Riemann sum

$$\sum_{i=0}^{n}(x_{i+1}-x_i)f(x_{i+1})$$

is $\int_0^1 f(t)P(t)\,dt$**, where** P **is a polynomial of degree** n**, independent of** f**, with** $0 \le P(t) \le 1$ **for** $0 \le t \le 1$**.**

Answer. We will show that the result holds with $P(t) = 1-(1-t)^n$.

Solution 1. We may drop the $i=n$ term in the Riemann sum, since $f(x_{n+1}) = f(1) = 0$. The volume of the region $0 < x_1 < \cdots < x_n < 1$ in $\mathbb{R}^n$ is

$$\int_0^1\int_0^{x_n}\cdots\int_0^{x_2} dx_1\,dx_2\,\cdots\,dx_n = 1/n!.$$

Therefore the expected value of the Riemann sum is $E = \left(\sum_{i=0}^{n-1} M_i\right)/(1/n!)$, where

$$\begin{aligned}
M_i &= \int_0^1\int_0^{x_n}\cdots\int_0^{x_2}(x_{i+1}-x_i)f(x_{i+1})\,dx_1\cdots dx_n\\
&= \int_0^1\int_0^{x_n}\cdots\int_0^{x_{i+1}}(x_{i+1}-x_i)\frac{x_i^{i-1}}{(i-1)!}f(x_{i+1})\,dx_i\,dx_{i+1}\cdots dx_n\\
&= \int_0^1\int_0^{x_n}\cdots\int_0^{x_{i+2}}\left(x_{i+1}\frac{x_{i+1}^i}{i!}-\frac{x_{i+1}^{i+1}}{(i+1)\cdot(i-1)!}\right)f(x_{i+1})\,dx_{i+1}\cdots dx_n\\
&= \int_0^1\int_0^{x_n}\cdots\int_0^{x_{i+2}}\frac{x_{i+1}^{i+1}}{(i+1)!}f(x_{i+1})\,dx_{i+1}\cdots dx_n\\
&\qquad\left(\text{since } \frac{1}{i}-\frac{1}{i+1}=\frac{1}{i(i+1)}\right)\\
&= \int_0^1\int_{x_{i+1}}^1\int_{x_{i+1}}^{x_n}\int_{x_{i+1}}^{x_{i+3}}\frac{x_{i+1}^{i+1}}{(i+1)!}f(x_{i+1})\,dx_{i+2}\cdots dx_n\,dx_{i+1}\\
&= \int_0^1\frac{(1-x_{i+1})^{n-(i+1)}}{(n-(i+1))!}\frac{x_{i+1}^{i+1}}{(i+1)!}f(x_{i+1})\,dx_{i+1}\\
&= \frac{1}{n!}\int_0^1\binom{n}{i+1}(1-x_{i+1})^{n-(i+1)}x_{i+1}^{i+1}f(x_{i+1})\,dx_{i+1}.
\end{aligned}$$

Therefore

$$\begin{aligned}
E &= \int_0^1\left(\sum_{i=0}^{n-1}\binom{n}{i+1}(1-x_{i+1})^{n-(i+1)}x_{i+1}^{i+1}\right)f(x_{i+1})\,dx_{i+1}\\
&= \int_0^1\left(\sum_{j=1}^{n}\binom{n}{j}(1-t)^{n-j}t^j\right)f(t)\,dt \qquad (\text{we set } j=i+1 \text{ and } t=x_{i+1})\\
&= \int_0^1(1-(1-t)^n)f(t)\,dt,
\end{aligned}$$

since the sum is the binomial expansion of $((1-t)+t)^n$ except that the $j=0$ term, which is $(1-t)^n$, is missing. Hence $E = \int_0^1 f(t)P(t)\,dt$ where $P(t) = 1-(1-t)^n$. Clearly $0 \le P(t) \le 1$ for $0 \le t \le 1$. ∎

Solution 2. Fix $n \geq 0$ and let $(y_1, \dots, y_n)$ be chosen uniformly from $[0,1]^n$. Fix $t \in [0,1]$, and let $E_n(t)$ denote the expected value of

$$R_n = t - \max\left(\{0\} \cup (\{y_1, \dots, y_n\} \cap [0,t])\right).$$

Let $(x_1, \dots, x_n)$ be the permutation of $(y_1, \dots, y_n)$ such that $x_1 \leq \cdots \leq x_n$. The distribution of $(x_1, \dots, x_n)$ equals the distribution in the problem. Define

$$S = \sum_{i=0}^{n-1} (y_{i+1} - z_{i+1}) f(y_{i+1})$$

where z_{i+1} is the largest y_j less than y_{i+1}, or 0 if no such y_j exists. The terms in S are a permutation of those in the original Riemann sum (ignoring the final term in the Riemann sum, which is zero), so the expected values of the Riemann sum and of S are equal. Each term in S has expected value $\int_0^1 E_{n-1}(t) f(t)\,dt$, since the expected value of $y_{i+1} - z_{i+1}$ *conditioned on the event* $y_{i+1} = t$ *for some* $t \in [0,1]$ is the definition of $E_{n-1}(t)$ using $(y_j)_{j \neq i+1} \in [0,1]^{n-1}$. Therefore the expected value of the Riemann sum is $n \int_0^1 E_{n-1}(t) f(t)\,dt$.

It remains to determine $E_n(t)$ for $n \geq 0$. Let y_{n+1} be chosen uniformly in $[0,1]$, independently of the $y_1, \dots, y_n$ in the definition of $E_n(t)$. Then $E_n(t)$ equals the probability that y_{n+1} is in $[0,t]$ and is closer to t than any other y_j, since this probability conditioned on a choice of $y_1, \dots, y_n$ equals R_n. On the other hand, this probability equals $(1-(1-t)^{n+1})/(n+1)$, since the probability that at least one of $y_1, \dots, y_{n+1}$ lies in $[0,t]$ equals $1-(1-t)^{n+1}$, and conditioned on this, the probability that the y_j in $[0,t]$ closest to t is y_{n+1} is $1/(n+1)$, since all possible indices for this closest y_j are equally likely. Thus $E_n(t) = (1-(1-t)^{n+1})/(n+1)$, and we may take $P(t) = nE_{n-1}(t) = 1-(1-t)^n$. ∎

Literature note. For more on Riemann sums, see [Spv, Ch. 13, Appendix 1].

The Fifty-First William Lowell Putnam Mathematical Competition
December 1, 1990

A1. (150, 9, 1, 0, 0, 0, 0, 0, 1, 1, 6, 33)
Let

$$T_0 = 2, \ \ T_1 = 3, \ \ T_2 = 6,$$

and for $n \geq 3$,

$$T_n = (n+4)T_{n-1} - 4nT_{n-2} + (4n-8)T_{n-3}.$$

The first few terms are

$$2, \ 3, \ 6, \ 14, \ 40, \ 152, \ 784, \ 5168, \ 40576, \ 363392.$$

Find, with proof, a formula for T_n **of the form** $T_n = A_n + B_n$**, where** (A_n) **and** (B_n) **are well-known sequences.**

Answer. We have $T_n = n! + 2^n$.

Motivation. The hardest part of this problem is guessing the formula. There are not many "well-known sequences" to guess. Observe that the terms are becoming divisible by high powers of 2 (but not any other prime), and the ratio of the last two terms given is roughly 8, and the ratio of the previous two is roughly 7.

Solution. The formula $T_n = n! + 2^n$ can be verified by induction.

Alternatively, set $t_n = n! + 2^n$. Clearly $t_0 = 2 = T_0$, $t_1 = 3 = T_1$ and $t_2 = 6 = T_2$. Also,

$$t_n - nt_{n-1} = 2^n - n2^{n-1}.$$

Now 2^n and $n2^{n-1}$ are both solutions of the linear recursion

$$f_n - 4f_{n-1} + 4f_{n-2} = 0; \tag{1}$$

this follows from direct substitution. Since $t_n - nt_{n-1}$ is a linear combination of solutions to (1), it must also be a solution. Hence

$$(t_n - nt_{n-1}) - 4(t_{n-1} - (n-1)t_{n-2}) + 4(t_{n-2} - (n-2)t_{n-3}) = 0,$$

or equivalently,

$$t_n = (n+4)t_{n-1} - 4nt_{n-2} + (4n-8)t_{n-3}.$$

Thus $t_n = T_n$, because they are identical for $n = 0, 1, 2$ and satisfy the same third-order recursion (1) for $n \geq 3$. ∎

Remark. Let $\left\{t_n^{(1)}\right\}_{n\geq 1}, \left\{t_n^{(2)}\right\}_{n\geq 1}, \dots, \left\{t_n^{(m)}\right\}_{n\geq 1}$ be sequences, each of one of the following forms:

(i) $\{\alpha^n\}_{n\geq 1}$ for some $\alpha \in \mathbb{C}$,

(ii) $\{P(n)\}_{n\geq 1}$ for some polynomial $P \in \mathbb{C}[n]$,

(iii) $\{(an+b)!\}_{n\geq 1}$ for some integers $a, b \geq 0$.

Let $Q \in \mathbb{C}[x_1, \dots, x_m]$ be a polynomial, and define $u_n = Q\left(t_n^{(1)}, \dots, t_n^{(m)}\right)$ for $n \geq 1$. Then one can show that the sequences $\{u_n\}_{n\geq 1}, \{u_{n+1}\}_{n\geq 1}, \{u_{n+2}\}_{n\geq 1}, \dots$, thought of as functions of n, lie in a finitely generated $\mathbb{C}[n]$-submodule of the $\mathbb{C}[n]$-module of all sequences of complex numbers. Therefore $\{u_n\}$ satisfies a nontrivial linear recursion with polynomial coefficients. The reader may enjoy finding this recursion explicitly for sequences such as $(n!)^2 + 3^n$ or $2^n n! + F_n$ where F_n is the nth Fibonacci number, defined at the end of 1988A5. Problem 1984B1 [PutnamII, p. 44] is a variation on this theme:

Let n be a positive integer, and define

$$f(n) = 1! + 2! + \cdots + n!.$$

Find polynomials $P(x)$ and $Q(x)$ such that

$$f(n+2) = P(n)f(n+1) + Q(n)f(n),$$

for all $n \geq 1$.

A2. (63, 25, 16, 4, 0, 0, 0, 4, 5, 6, 21, 57)

Is $\sqrt{2}$ the limit of a sequence of numbers of the form $\sqrt[3]{n} - \sqrt[3]{m}$, $(n, m = 0, 1, 2, \ldots)$?

Answer. Yes. In fact, *every* real number r is a limit of numbers of the form $\sqrt[3]{n} - \sqrt[3]{m}$.

Solution 1. By the binomial expansion,

$$\sqrt[3]{n+1} - \sqrt[3]{n} = n^{1/3}\left(1 + \frac{1}{n}\right)^{1/3} - n^{1/3} = n^{1/3}\left(1 + O\left(\frac{1}{n}\right)\right) - n^{1/3} = O(n^{-2/3})$$

so $\sqrt[3]{n+1} - \sqrt[3]{n} \to 0$ as $n \to \infty$. (Alternatively, one could use

$$\sqrt[3]{n+1} - \sqrt[3]{n} = \frac{1}{\sqrt[3]{(n+1)^2} + \sqrt[3]{n(n+1)} + \sqrt[3]{n^2}} = O(n^{-2/3})$$

or

$$\sqrt[3]{n+1} - \sqrt[3]{n} = \frac{1}{3}\int_n^{n+1} x^{-2/3}\,dx = O(n^{-2/3}),$$

or the Mean Value Theorem applied to the difference quotient $\frac{\sqrt[3]{n+1} - \sqrt[3]{n}}{(n+1)-n}$.)

If $m > r^3$, then the numbers

$$0 = \sqrt[3]{m} - \sqrt[3]{m} < \sqrt[3]{m+1} - \sqrt[3]{m} < \cdots < \sqrt[3]{m+7m} - \sqrt[3]{m} = \sqrt[3]{m}$$

partition the interval $[0, \sqrt[3]{m}]$, containing r, in such a way that the largest subinterval is of size $O(m^{-2/3})$. By taking m sufficiently large, one can find a difference among these that is arbitrarily close to r. ■

Remark. We show more generally, that for any sequence $\{a_n\}$ with $a_n \to +\infty$ and $a_{n+1} - a_n \to 0$, the set $S = \{\, a_n - a_m : m, n \geq 0 \,\}$ is dense in $\mathbb{R}$. Given $r \geq 0$ and $\epsilon > 0$, fix m such that $|a_{M+1} - a_M| < \epsilon$ for all $M \geq m$. If n is the smallest integer $\geq m$ with $a_n \geq a_m + r$, then $a_n < a_m + r + \epsilon$, so $a_n - a_m$ is within ϵ of r. This shows that S is dense in $[0, \infty)$, and by symmetry S is dense also in $(-\infty, 0]$.

Remark. Let $f(x)$ be a function such that $f(x) \to +\infty$ and $f'(x) \to 0$ as $x \to +\infty$. The Mean Value Theorem shows that the hypotheses of the previous remark are satisfied by the sequence $a_n = f(n)$.

Solution 2. Fix $r \in \mathbb{R}$ and $\epsilon > 0$. Then for sufficiently large positive integers n,

$$(n+r)^3 - (n+r-\epsilon)^3 = 3n^2\epsilon + O(n) > 1,$$

so $(n+r-\epsilon)^3 \leq \lfloor (n+r)^3 \rfloor \leq (n+r)^3$, and $\sqrt[3]{\lfloor (n+r)^3 \rfloor}$ is within ϵ of $n+r$. Hence

$$\lim_{n\to\infty} \left(\sqrt[3]{\lfloor (n+r)^3 \rfloor} - \sqrt[3]{n} \right) = r. \blacksquare$$

Solution 3. As in Solution 1, $\sqrt[3]{n+1} - \sqrt[3]{n} \to 0$ as $n \to \infty$, so the set $S = \{\sqrt[3]{n} - \sqrt[3]{m}\}$ contains arbitrarily small positive numbers. Also S is closed under multiplication by positive integers k since $k(\sqrt[3]{n} - \sqrt[3]{m}) = \sqrt[3]{k^3 n} - \sqrt[3]{k^3 m}$. Any set with the preceding two properties is dense in $[0, \infty)$, because any finite open interval (a, b) in $[0, \infty)$ contains a multiple of any element of $S \cap (0, b-a)$. By symmetry S is dense in $(-\infty, 0]$ too. $\blacksquare$

Solution 4. Let $b_n = (n\sqrt[3]{5} \bmod 1) \in [0, 1]$. As in the remark following 1988B3, $\{b_n\}$ is dense in $[0, 1]$. Thus given $\epsilon > 0$, we can find n such that $n\sqrt[3]{5} - r$ is within ϵ of some integer $m \geq 0$. Then $\sqrt[3]{5n^3} - \sqrt[3]{m^3} = n\sqrt[3]{5} - m$ is within ϵ of r. $\blacksquare$

A3. (4, 4, 4, 0, 0, 0, 0, 0, 22, 36, 76, 55)

Prove that any convex pentagon whose vertices (no three of which are collinear) have integer coordinates must have area $\geq 5/2$.

Solution. A *lattice polygon* is a plane polygon whose vertices are *lattice points*, i.e., points with integer coordinates. The area of any convex lattice polygon has area equal to half an integer: this follows from Pick's Theorem mentioned in the second remark below; alternatively, by subdivision one can reduce to the case of a triangle, in which case the statement follows from the first remark below.

Consider a convex lattice pentagon $ABCDE$ of minimal area. Since the area is always half an integer, the minimum exists. If the interior of side AB contains a lattice point F, then $AFCDE$ is a convex lattice pentagon with smaller area, contradicting the choice of $ABCDE$. (As is standard, vertices are listed in order around each polygon.) Applying this argument to each side, we may assume that all boundary lattice points are vertices.

Separate the vertices into four classes according to the parity of their coordinates. By the Pigeonhole Principle, one class must contain at least two vertices. The midpoint M between two such vertices has integer coordinates. By the previous paragraph, these two vertices cannot form a side of the polygon. Also, the pentagon is convex, so M is in the interior of the pentagon. Connecting M to the vertices divides the polygon into 5 triangles, each of area at least $1/2$, so the whole polygon has area at least $5/2$. $\blacksquare$

Remark. The bound $5/2$ cannot be improved: the polygon with vertices $(0, 0)$, $(1, 0)$, $(2, 1)$, $(1, 2)$, $(0, 1)$ is convex and has area $5/2$ (see Figure 14).

Remark. The area of a plane triangle with vertices (x_1, y_1), (x_2, y_2), (x_3, y_3) equals

$$\frac{1}{2} \left| \det \begin{pmatrix} x_1 & y_1 & 1 \\ x_2 & y_2 & 1 \\ x_3 & y_3 & 1 \end{pmatrix} \right|$$

In particular, if $x_i, y_i \in \mathbb{Z}$, the area is half an integer.

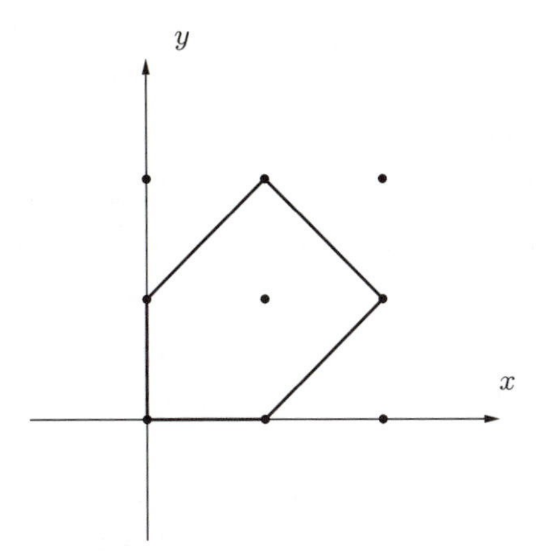

FIGURE 14.
A convex pentagon with area 5/2 whose vertices have integer coordinates.

For a formula for the volume of an n-dimensional simplex in $\mathbb{R}^n$, see Solution 5 to 1993B5.

Remark (Pick's Theorem). Given a lattice polygon, let i be the number of internal lattice points and let b be the number of boundary lattice points. Pick's Theorem [Lar1, p. 68] states that the area of the polygon equals $i + b/2 - 1$.

In the previous solution, we could have used Pick's Theorem in two places: in the first paragraph to prove that a lattice polygon has half-integer area (even if it is not convex), and as a substitute for the last sentence, using $i \geq 1$ and $b = 5$.

Related question. Problem 1981A6 [PutnamII, p. 37] is similar:

> Suppose that each of the vertices of $\triangle ABC$ is a lattice point in the (x, y)-plane and that there is exactly one lattice point P in the *interior* of the triangle. The line AP is extended to meet BC at E. Determine the largest possible value for the ratio of the lengths of segments $|AP|/|PE|$.

Here is a solution to 1981A6 using Pick's Theorem. (See [PutnamII, p. 118] for a non-Pick solution.)

We may reduce to the case where BC has no lattice points on it other than B, C, and possibly E, by replacing the base BC with the shortest segment along it with lattice endpoints and containing E in its interior.

Case 1: E is a lattice point. Then the reflection of E across P is also a lattice point, so it must coincide with A, and $|AP|/|PE| = 1$.

Case 2: E is not a lattice point. Without loss of generality (by applying an affine transformation preserving the lattice), we may assume $B = (0, 0)$ and $C = (1, 0)$. If $A = (x, y)$, $y > 0$, then $x \neq 0, 1$, and by Pick's Theorem,

$$\begin{aligned} y &= 2 + \#\{\text{boundary lattice points}\} - 2 \\ &= d + e + 1 \end{aligned}$$

where $d = \gcd(x, y)$, $e = \gcd(1 - x, y)$. Since d and e are relatively prime, de divides y, so $de \leq d + e + 1$, or equivalently $(d - 1)(e - 1) \leq 2$. We have several subcases:

- If $d = 1$, then $y = 2 + e$ is divisible by e, so $e = 1$ or 2. If $e = 1$, then $y = 2$, and the ratio $|AP|/|PE|$ is 2. If $e = 2$, then $y = 3$, and the ratio is 3.
- If $e = 1$, then essentially the same argument gives a ratio of 2 or 3.
- The case $d = e = 2$ is not allowed, since d and e are coprime.
- If $d = 3$ and $e = 2$, then $y = 6$, giving a ratio of 5.
- If $d = 2$ and $e = 3$, then the ratio is also 5.

Hence the maximum is 5. This argument can be refined to show that equality is achieved only for one triangle (up to automorphisms of the plane preserving lattice points).

Remark. The group of linear transformations of the plane preserving the lattice points and fixing the origin is the group $\mathrm{GL}_2(\mathbb{Z})$ defined in the introduction. It is important in number theory and related fields.

Remark. Fix integers $d \geq 2$ and $k \geq 1$. It follows easily from [He] that up to the action of $\mathrm{GL}_d(\mathbb{Z})$ and translation by lattice points, there are only finitely many convex lattice polytopes in $\mathbb{R}^n$ having exactly k interior lattice points. For $d = 2$ and $k = 1$, we get 16 polygons; see [PRV] for a figure showing all of them. Five of these 16 are triangles, and checking each case gives another proof that $|AP|/|PE| \leq 5$, and that equality is possible for only one of the five triangles.

A4. (44, 7, 6, 6, 0, 0, 0, 0, 32, 15, 30, 61)

Consider a paper punch that can be centered at any point of the plane and that, when operated, removes from the plane precisely those points whose distance from the center is irrational. How many punches are needed to remove every point?

Answer. Three punches are needed.

Solution. Punches at two points P and Q are not enough to remove all points, because if r is any rational number exceeding $PQ/2$, the circles of radius r centered at P and Q intersect in at least one point R, and R is not removed by either punch. We next show that three carefully chosen punches suffice.

Proof 1: Existential. Punch twice, at distinct centers. Since each punch leaves countably many circles, and any two distinct circles intersect in at most two points, the two punches leave behind a countable set. Consider all circles with rational radii centered at points of this set. Their intersections with a fixed line L form a countable set S. A point of $L - S$ is at an irrational distance from all unpunched points; apply the third punch there. ■

Remark. The fact that the plane is not a countable union of circles can also be deduced from measure theory: a circle (without its interior) in the plane has measure zero, and a countable union of measure zero sets still has measure zero, but the entire plane has infinite measure. See [Ru] for more on measure theory.

Proof 2: Constructive. Choose $\alpha \in \mathbb{R}$ such that α^2 is irrational, for example $\alpha = \sqrt[3]{2}$. Use punches centered at $A = (-\alpha, 0)$, $B = (0, 0)$, and $C = (\alpha, 0)$. If $P = (x, y)$ is any point,

$$AP^2 + CP^2 - 2BP^2 = (x+\alpha)^2 + y^2 + (x-\alpha)^2 + y^2 - 2(x^2+y^2) = 2\alpha^2$$

is irrational, so AP, BP, CP cannot all be rational. Hence all P get removed. ■

Remark. The motivation for taking $AP^2 + CP^2 - 2BP^2$ is that it is the linear combination which eliminates the terms involving x or y.

Remark. Both proofs easily generalize to prove the same result where the punch removes only those points whose distance from the center is *transcendental.* Recall that a real or complex number α is said to be *algebraic* if α is a zero of a nonzero polynomial with rational coefficients, and α is said to be *transcendental* otherwise. In Proof 1, observe that the set of real algebraic numbers is countable. In Proof 2, simply take α transcendental.

Remark. Essentially the same question appeared as [New, Problem 28].

Related question. There are many interesting questions concerning distances between points in a subset of the plane. For example, for any set of n points, in the plane, Erdős [Er], [Hon2, Ch. 12] proved that

- the number of different distances produced must be at least $\sqrt{n-3/4} - 1/2$,
- the smallest distance produced cannot occur more often than $3n-6$ times,
- the greatest distance produced cannot occur more often than n times,
- no distance produced can occur more than $2^{-1/2}n^{3/2} + n/4$ times.

Also, Problem 5 on the 1987 International Mathematical Olympiad [IMO87] asks

> Let n be an integer greater than or equal to 3. Prove that there is a set of n points in the plane such that the distance between any two points is irrational and each set of three points determines a nondegenerate triangle with rational area.

A5. (29, 5, 0, 0, 0, 0, 0, 0, 1, 0, 58, 108)

If A and B are square matrices of the same size such that ABAB = 0, does it follow that BABA = 0?

Answer. No.

Solution 1. Direct multiplication shows that the 3×3 matrices

$$\mathbf{A} = \begin{pmatrix} 0 & 0 & 1 \\ 0 & 0 & 0 \\ 0 & 1 & 0 \end{pmatrix}, \quad \mathbf{B} = \begin{pmatrix} 0 & 0 & 1 \\ 1 & 0 & 0 \\ 0 & 0 & 0 \end{pmatrix} \tag{1}$$

give a counterexample. ■

Solution 2. A more enlightening way to construct a counterexample is to use a transition diagram, as in the following example. Let e_1, e_2, e_3, e_4 be a basis of a four-dimensional vector space. Represent the matrices as in Figure 15. For example, the arrow from e_4 to e_3 labelled $\mathbf{B}$ indicates that $\mathbf{B}e_4 = e_3$; the arrow from e_1 to 0 indicates that $\mathbf{A}e_1 = 0$. Then it can be quickly checked that $\mathbf{ABAB}$ annihilates the four basis vectors, but $\mathbf{BABA}e_4 = e_1$. (Be careful with the order of multiplication when checking!) ■

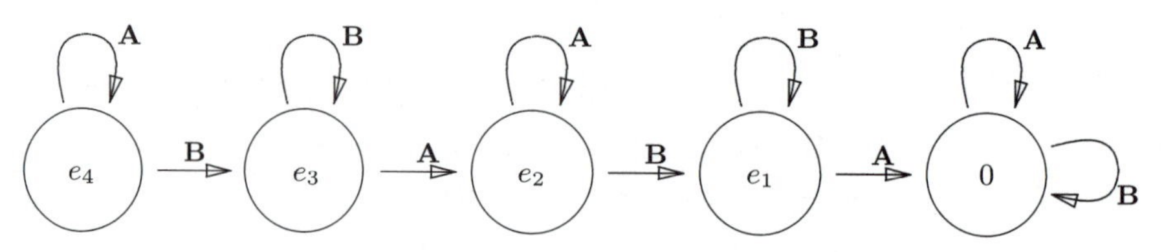

FIGURE 15.
Schematic representation of counterexample in Solution 2 to 1990A5. Alternatively, a transition diagram for a finite automaton.

Remark. The counterexample of Solution 1 also can be obtained from a transition diagram.

Remark. There are no 1×1 or 2×2 counterexamples. The 1×1 case is clear. For the 2×2 case, observe that $\mathbf{ABAB} = \mathbf{0}$ implies $\mathbf{B}(\mathbf{ABAB})\mathbf{A} = \mathbf{0}$, and hence $\mathbf{BA}$ is nilpotent. But if a 2×2 matrix $\mathbf{M}$ is nilpotent, its characteristic polynomial is x^2, so $\mathbf{M}^2 = \mathbf{0}$ by the Cayley-Hamilton Theorem [Ap2, Theorem 7.8]. Thus $\mathbf{BABA} = \mathbf{0}$.

Remark. For any $n \geq 3$, there exist $n \times n$ counterexamples: enlarge the matrices in (1) by adding rows and columns of zeros.

Stronger result. Here we present a conceptual construction of a counterexample, requiring essentially no calculations. Define a *word* to be a finite sequence of $\mathbf{A}$'s and $\mathbf{B}$'s. (The empty sequence $\emptyset$ is also a word.) Let S be a finite set of words containing $\mathbf{BABA}$ and its "right subsequences" $\mathbf{ABA}$, $\mathbf{BA}$, $\mathbf{A}$, $\emptyset$, but not containing any word having $\mathbf{ABAB}$ as a subsequence. Consider a vector space with basis corresponding to these words (i.e., $e_{\mathbf{BABA}}$, $e_{\mathbf{ABA}}$, $e_\emptyset$, etc.). Let $\mathbf{A}$ be the linear transformation mapping e_w to $e_{\mathbf{A}w}$ if $\mathbf{A}w \in S$ and to 0 otherwise. Define a linear transformation $\mathbf{B}$ similarly. Then $\mathbf{ABAB} = 0$ but $\mathbf{BABA} \neq 0$. (This gives a very general way of dealing with any problem of this type.)

Remark. To help us find a counterexample, we imposed the restriction that each of $\mathbf{A}$ and $\mathbf{B}$ maps each standard basis vector e_i to some e_j or to 0. With this restriction, the problem can be restated in terms of automata theory:

> Does there exist a finite automaton with a set of states $\Sigma = \{0, e_1, e_2, \ldots, e_n\}$ in which all states are initial states and all but 0 are final states, and a set of two productions $\{\mathbf{A}, \mathbf{B}\}$ each mapping 0 to 0, such that the language it accepts contains $\mathbf{ABAB}$ but not $\mathbf{BABA}$?

See Chapter 3 of [Sa] for terminology. The language accepted by such a finite automaton is defined as the set of words in $\mathbf{A}$ and $\mathbf{B}$ that correspond to a sequence of productions leading from some initial state to some final state. Technically, since our finite automaton does not have a unique initial state, it is called *nondeterministic*, even though each production maps any given state to a unique state. (Many authors [HoU, p. 20] do not allow multiple initial states, even in nondeterministic finite automata; we could circumvent this by introducing a new artificial initial state, with a new nondeterministic production mapping it to the desired initial states.) One theorem of automata theory [Sa, Theorem 3.3] is that any language accepted by a nondeterministic finite automaton is also the language accepted by some deterministic finite automaton.

Many lexical scanners, such as the UNIX utility `grep` [Hu], are based on the theory of finite automata. See the remark in 1989A6 for the appearance of automata theory in a very different context.

A6. (6, 6, 54, 0, 0, 0, 0, 4, 0, 45, 85)

If X is a finite set, let $|X|$ denote the number of elements in X. Call an ordered pair (S,T) of subsets of $\{1,2,\ldots,n\}$ *admissible* if $s > |T|$ for each $s \in S$, and $t > |S|$ for each $t \in T$. How many admissible ordered pairs of subsets of $\{1,2,\ldots,10\}$ are there? Prove your answer.

Answer. The number of admissible ordered pairs of subsets of $\{1,2,\ldots,10\}$ equals the 22nd Fibonacci number $F_{22} = 17711$.

Solution 1. Let $A_{m,n}$ be the set of admissible pairs (S,T) with $S \subseteq \{1,2,\ldots,m\}$ and $T \subseteq \{1,2,\ldots,n\}$, and let $a_{m,n} = |A_{m,n}|$. Suppose $m \geq n \geq 1$. Then $A_{m-1,n} \subseteq A_{m,n}$. We now show that the maps

$$\begin{aligned} A_{m,n} - A_{m-1,n} &\leftrightarrow A_{m-1,n-1} \\ (S,T) &\mapsto (S-\{m\},\{t-1 : t \in T\}) \\ (U \cup \{m\},\{v+1 : v \in V\}) &\leftarrow\!\shortmid (U,V) \end{aligned}$$

are well-defined inverse bijections.

If $(S,T) \in A_{m,n} - A_{m-1,n}$, then $m \in S$. Let $S' = S - \{m\}$ and $T^- = \{t-1 : t \in T\}$. Then $|S| \geq 1$, and $t > |S| \geq 1$ for all $t \in T$, so $T^- \subseteq \{1,2,\ldots,n-1\}$. Since (S,T) is admissible, each element of S' is greater than $|T| = |T^-|$. Also, each element of T is greater than $|S|$, so each element of T^- is greater than $|S| - 1 = |S'|$. Hence $(S',T^-) \in A_{m-1,n-1}$.

If $(U,V) \in A_{m-1,n-1}$, let $U' = U \cup \{m\} \subseteq \{1,2,\ldots,m\}$ and $V^+ = \{v+1 : v \in V\} \subseteq \{1,2,\ldots,n\}$. Since (U,V) is admissible, each element of U is greater than $|V|$, but $m \geq n > |V|$ also, so each element of U' is greater than $|V|$. Moreover, each element of V is greater than $|U|$, so each element of V^+ is greater than $|U|+1 = |U'|$. Hence $(U',V^+) \in A_{m,n} - A_{m-1,n}$.

Composing the two maps just defined in either order gives the identity, so both are bijections. Hence $a_{m,n} = a_{m-1,n} + a_{m-1,n-1}$ for $m \geq n \geq 1$. In particular, $a_{n,n} = a_{n,n-1} + a_{n-1,n-1}$ (because $a_{i,j} = a_{j,i}$), and $a_{n,n-1} = a_{n-1,n-1} + a_{n-1,n-2}$, so each term of

$$a_{0,0},\ a_{1,0},\ a_{1,1},\ a_{2,1},\ a_{2,2},\ a_{3,2},\ a_{3,3},\ \ldots$$

is the sum of the two preceding terms. Starting from $a_{0,0} = 1$ and $a_{1,0} = 2$, we find that the 21st term in the sequence

$$1,\ 2,\ 3,\ 5,\ 8,\ 13,\ 21,\ 34,\ 55,\ 89,\ 144,\ 233,\ 377,$$

$$610,\ 987,\ 1597,\ 2584,\ 4181,\ 6765,\ 10946,\ 17711,\ldots$$

is $a_{10,10} = 17711$. (The sequence is the Fibonacci sequence defined at the end of 1988A5, but starting with $F_2 = 1$.) ∎

Solution 2 (Jeremy Rouse). Let $a_{m,n}$ be as in Solution 1. If S is an i-element subset of $\{j+1, j+2, \ldots, m\}$ and T is a j-element subset of $\{i+1, i+2, \ldots, n\}$, then (S,T) is an admissible pair; conversely, each admissible pair (S,T) with $S \subseteq \{1,2,\ldots,m\}$, $T \subseteq \{1,2,\ldots,n\}$, $|S| = i$, and $|T| = j$ arises in this way. Hence $a_{m,n} = \sum_{i,j} \binom{m-j}{i}\binom{n-i}{j}$, where the sum ranges over pairs of nonnegative integers (i,j) satisfying $i+j \le \min\{m,n\}$ (so that the binomial coefficients are nonzero). Let F_n be the nth Fibonacci number. We will give a bijective proof that

$$\sum_{i,j} \binom{n-j}{i}\binom{n-i}{j} = F_{2n+2} \qquad (1)$$

for all $n \ge 0$.

For $m \ge 1$, let $\mathcal{R}_m$ mean "$1 \times m$ rectangle," and let N_m denote the number of ways to tile an $\mathcal{R}_m$ with 1×1 squares and 1×2 dominos (rectangles). We now prove (1) by showing that both sides equal N_{2n+1}. A tiling of an $\mathcal{R}_m$ ends either with a square or a domino, so $N_m = N_{m-1} + N_{m-2}$ for $m \ge 3$. Together with $N_1 = 1$ and $N_2 = 2$, this proves $N_m = F_{m+1}$ by induction. In particular N_{2n+1} equals F_{2n+2}, the right-hand side of (1).

On the other hand, if we start with a *pair* of tilings, one a tiling of an $\mathcal{R}_{n+i-j}$ by $n-i-j$ squares and i dominos, and the other a tiling of an $\mathcal{R}_{n+j-i}$ by $n-i-j$ squares and j dominos, we may form a tiling of an $\mathcal{R}_{2n+1}$ by appending the two, with a square inserted in between. Conversely, any tiling of an $\mathcal{R}_{2n+1}$ arises from such a pair: every tiling of an $\mathcal{R}_{2n+1}$ contains an odd number of squares, so there is a "median" square, and the pieces to the left and right of this square constitute a pair of tilings. The number of such pairs for fixed i and j equals $\binom{n-j}{i}\binom{n-i}{j}$, so N_{2n+1} equals $\sum_{i,j}\binom{n-j}{i}\binom{n-i}{j}$, the left-hand side of (1).

This proves (1). The desired value $a_{10,10} = F_{22}$ is then found by calculating $F_0, F_1, \ldots, F_{22}$ successively, using $F_{n+2} = F_{n+1} + F_n$. ■

B1. (114, 2, 52, 0, 0, 0, 0, 11, 5, 3, 10, 4)

Find all real-valued continuously differentiable functions f on the real line such that for all x

$$(f(x))^2 = \int_0^x \left((f(t))^2 + (f'(t))^2\right) dt + 1990.$$

Answer. There are two such functions, namely $f(x) = \sqrt{1990}e^x$, and $f(x) = -\sqrt{1990}e^x$.

Solution. For a given f, the functions on the left- and right-hand sides are equal if and only if their values at 0 are equal, i.e., $f(0)^2 = 1990$, and their derivatives are equal for all x, i.e.,

$$2f(x)f'(x) = (f(x))^2 + (f'(x))^2 \qquad \text{for all } x.$$

The latter condition is equivalent to each of the following: $(f(x) - f'(x))^2 = 0$, $f'(x) = f(x)$, $f(x) = Ce^x$ for some constant C. Combining this condition with $f(0)^2 = 1990$ yields $C = \pm\sqrt{1990}$, so the desired functions are $f(x) = \pm\sqrt{1990}e^x$. ■

B2. (23, 5, 4, 9, 0, 0, 3, 0, 0, 32, 125)

Prove that for $|x| < 1$, $|z| > 1$,

$$1+\sum_{j=1}^{\infty}(1+x^j)\frac{(1-z)(1-zx)(1-zx^2)\cdots(1-zx^{j-1})}{(z-x)(z-x^2)(z-x^3)\cdots(z-x^j)}=0.$$

Solution. Let $S_0 = 1$, and for $n \geq 1$, let

$$S_n = 1+\sum_{j=1}^{n}(1+x^j)\frac{(1-z)(1-zx)(1-zx^2)\cdots(1-zx^{j-1})}{(z-x)(z-x^2)(z-x^3)\cdots(z-x^j)}.$$

Since $S_1 = (1-zx)/(z-x)$ and $S_2 = (1-zx)(1-zx^2)/(z-x)(z-x^2)$, we suspect that

$$S_n = \frac{(1-zx)(1-zx^2)\cdots(1-zx^n)}{(z-x)(z-x^2)(z-x^3)\cdots(z-x^n)},$$

which is easily proved by induction.

It remains to prove that $\lim_{n\to\infty} S_n = 0$. If $S_n = 0$ for some n, then $S_N = 0$ for all $N \geq n$, so $\lim_{n\to\infty} S_n = 0$. Otherwise

$$\frac{S_{n+1}}{S_n} = \frac{1-zx^{n+1}}{z-x^{n+1}} \to \frac{1}{z}$$

as $n \to \infty$, since $x^{n+1} \to 0$. By the Ratio Test, $\lim_{n\to\infty} S_n = 0$. ∎

B3. (97, 7, 4, 2, 0, 0, 0, 0, 12, 2, 54, 23)

Let S be a set of 2×2 integer matrices whose entries a_{ij} (1) are all squares of integers, and, (2) satisfy $a_{ij} \leq 200$. Show that if S has more than 50387 ($= 15^4 - 15^2 - 15 + 2$) elements, then it has two elements that commute.

Solution. Let U be the set of 2×2 matrices satisfying (1) and (2). Let D be the set of diagonal matrices in U, and let J be the set of multiples of $\begin{pmatrix}1 & 1\\ 1 & 1\end{pmatrix}$ in U. The numbers less than or equal to 200 that are squares of integers are the 15 numbers 0^2, 1^2, ... , 14^2, so $|U| = 15^4$, $|D| = 15^2$, and $|J| = 15$. Now

(i) any two matrices from D commute,

(ii) any two matrices from J commute, and

(iii) $\begin{pmatrix}1 & 1\\ 0 & 1\end{pmatrix}$ and $\begin{pmatrix}1 & 4\\ 0 & 1\end{pmatrix}$ commute.

Suppose that no two elements of S commute. Write

$$S = \big(S\cap(D\cup J)\big)\cup\big(S\cap(D\cup J)^c\big).$$

(Here X^c denotes the complement of X.) By (i) and (ii), S can contain at most one element of D and at most one element of J, so $|S\cap(D\cup J)| \leq 2$. By (iii),

$$\begin{aligned}|S\cap(D\cup J)^c| &< |U\cap(D\cup J)^c|\\ &= |U|-|D|-|J|+|D\cap J|\\ &= 15^4-15^2-15+1.\end{aligned}$$

Hence $|S| \leq 2 + (15^4 - 15^2 - 15) = 50387$. ∎

Remark. The number 50387 is far from the best possible. Liu and Schwenk have shown, using an inclusion-exclusion argument, that the maximum number of elements in U in which no two elements commute is 32390 [LS]. Because the bound given in the problem is so far from being optimal, there are many possible solutions.

B4. (9, 5, 2, 1, 0, 0, 0, 0, 3, 2, 63, 116)

Let G be a finite group of order n generated by a and b. Prove or disprove: there is a sequence

$$g_1, g_2, g_3, \ldots, g_{2n}$$

such that

(1) every element of G occurs exactly twice, and

(2) g_{i+1} equals $g_i a$ or $g_i b$, for $i = 1, 2, \ldots, 2n$. (Interpret g_{2n+1} as g_1.)

Solution. We use graph theory terminology; see the remark below. Construct a directed multigraph $\mathcal{D}$ whose vertices are the elements of G, and whose arcs are indexed by $G \times \{a, b\}$, such that the arc corresponding to the pair (g, x) goes from vertex g to vertex gx. (See Figure 16 for an example, with G equal to the symmetric group S_3, a equal to the transposition (12), and b equal to the 3-cycle (123).) At the vertex g, there are two arcs going out (to ga and to gb), and two arcs coming in (from ga^{-1} and from gb^{-1}). Also, $\mathcal{D}$ is weakly connected, since a and b generate G. Hence, by the first theorem in the remark below, $\mathcal{D}$ has an Eulerian circuit. Take g_1, g_2, ... , g_m to be the the startpoints of the arcs in this circuit, in order. Each element of G occurs exactly twice in this sequence, since each vertex of $\mathcal{D}$ has outdegree 2; in particular $m = 2n$. Also, for $1 \le i \le 2n$, the element g_{i+1} equals either $g_i a$ or $g_i b$, because the two outgoing arcs from g_i end at $g_i a$ and $g_i b$. ∎

Remark (Eulerian paths and circuits). A *directed multigraph* $\mathcal{D}$ consists of a set V (whose elements are called *vertices*), and a set E (whose elements are called *arcs* or sometimes *edges* or *directed edges*), with a map $E \to V \times V$ (thought of as sending an arc to the pair consisting of its startpoint and the endpoint). Typically one draws each element of V as a point, and each arc of E as an arc from the startpoint to the endpoint, with an arrow to indicate the direction. What makes it a multigraph is that for some vertices $v, w \in V$, there may be more than one arc from v to w. Also, there may be *loops*: arcs from a vertex to itself. Call $\mathcal{D}$ *finite* if V and E are both finite sets.

The *outdegree* of a vertex v is the number of arcs in E having v as startpoint. Similarly the *indegree* of v is the number of arcs in E having v as endpoint. If $v, w \in V$ then a *path* from v to w in $\mathcal{D}$ is a finite sequence of arcs in E, not necessarily distinct, such that the startpoint of the first arc is v, the endpoint of each arc (other than the last) is the startpoint of the next arc, and the endpoint of the last arc is w. Such a path is called a *circuit* or *cycle* if $v = w$. An *Eulerian path* in $\mathcal{D}$ is a path in which each arc in E occurs exactly once. An *Eulerian circuit* is a circuit in which each arc in E occurs exactly once. We say that $\mathcal{D}$ is *strongly connected* if for every two distinct vertices $v, w \in V$, there is a path from v to w in $\mathcal{D}$. On the other hand, $\mathcal{D}$ is *weakly connected* if for every two distinct vertices $v, w \in V$, there is a path from v to w in

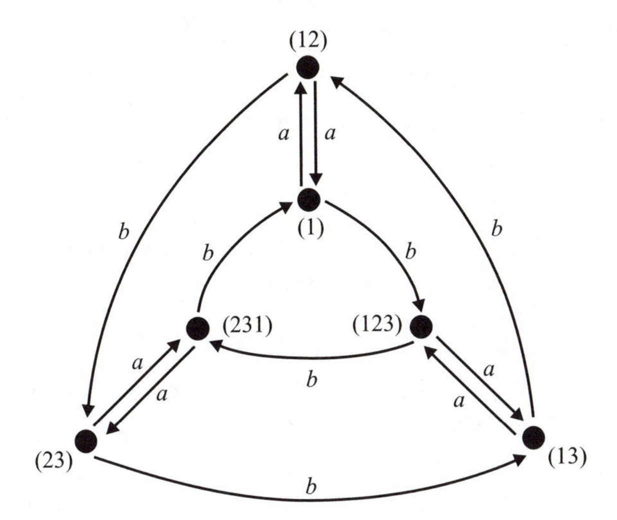

FIGURE 16.
Cayley graph of the group S_3 with generators $a = (12)$ and $b = (123)$

the corresponding undirected graph, that is, a path consisting of arcs some of which may be the reverses of the arcs in E.

Then one can prove the following two theorems.

- A finite directed multigraph with at least one arc has an Eulerian circuit if and only if it is weakly connected and the indegree and outdegree are equal at each vertex.
- A finite directed multigraph with at least one arc has an Eulerian path but not an Eulerian circuit if and only if it is weakly connected and the indegree and outdegree are equal at each vertex, except at one vertex at which indegree is 1 larger than outdegree, and one other vertex at which outdegree is 1 larger than indegree.

See Chapter 7 of [Ros2], especially §7.4 and §7.5.

Remark. The directed multigraph constructed in the solution is called the *Cayley digraph* or *Cayley diagram* associated to G and its set of generators $\{a, b\}$. See [Fr, pp. 87–91].

B5. (16, 11, 15, 12, 0, 0, 0, 11, 5, 5, 48, 78)

Is there an infinite sequence $a_0, a_1, a_2, \ldots$ of nonzero real numbers such that for $n = 1, 2, 3, \ldots$ the polynomial

$$p_n(x) = a_0 + a_1 x + a_2 x^2 + \cdots + a_n x^n$$

has exactly n distinct real roots?

Answer. Yes, such an infinite sequence exists.

Solution 1. Take $a_0 = 1$, $a_1 = -1$, and for $n \geq 1$ construct a_{n+1} inductively as follows. Suppose $p_n(x)$ has n distinct real zeros: $x_1 < x_2 < \cdots < x_n$. Choose $\alpha_0, \ldots,$ α_n so that

$$\alpha_0 < x_1 < \alpha_1 < \cdots < x_n < \alpha_n.$$

Then the signs of $p_n(\alpha_0), p_n(\alpha_1), \ldots, p_n(\alpha_n)$ alternate. Define $a_{n+1} = -\epsilon \operatorname{sgn}(p_n(\alpha_n))$, where ϵ is positive and small enough that

$$\operatorname{sgn}\left(p_{n+1}(\alpha_i)\right) = \operatorname{sgn}\left(p_n(\alpha_i)\right)$$

for all i. Let

$$p_{n+1}(x) = p_n(x) + a_{n+1}x^{n+1}.$$

By the Intermediate Value Theorem, p_{n+1} has a zero between α_i and α_{i+1} for $0 \leq i \leq n-1$, and a zero greater than α_n since

$$\operatorname{sgn}\left(p_{n+1}(\alpha_n)\right) \neq \lim_{x\to\infty} \operatorname{sgn}\left(p_{n+1}(x)\right).$$

Because $p_{n+1}(x)$ is of degree $n+1$, it cannot have more than these $n+1$ zeros, so $p_{n+1}(x)$ has $n+1$ distinct real zeros, as desired. ■

Solution 2. For $n \geq 0$, let $a_n = (-1)^n 10^{-n^2}$. For $0 \leq k \leq n$,

$$\begin{aligned}
(-1)^k 10^{-k^2} p_n(10^{2k}) &= \sum_{i=0}^{n} (-1)^{i-k} 10^{-(i-k)^2} \\
&= \sum_{j=-k}^{n-k} (-1)^j 10^{-j^2} \\
&> 1 - 2\sum_{j=1}^{\infty} 10^{-j^2} \\
&> 0,
\end{aligned}$$

so $p_n(1), p_n(10^2), p_n(10^4), \ldots, p_n(10^{2n})$ alternate in sign. By the Intermediate Value Theorem, it follows that $p_n(x)$ has at least n distinct real zeros. Since $p_n(x)$ has degree n, there cannot be more than n zeros. ■

Remark. Solution 2 is motivated by the theory of Newton polygons for polynomials with coefficients in the field $\mathbb{Q}_p$ of p-adic numbers. Let $|\cdot|_p$ denote the p-adic absolute value on $\mathbb{Q}_p$. If $\{a_i\}_{i\geq 0}$ is a sequence of nonzero p-adic numbers such that the lower convex hull of $\{\,(i, -\ln |a_i|_p) : 0 \leq i \leq n\,\}$ consists of n segments with different slopes, then $\sum_{i=0}^{n} a_i x^i$ has n distinct zeros in $\mathbb{Q}_p$; in particular this holds for $a_i = p^{i^2}$. The analogous statement over $\mathbb{R}$, with $|\cdot|_p$ replaced by the standard absolute value, is not true in general, but it is true if the differences between the slopes are sufficiently large relative to n. For an introduction to p-adic numbers and Newton polygons, see [Kob].

B6. (5, 0, 0, 0, 0, 0, 0, 0, 38, 6, 29, 123)

Let S be a nonempty closed bounded convex set in the plane. Let K be a line and t a positive number. Let L_1 and L_2 be support lines for S parallel to K, and let $\overline{L}$ be the line parallel to K and midway between L_1 and L_2. Let $B_S(K,t)$ be the band of points whose distance from $\overline{L}$ is at most $(t/2)w$, where w is the distance between L_1 and L_2. What is the smallest t such that

$$S \cap \bigcap_K B_S(K,t) \neq \emptyset$$

for all S? (K runs over all lines in the plane.)

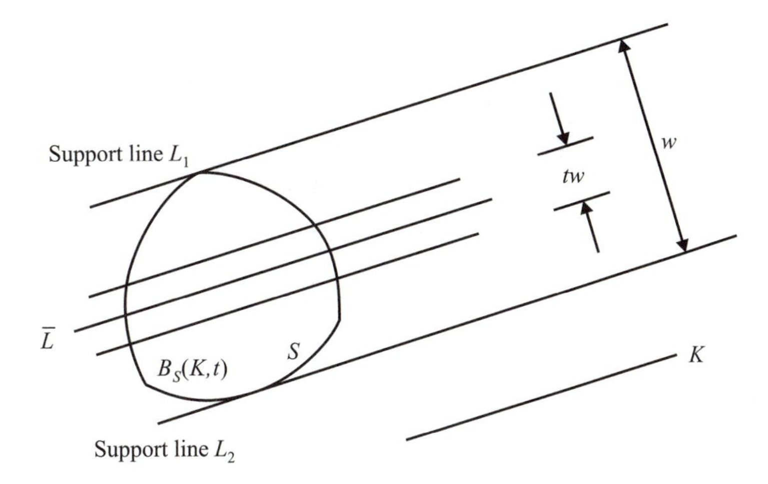

Answer. The smallest t is $1/3$.

Motivation. To approach this problem, it is natural to experiment with simple shapes. The triangle suggests that the answer should be $t = 1/3$, and that for any $t \geq 1/3$, the intersection

$$S \cap \bigcap_K B_S(K,t)$$

should contain the centroid, and is hence nonempty.

Solution 1. We first show that the intersection can be empty for $t < 1/3$. Suppose that S is a triangle. Dissect the triangle into 9 congruent subtriangles, as shown in Figure 17. If K is parallel to one of the sides of the triangle, $S \cap B_S(K,t)$ is contained in the three subtriangles in the "middle strip" (and does not meet the boundary of the strip). Hence if $t < 1/3$, and K_1, K_2, K_3 are parallel to the three sides of the triangle,

$$S \cap \bigcap_{i=1}^{3} B_S(K_i,t)$$

is empty. This is illustrated in Figure 18. (Also, if $t = 1/3$, then this intersection consists of just the centroid. This motivates the rest of the solution.)

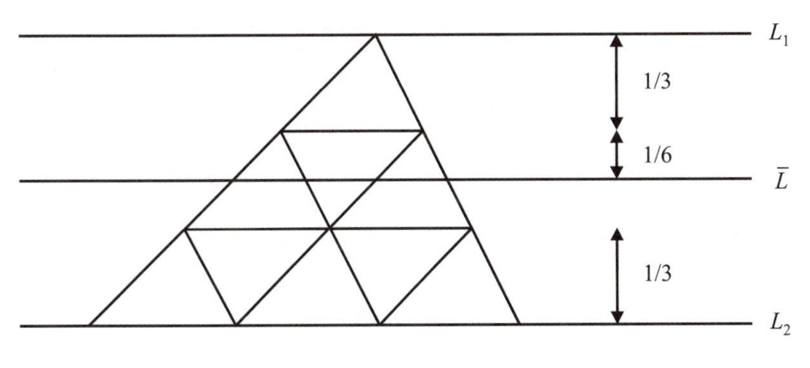

FIGURE 17.

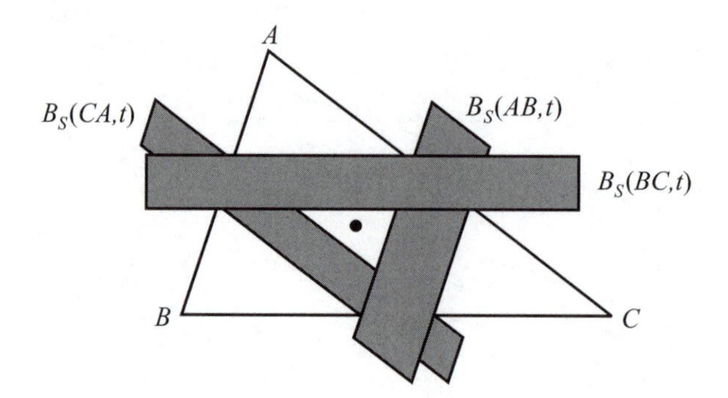

FIGURE 18.
$S \cap \bigcap_K B_S(K,t)$ may be empty for $t < 1/3$.

Recall that the *centroid* of a measurable region S in $\mathbb{R}^2$ is the unique point P such that $\int_{Q\in S} \overrightarrow{PQ}\, dA = 0$. Equivalently, the coordinates of the centroid $(\overline{x}, \overline{y})$ are given by $\overline{x} = \int_S x\, dA / \int_S 1\, dA$, and $\overline{y} = \int_S y\, dA / \int_S 1\, dA$. If S is convex, then the centroid lies within S.

We now show that the intersection of the problem is nonempty for $t \geq 1/3$ for any S, by showing that each strip $B_S(K,t)$ contains the centroid of S. By symmetry, it suffices to show that the centroid of S is at most $2/3$ of the distance from L_1 to L_2. Think of L_1 as the upper support line. (See Figure 19.) Let P_i be a point of contact of L_i with S, for $i = 1, 2$. For a variable point Q to the left of P_2 on L_2 (possibly $Q = P_2$), let $\mathcal{A}$ be the intersection of S with the open half-plane to the left of $\overleftrightarrow{QR_1}$, and let $\mathcal{B}$ be the part of (possibly degenerate) $\triangle QP_1P_2$ lying *outside* S. As Q moves to a nearby point Q', Area($\mathcal{A}$) and Area($\mathcal{B}$) each change by at most Area($\triangle QQ'P_1$), which can be made arbitrarily small by choosing Q' close to Q; hence Area($\mathcal{A}$) and Area($\mathcal{B}$) vary continuously as functions of Q. The difference $\delta(Q) = \text{Area}(\mathcal{A}) - \text{Area}(\mathcal{B})$ is also a continuous function of Q. At $Q = P_2$, Area($\mathcal{A}$) ≥ 0 and Area($\mathcal{B}$) $= 0$, so $\delta(P_2) \geq 0$. But as Q tends to infinity along L_1, Area($\mathcal{A}$) is bounded by Area(S) and Area($\mathcal{B}$) grows without bound, so $\delta(Q) < 0$ for some Q. By the Intermediate Value Theorem, there is some position of Q for which $\delta(Q) = 0$, i.e., for which Area($\mathcal{A}$) = Area($\mathcal{B}$). Fix such a Q.

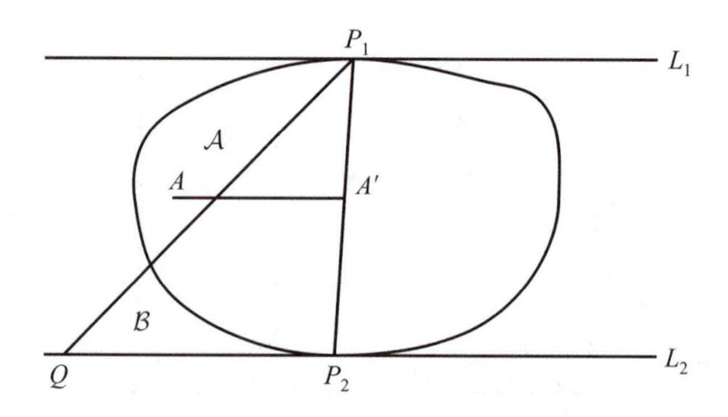

FIGURE 19.

We claim that if $A \in \mathcal{A}$ and $B \in \mathcal{B}$, then B lies below A. To show this, let A' be the intersection of $\overline{P_1P_2}$ with the horizontal line through A. Since S is convex, $A' \in \overline{P_1P_2} \subseteq S$ and $\triangle AA'P_1 \subseteq S$. Then $B \notin \triangle AA'P_1$ by the definition of $\mathcal{B}$. But $\triangle AA'P_1$ contains all points of $\triangle QP_1P_2$ lying above or at the same level as A, so B must lie below A.

Let $\tilde{S}$ denote the region obtained from S by removing $\mathcal{A}$ and adding $\mathcal{B}$, and performing the corresponding operations to the right of $\overline{P_1P_2}$. By the previous paragraph, the centroid of $\tilde{S}$ lies at least as low as the centroid of S. But $\tilde{S}$ is a triangle, with base on L_2 and opposite vertex at P_1, so the centroid of $\tilde{S}$ lies exactly $2/3$ of the way from L_1 to L_2. Hence the centroid of S lies at most $2/3$ of the way from L_1 to L_2.

Thus the minimal t for which the intersection is nonempty is $1/3$. ∎

Solution 2. As in the first paragraph of Solution 1, $t \geq 1/3$. We now show that $t = 1/3$ works, by proving that the centroid of S is in $B_S(K,t)$ for all K. Without loss of generality, we may rotate, rescale, and translate to assume that L_2 is the x-axis and L_1 is the line $y = 3$. Let P be a point where S meets L_2. Let A be the area of S. It suffices to show that the centroid (x, y) satisfies $y \leq 2$, since then $y \geq 1$ by symmetry.

Partially cover S with nonoverlapping inscribed triangles each having one vertex at P, as in Figure 20, and let ϵA be the area of the part of S not covered. Each triangle has vertex y-coordinates 0, y_1, y_2 where $0 \leq y_1, y_2 \leq 3$, so the centroid of the triangle has y-coordinate at most 2. Let $\overline{y}_\triangle$ denote the y-coordinate of the centroid of the triangle-tiled portion of S, let $\overline{y}_\epsilon$ denote the y-coordinate of the centroid of the remainder, and let $\overline{y}_S$ denote the y-coordinate of the centroid of S. Then $\overline{y}_\triangle \leq 2$, $\overline{y}_\epsilon \leq 3$, and

$$\begin{aligned} A\overline{y}_S &= \epsilon A \overline{y}_\epsilon + (A - \epsilon A)\overline{y}_\triangle \\ \overline{y}_S &= \epsilon \overline{y}_\epsilon + (1-\epsilon)\overline{y}_\triangle \\ &\leq 3\epsilon + 2(1-\epsilon) \\ &\leq 2 + \epsilon. \end{aligned}$$

But ϵ can be made arbitrarily small by choosing the triangles appropriately, so $\overline{y}_S \leq 2$, as desired. ∎

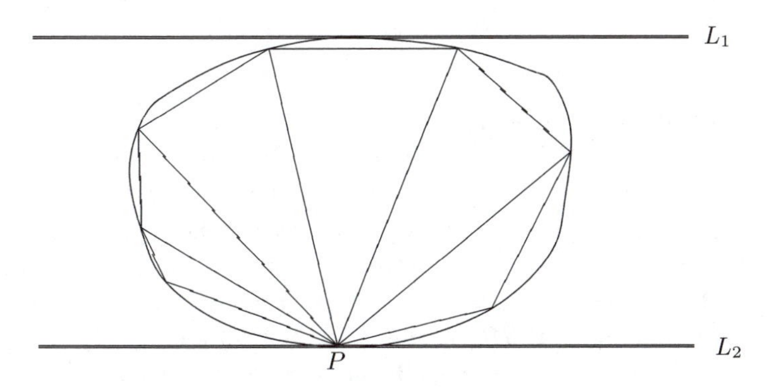

FIGURE 20.
Partially covering S by triangles.

Remark. We sketch a justification of the last sentence. Using the convexity of S, one can prove that there exist continuous functions $f(x) \leq g(x)$ defined on an interval $[a, b]$ such that S is the region between the graphs of f and g on $[a, b]$. The approximations to $A = \int_a^b g(x)\,dx - \int_a^b f(x)\,dx$ given by the Trapezoid Rule can be made arbitrarily close to A by taking sufficiently fine subdivisions of $[a, b]$. We may assume that the x-coordinate of P is one of the sample points; then the approximation is represented by the area of an inscribed polygon with one vertex at P. Cut the polygon into triangles by connecting P to the other vertices with line segments.

Remark. A more analytic approach to proving that the centroid is in the central 1/3 of the strip is the following. Choose coordinates so that L_1 and L_2 are the lines $y = 1$ and $y = 0$ respectively. Let $f(t)$ be the length of the intersection of S and the line $y = t$. Convexity of S implies that $f(t)$ is a nonnegative concave-down continuous function on $[0, 1]$, and the desired result would follow from

$$\int_0^1 tf(t)\,dt \geq \frac{1}{3}\int_0^1 f(t)\,dt. \tag{1}$$

(Geometrically, this states that the centroid is at least 1/3 of the way from L_2 to L_1. Along with the corresponding statement with the roles of L_1 and L_2 reversed, this shows that the centroid is in $B_S(K, 1/3)$.)

For any function f *with continuous second derivative*, integration by parts twice yields

$$\begin{aligned} \int_0^1 \left(t - \frac{1}{3}\right) f(t)\,dt &= \left(\frac{t^2}{2} - \frac{t}{3}\right) f(t)\bigg|_0^1 - \int_0^1 \left(\frac{t^2}{2} - \frac{t}{3}\right) f'(t)\,dt \qquad (2)\\ &= \frac{f(1)}{6} - \left(\frac{t^3}{6} - \frac{t^2}{6}\right) f'(t)\bigg|_0^1 + \int_0^1 \left(\frac{t^3}{6} - \frac{t^2}{6}\right) f''(t)\,dt \\ &= \frac{f(1)}{6} + \int_0^1 \left(\frac{t^3}{6} - \frac{t^2}{6}\right) f''(t)\,dt. \end{aligned}$$

If in addition $f(1) \geq 0$ and f is concave-down, then this implies (1) since the final integrand is everywhere nonnegative.

To prove

$$\int_0^1 \left(t - \frac{1}{3}\right) f(t)\,dt \geq \frac{f(1)}{6} \tag{3}$$

for *all* concave-down continuous functions, including those that are not twice differentiable, it remains to prove that any concave-down continuous function f on $[0, 1]$ is a uniform limit of concave-down functions with continuous second derivatives. Adjusting f by a linear function, we may assume $f(0) = f(1) = 0$. By continuity of f at 0 and 1, the function $f(t)$ is the uniform limit of the concave-down continuous functions $\min\{f(t), ct, c(1-t)\}$ on $[0, 1]$ as $c \to +\infty$. Hence we may replace f by such an approximation to assume that $f(t) \leq \min\{ct, c(1-t)\}$ for some $c > 0$. We can then extend f to a concave-down continuous function on $\mathbb{R}$ by setting $f(t) = ct$ for $t < 0$ and $f(t) = c(1-t)$ for $t > 1$. We now show that it is the uniform limit of concave-down smooth functions. Define a sequence of smooth nonnegative functions δ_n supported on $[-1/n, 1/n]$ with $\int_{-\infty}^{\infty} \delta_n(t)\,dt = 1$, and define the convolution

$$f_n(t) = \int_{-\infty}^{\infty} f(u)\delta_n(t-u)\,du.$$

Then f is the uniform limit of f_n on $[0,1]$, and f_n is smooth. Finally, f_n is also concave-down, because the convolution can be viewed as a weighted average of translates of f.

Remark (Eric Wepsic). Alternatively, one can prove (3) for all concave-down continuous functions by discretizing (2). For $n \geq 1$, define

$$A_n = \sum_{i=1}^{n} \left(\frac{i}{n} - \frac{1}{3}\right) f\left(\frac{i}{n}\right) \frac{1}{n}, \quad \text{and}$$

$$B_n = \frac{f(1)}{6} + \sum_{i=1}^{n} g\left(\frac{i}{n}\right) \left(\frac{f((i+1)/n) - 2f(i/n) + f((i-1)/n)}{1/n^2}\right) \frac{1}{n},$$

where $g(t) = t^3/6 - t^2/6$. (These are supposed to be approximations to the two ends of (2). The ratio

$$\frac{f((i+1)/n) - 2f(i/n) + f((i-1)/n)}{1/n^2}$$

is an approximation of $f''(i/n)$ in the same spirit as the formula in Proof 4 of the lemma in 1992A4.)

It is not quite true that $A_n = B_n$, but we can bound the difference. First collect terms in B_n with the same value of f, and use $g(0) = g(1) = 0$, to obtain

$$B_n = \sum_{i=1}^{n} b_{in}\, f(i/n)$$

with

$$b_{in} = \begin{cases} n\left(g\left(\frac{1}{n}\right) - g\left(\frac{0}{n}\right) + 0\right), & \text{if } i = 0 \\ n\left(g\left(\frac{i+1}{n}\right) - 2g\left(\frac{i}{n}\right) + g\left(\frac{i-1}{n}\right)\right), & \text{if } 1 \leq i \leq n-1 \\ \frac{1}{6} + n\left(0 - g\left(\frac{n}{n}\right) + g\left(\frac{n-1}{n}\right)\right), & \text{if } i = n. \end{cases}$$

Since g is infinitely differentiable, Taylor's Theorem [Spv, Ch. 19, Theorem 4] centered at x shows

$$g\left(x + \frac{1}{n}\right) - 2g(x) + g\left(x - \frac{1}{n}\right) = \frac{g''(x)}{n^2} + O\left(\frac{1}{n^3}\right)$$

and

$$g\left(x \pm \frac{1}{n}\right) - g(x) = \pm\frac{g'(x)}{n} + O\left(\frac{1}{n^2}\right)$$

as $n \to \infty$, where the constants implied by the O's are uniform for $x \in [0,1]$. Thus

$$b_{in} = \begin{cases} g'(0) + O\left(\frac{1}{n}\right), & \text{if } i = 0 \\ \frac{g''(i/n)}{n} + O\left(\frac{1}{n^2}\right), & \text{if } 1 \le i \le n-1 \\ \frac{1}{6} - g'(1) + O\left(\frac{1}{n}\right), & \text{if } 1 \le i \le n-1 \end{cases}$$

$$= \begin{cases} O\left(\frac{1}{n}\right), & \text{if } i = 0 \\ \left(\frac{i}{n} - \frac{1}{3}\right)\frac{1}{n} + O\left(\frac{1}{n^2}\right), & \text{if } 1 \le i \le n-1 \\ O\left(\frac{1}{n}\right), & \text{if } 1 \le i \le n-1. \end{cases}$$

where again the implied constants are uniform. Except for the O's, these are the same as the coefficients of $f(i/n)$ in A_n. Thus, if $M = \sup\{\,|f(t)| : t \in [0,1]\,\}$, then

$$A_n - B_n = O(1/n)M + \left(\sum_{i=1}^{n-1} O(1/n^2)M\right) + O(1/n)M = O(1/n).$$

In particular, $\lim_{n\to\infty}(A_n - B_n) = 0$. If f is concave-down, then for all i, $f((i+1)/n) - 2f(i/n) + f((i-1)/n) \le 0$ and $g(i/n) \le 0$, so the definition of B_n implies $B_n \ge f(1)/6$ for all n. If f is continuous, then A_n is the nth Riemann sum for $\int_0^1 (t - 1/3) f(t)\,dt$, so $\lim_{n\to\infty} A_n = \int_0^1 (t - 1/3) f(t)\,dt$. Combining the three previous sentences yields (3), whenever f is concave-down and continuous.

Remark. A similar result is that for every compact convex set S in the plane, there exists at least one point $P \in S$ such that every chord AB of S containing P satisfies

$$1/2 \le \frac{AP}{BP} \le 2.$$

This result can be generalized to an arbitrary number of dimensions [Berg, Corollary 11.7.6].

The Fifty-Second William Lowell Putnam Mathematical Competition December 7, 1991

A1. (189, 0, 3, 0, 0, 0, 0, 0, 0, 1, 20, 0)

A 2×3 rectangle has vertices at $(0,0)$, $(2,0)$, $(0,3)$, and $(2,3)$. It rotates 90° clockwise about the point $(2,0)$. It then rotates 90° clockwise about the point $(5,0)$, then 90° clockwise about the point $(7,0)$, and finally, 90° clockwise about the point $(10,0)$. (The side originally on the x-axis is now back on the x-axis.) Find the area of the region above the x-axis and below the curve traced out by the point whose initial position is $(1,1)$.

Answer. The area of the region is $7\pi/2 + 6$.

Solution.

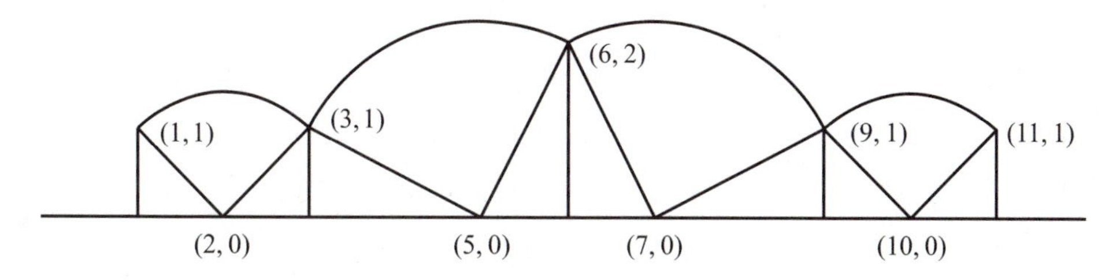

FIGURE 21.

The point $(1,1)$ rotates around $(2,0)$ to $(3,1)$, then around $(5,0)$ to $(6,2)$, then around $(7,0)$ to $(9,1)$, then around $(10,0)$ to $(11,1)$. (See Figure 21.) The area of concern consists of four 1×1 right triangles of area $1/2$, four 1×2 triangles of area 1, two quarter circles of area $(\pi/4)(\sqrt{2})^2 = \pi/2$, and two quarter circles of area $(\pi/4)(\sqrt{5})^2 = 5\pi/4$, for a total area of $7\pi/2 + 6$. ■

A2. (150, 17, 2, 1, 0, 0, 0, 0, 3, 5, 15, 20)

Let $\mathbf{A}$ and $\mathbf{B}$ be different $n \times n$ matrices with real entries. If $\mathbf{A}^3 = \mathbf{B}^3$ and $\mathbf{A}^2\mathbf{B} = \mathbf{B}^2\mathbf{A}$, can $\mathbf{A}^2 + \mathbf{B}^2$ be invertible?

Answer. No.

Solution. We have

$$(\mathbf{A}^2 + \mathbf{B}^2)(\mathbf{A} - \mathbf{B}) = \mathbf{A}^3 - \mathbf{B}^3 - \mathbf{A}^2\mathbf{B} + \mathbf{B}^2\mathbf{A} = \mathbf{0},$$

and $\mathbf{A} - \mathbf{B} \neq \mathbf{0}$, so $\mathbf{A}^2 + \mathbf{B}^2$ is not invertible. ■

A3. (42, 35, 29, 0, 0, 0, 0, 0, 6, 5, 63, 33)

Find all real polynomials $p(x)$ of degree $n \geq 2$ for which there exist real numbers $r_1 < r_2 < \cdots < r_n$ such that

(i) $p(r_i) = 0, \quad i = 1, 2, \ldots, n,$ **and**

(ii) $p'\left(\frac{r_i + r_{i+1}}{2}\right) = 0 \quad i = 1, 2, \ldots, n-1,$

where $p'(x)$ denotes the derivative of $p(x)$.

Answer. The real polynomials with the required property are exactly those that are of degree 2 with 2 distinct real zeros.

Solution.

All degree 2 polynomials with 2 distinct real zeros work. If $p(x) = ax^2 + bx + c$ has two real zeros $r_1 < r_2$ (i.e., $b^2 - 4ac > 0$), then

$$p(x) = a(x - r_1)(x - r_2);$$

comparing coefficients of x, we get $-b/a = r_1 + r_2$, from which

$$p'(x) = 2a(x - (r_1 + r_2)/2).$$

Geometrically, this is clearer: $y = p(x)$ is a parabola, symmetric about some vertical axis $x = d$, and $p'(d) = 0$. The zeros $x = r_1$, $x = r_2$ must also be symmetric about the axis, so $d = (r_1 + r_2)/2$.

No polynomial of higher degree works. Suppose $r_1 < \cdots < r_n$ are all real and $n > 2$, so

$$p(x) = a(x - r_1)(x - r_2)\cdots(x - r_n).$$

Let $r = (r_{n-1} + r_n)/2$.

From here, we can follow two (similar) approaches.

Approach 1 uses the following exercise: If $p(x)$ is a degree n polynomial with zeros $r_1, \ldots, r_n$, then

$$p'(x) = p(x)\left(\frac{1}{x - r_1} + \cdots + \frac{1}{x - r_n}\right) \tag{1}$$

for $x \notin \{r_1, \ldots, r_n\}$. (This can be shown using the product rule for derivatives, or more directly by computing $\frac{d}{dx} \ln p(x)$ in two ways. This useful equation also comes up in 1992A2.) Since $p(r) \neq 0$,

$$\begin{aligned}
\frac{p'(r)}{p(r)} &= \frac{1}{r - r_1} + \cdots + \frac{1}{r - r_{n-2}} + \frac{1}{r - r_{n-1}} + \frac{1}{r - r_n} \\
&= \frac{1}{r - r_1} + \cdots + \frac{1}{r - r_{n-2}} \qquad (\text{since } r - r_n = -(r - r_{n-1})) \\
&> 0,
\end{aligned}$$

so $p'(r) \neq 0$.

Approach 2 is the following: let $q(x) = (x - r_1)\cdots(x - r_{n-2})$ so

$$p(x) = (x - r_1)(x - r_2) \cdot q(x),$$

and apply the product rule to obtain

$$p'(x) = a \cdot 2(x - r)q(x) + a(x - r_{n-1})(x - r_n)q'(x).$$

Rolle's Theorem (see remark below) implies that all the zeros of $q'(x)$ lie between r_1 and r_{n-2}. Hence $(r_{n-1} + r_n)/2$ is not a zero of $q'(x)$, so $p(x)$ does not satisfy the hypotheses of the problem. ∎

Remark. *Rolle's Theorem* is the following:

Rolle's Theorem. *Let $[a,b]$ be a closed interval in $\mathbb{R}$. Let $f(t)$ be a function that is continuous on $[a,b]$ and differentiable on (a,b), and suppose that $f(a) = f(b)$. Then there exists $c \in (a,b)$ such that $f'(c) = 0$.*

In particular, if $f(t) \in \mathbb{R}[t]$ is a polynomial of degree n with n distinct real zeros, then the real zeros of $f'(t)$ are contained in the interval spanned by the zeros of $f(t)$. The previous sentence has a generalization to complex polynomials, known as *Lucas' Theorem*, or sometimes as the *Gauss-Lucas Theorem*:

Lucas' Theorem. *If $p(z) \in \mathbb{C}[z]$ is a polynomial of degree at least 1, then the zeros of $p'(z)$ are contained in the convex hull of the set of zeros of $p(z)$.*

See [Mar] for this and many related results, as well as the following open question:

> Let $p(z) \in \mathbb{C}[z]$ be a polynomial whose complex zeros all lie in the disc $|z| \leq 1$, and let z_1 be any one of the zeros. Must $p'(z)$ have a zero in the disc $|z - z_1| \leq 1$?

B. Sendov conjectured in 1962 that the answer is yes, but in the literature it is sometimes called the "Ilyeff Conjecture," because of a misattribution. The statement has been proved for $\deg p \leq 8$ [BX].

Related question. Let $n > 2$, and let $p(x) \in \mathbb{R}[x]$ be a polynomial of degree n with n distinct real zeros $r_1 < \cdots < r_n$. Rolle's Theorem implies that $p'(x)$ has $n-1$ distinct real zeros, interlaced between the zeros of $p(x)$. Prove that the largest real zero of $p'(x)$ is closer to r_n than to r_{n-1}, and that the smallest real zero of $p'(x)$ is closer to r_1 than to r_2.

See [An] for related ideas.

A4. (86, 33, 43, 0, 0, 0, 0, 0, 12, 3, 21, 15)

Does there exist an infinite sequence of closed discs $D_1, D_2, D_3, \ldots$ in the plane, with centers $c_1, c_2, c_3, \ldots$, respectively, such that

(i) the c_i have no limit point in the finite plane,

(ii) the sum of the areas of the D_i is finite, and

(iii) every line in the plane intersects at least one of the D_i?

Answer. Yes, such a sequence of closed discs exists.

Solution. Let $a_i = 1/i$ for $i \geq 1$ (or choose any other sequence of positive numbers a_i satisfying $\sum_{i=1}^{\infty} a_i = \infty$ and $\sum_{i=1}^{\infty} a_i^2 < \infty$). For $n \geq 1$, let $A_n = a_1 + a_2 + \cdots + a_n$. Let U be the union of the discs of radius a_n centered at $(A_n, 0)$, $(-A_n, 0)$, $(0, A_n)$, $(0, -A_n)$, for all $n \geq 1$. Then U covers the two coordinate axes, and has finite total area. Every line in the plane meets at least one axis, and hence meets U. Finally, the centers have no limit point, since every circle C centered at the origin encloses at most finitely many centers: if C has radius R, we can choose n such that $A_n > R$, and then less than $4n$ centers lie inside C. ■

Motivation. The essential idea is as follows. This construction covers a one-dimensional set (the union of the two axes). Any line must meet this set, and it is possible to cover it with an infinite number of circles of finite area. Then a little care must be taken to make sure the centers do not have a limit point. Another one-dimensional set that would work is the union of the circles of integer radius centered at the origin.

Related question. *Gabriel's Horn* is the surface of revolution G obtained by revolving the graph of $y = 1/x$, $x \geq 1$ around the x-axis. Prove that G encloses a finite volume but has infinite surface area. (This leads to the following fun "paradox": one cannot paint the inside of the horn because it has infinite surface area, yet one can paint it by filling it with a finite amount of paint and then emptying it!)

A5. (23, 4, 5, 0, 0, 0, 0, 0, 3, 6, 82, 90)

Find the maximum value of

$$\int_0^y \sqrt{x^4 + (y - y^2)^2}\, dx$$

for $0 \leq y \leq 1$.

Answer. The maximum value of the integral is $1/3$.

Motivation. To find the maximum of $I(y)$, one naturally looks at $I'(y)$; if it is never 0, then the maximum must occur at an endpoint of the interval in question.

Solution 1. For $0 \leq y \leq 1$ let $I(y) = \int_0^y \sqrt{x^4 + (y - y^2)^2} dx$.

Note that $I(1) = \int_0^1 x^2 dx = 1/3$. We will prove $I'(y) > 0$ for $0 < y < 1$. Since $I(y)$ is continuous on $[0, 1]$, this will prove that $1/3$ is the maximum.

By the remark following this solution,

$$I'(y) = \sqrt{y^4 + (y - y^2)^2} + (y - y^2)(1 - 2y) \int_0^y \frac{1}{\sqrt{x^4 + (y - y^2)^2}} dx,$$

so we must check that

$$\sqrt{2y^2 - 2y + 1} > (1 - y)(2y - 1) \int_0^y \frac{1}{\sqrt{x^4 + (y - y^2)^2}} dx. \tag{1}$$

If $y < 1/2$, (1) holds because the right side is negative. If $y \geq 1/2$, (1) would follow from

$$\begin{aligned}\sqrt{2y^2 - 2y + 1} > (1 - y)(2y - 1) &\int_0^y \frac{1}{\sqrt{(y - y^2)^2}} dx \\ &= (1 - y)(2y - 1)y/(y - y^2) \\ &= 2y - 1,\end{aligned}$$

which is equivalent to $2y > 2y^2$, hence true, since $0 < y < 1$. ■

Remark. We discuss how to differentiate certain integrals depending on a parameter. Fix $M > 0$, and define the triangle

$$T = \{\, (x, y) \in \mathbb{R}^2 : 0 < y < M, 0 \leq x \leq y \,\}.$$

Suppose that $f(x, y)$ is a continuous function on T such that the partial derivative f_y (alternative notation for $\partial f / \partial y$) exists on the interior of T and extends to a continuous function on T. For $0 < y < M$, let $I(y) = \int_0^y f(x, y)\, dx$. We wish to compute $I'(y)$ for $0 < y < M$.

We will express $I'(y)$ in terms of partial derivatives of the auxiliary function $J(t, u) = \int_0^t f(x, u)\, dx$ defined for $(t, u) \in T$. By the Fundamental Theorem of Calculus, $J_t(t, u) = f(t, u)$ on T. (If $t = 0$ or $t = u$, we consider J_t as only a one-sided derivative.)

Since f_y extends to a continuous function on T, we may differentiate under the integral sign to obtain $J_u(t,u) = \int_0^t f_y(x,u)\,dx$ on T. Then by the Multivariable Chain Rule [Ru, p. 214],

$$\begin{aligned} I'(y) &= \frac{dJ(y,y)}{dy} \\ &= J_t(y,y)\frac{\partial y}{\partial y} + J_u(y,y)\frac{\partial y}{\partial y} \\ &= f(y,y) + \int_0^y f_y(x,y)\,dx. \end{aligned}$$

The same method can be used to differentiate integrals of the form

$$I(y) = \int_{g(y)}^{h(y)} f(x,y)\,dx.$$

Solution 2. If $u, v \geq 0$, then $\sqrt{u^2+v^2} \leq \sqrt{u^2+2uv+v^2} = u+v$. Taking $u = x^2$ and $v = y - y^2$ we obtain

$$\begin{aligned} \int_0^y \sqrt{x^4 + (y-y^2)^2}\,dx &\leq \int_0^y \left(x^2 + (y-y^2)\right)\,dx \\ &= y^2 - \frac{2}{3}y^3 \\ &\leq \frac{1}{3}, \end{aligned}$$

with equality everywhere if $y = 1$: the last inequality uses the positivity of $\frac{d}{dy}\left(y^2 - \frac{2}{3}y^3\right) = 2y(1-y)$ on $(0,1)$. Thus the maximum value is $1/3$. ■

A6. (8, 21, 8, 1, 0, 0, 0, 0, 6, 7, 40, 122)

Let $A(n)$ denote the number of sums of positive integers $a_1 + a_2 + \cdots + a_r$ which add up to n with $a_1 > a_2 + a_3$, $a_2 > a_3 + a_4$, $\ldots$, $a_{r-2} > a_{r-1} + a_r$, $a_{r-1} > a_r$. Let $B(n)$ denote the number of $b_1 + b_2 + \cdots + b_s$ which add up to n, with

(i) $b_1 \geq b_2 \geq \cdots \geq b_s$,

(ii) each b_i is in the sequence $1, 2, 4, \ldots, g_j, \ldots$ defined by $g_1 = 1$, $g_2 = 2$, and $g_j = g_{j-1} + g_{j-2} + 1$, and

(iii) if $b_1 = g_k$ then every element in $\{1, 2, 4, \ldots, g_k\}$ appears at least once as a b_i.

Prove that $A(n) = B(n)$ for each $n \geq 1$.

(For example, $A(7) = 5$ because the relevant sums are 7, $6+1$, $5+2$, $4+3$, $4+2+1$, and $B(7) = 5$ because the relevant sums are $4+2+1$, $2+2+2+1$, $2+2+1+1+1$, $2+1+1+1+1+1$, $1+1+1+1+1+1+1$.)

Solution 1. First, given a sum counted by $A(n)$, we construct a "tableau" of Fibonacci numbers. Start with two rows, the lower row with a_r ones and the upper with a_{r-1} ones, as shown below.

$$\begin{array}{rccccccc} a_{r-1}: & 1 & 1 & 1 & 1 & 1 & 1 & 1 \\ a_r: & 1 & 1 & 1 & 1 & 1 \end{array}$$

The top row is longer than the bottom row since $a_{r-1} > a_r$.

Since $a_{r-2} > a_{r-1} + a_r$, we may add a new row on top, such that each new entry directly above two entries is the sum of those two, each entry above a single entry is equal to that entry, and the rest are enough ones (at least one) to make the sum of the new row equal a_{r-2}.

$$\begin{array}{rcccccccccc} a_{r-2}: & 2 & 2 & 2 & 2 & 2 & 1 & 1 & 1 & 1 & 1 \\ a_{r-1}: & 1 & 1 & 1 & 1 & 1 & 1 & 1 \\ a_r: & 1 & 1 & 1 & 1 & 1 \end{array}$$

Next, $a_{r-3} > a_{r-2} + a_{r-1}$, so we can add another row on top, such that each entry above at least two entries is the sum of the two below it, each entry above one entry is equal to that entry, and the rest are enough ones (at least one) to make the row sum a_{r-3}, as shown below.

$$\begin{array}{rccccccccccc} a_{r-3}: & 3 & 3 & 3 & 3 & 3 & 2 & 2 & 1 & 1 & 1 & 1 \\ a_{r-2}: & 2 & 2 & 2 & 2 & 2 & 1 & 1 & 1 & 1 & 1 \\ a_{r-1}: & 1 & 1 & 1 & 1 & 1 & 1 & 1 \\ a_r: & 1 & 1 & 1 & 1 & 1 \end{array}$$

Continue this construction to make a tableau of r rows, in which the total of all entries is n. Let s be the number of columns, and let b_i denote the sum of the ith column. Then $b_i = F_1 + F_2 + \cdots + F_j$, where j is the length of the ith column. The sequence $g'_j = F_1 + F_2 + \cdots + F_j$ satisfies $g'_1 = 1 = g_1$, $g'_2 = 2 = g_2$ and for $j \geq 3$,

$$\begin{aligned} g'_{j-1} + g'_{j-2} + 1 = F_1 + F_2 + \cdots + F_{j-1} & \\ + F_1 + \cdots + F_{j-2} + 1 & \\ = F_2 + F_3 + \cdots + F_j \quad + F_1 & \qquad (\text{since } F_1 = F_2 = 1) \\ = g'_j, & \end{aligned}$$

so $g'_j = g_j$ for all j. Thus the b_i satisfy condition (i) in the definition of $B(n)$. The sum of the b_i equals the sum of all entries of the tableau, namely n. Each row is at least one longer than the next row, so each column is either equal in length or one longer than the next column. Thus the b_i satisfy conditions (ii) and (iii) as well.

Conversely, given a sum $b_1 + \cdots + b_s$ counted by $B(n)$, construct a (top and left justified) tableau in which the entries of the ith column from the top down are F_k, $F_{k-1}, \ldots, F_1$, where k is the positive integer such that $b_i = g_k$; then let a_i denote the sum of the ith row.

It is straightforward to check that these constructions define inverse bijections between the partitions counted by $A(n)$ and those counted by $B(n)$. Thus $A(n) = B(n)$ for all n. ■

Solution 2. Let $A(n,r)$ denote the number of sums counted by $A(n)$ with exactly r terms. Let $B(n,r)$ denote the number of sums counted by $B(n)$ in which $b_1 = g_r$.

Fix $r \geq 1$. Form the $r \times r$ matrix

$$M = (m_{ij}) = \begin{pmatrix} 1 & -1 & -1 & 0 & \cdots & 0 \\ 0 & 1 & -1 & -1 & \cdots & 0 \\ 0 & 0 & 1 & -1 & \cdots & 0 \\ 0 & 0 & 0 & 1 & \cdots & 0 \\ \vdots & \vdots & \vdots & \vdots & \ddots & \vdots \\ 0 & 0 & 0 & 0 & \cdots & 1 \end{pmatrix}$$

with

$$m_{ij} = \begin{cases} 1 & \text{if } j - i = 0, \\ -1 & \text{if } j - i = 1 \text{ or } j - i = 2, \\ 0 & \text{otherwise.} \end{cases}$$

Let $\mathbf{a}$ and $\mathbf{c}$ denote column vectors $(a_1, a_2, \ldots, a_r)$ and $(c_1, c_2, \ldots, c_r)$, respectively. Since M is an integer matrix with determinant 1, the relation $M\mathbf{a} = \mathbf{c}$ is a bijection between the set of $\mathbf{a}$ in $\mathbb{Z}^r$ and the set of $\mathbf{c}$ in $\mathbb{Z}^r$. Under this bijection, the inequalities $a_1 > a_2 + a_3, \ldots, a_{r-2} > a_{r-1} + a_r, a_{r-1} > a_r, a_r > 0$ correspond to $c_1 > 0, \ldots, c_{r-2} > 0, c_{r-1} > 0, c_r > 0$. Also, applying the identity

$$\begin{pmatrix} g_1 & g_2 \cdots & g_r \end{pmatrix} M = \begin{pmatrix} 1 & 1 & \cdots & 1 \end{pmatrix}$$

to $\mathbf{a}$ shows that the condition $\sum a_i = n$ corresponds to $\sum g_i c_i = n$. Thus $A(n,r)$, which counts the number of $\mathbf{a} \in \mathbb{Z}^r$ satisfying the inequalities and sum condition, equals the number of $\mathbf{c} \in \mathbb{Z}^r$ satisfying $c_i > 0$ for all i and $\sum g_i c_i = n$. Such a $\mathbf{c}$ may be matched with the sequence $b_1, b_2, \ldots, b_s$ consisting of c_r copies of $g_r, \ldots,$ and c_1 copies of g_1; the b-sequences arising in this way are exactly the sequences summing to n that satisfy conditions (i)–(iii) of the problem and $b_1 = g_r$. Hence the number of such $\mathbf{c}$ equals $B(n,r)$.

Thus $A(n,r) = B(n,r)$. Summing over r yields $A(n) = B(n)$. ∎

Related question. The most famous result along these lines is the following theorem of Euler [NZM, Theorem 10.2]:

Let $O(n)$ be the number of (nonincreasing) partitions of a positive integer n into odd parts, and let $D(n)$ be the number of (nonincreasing) partitions of n into different parts. Then $O(n) = D(n)$ for all n.

B1. (192, 6, 2, 0, 6, 0, 0, 0, 0, 5, 0, 2)

For each integer $n \geq 0$, let $S(n) = n - m^2$, where m is the greatest integer with $m^2 \leq n$. Define a sequence $(a_k)_{k=0}^{\infty}$ by $a_0 = A$ and $a_{k+1} = a_k + S(a_k)$ for $k \geq 0$. For what positive integers A is this sequence eventually constant?

Answer. This sequence is eventually constant if and only if A is a perfect square.

Solution. If a_k is a perfect square, then $a_{k+1} = a_k$, and the sequence is constant thereafter.

Conversely, if a_k is *not* a perfect square, then suppose $r^2 < a_k < (r+1)^2$. Then $S(a_k) = a_k - r^2$ is in the interval $[1, 2r]$, so $a_{k+1} = r^2 + 2S(a_k)$ is greater than r^2 but less than $(r+2)^2$, and not equal to $(r+1)^2$ by parity. Thus a_{k+1} is also not a perfect square, and is greater than a_k. Hence if A is not a perfect square, then no a_k is a perfect square, and the sequence diverges to infinity. ■

B2. (93, 30, 8, 0, 0, 0, 0, 0, 7, 1, 57, 17)

Suppose f and g are nonconstant, differentiable, real-valued functions on $\mathbb{R}$. Furthermore, suppose that for each pair of real numbers x and y,

$$f(x+y) = f(x)f(y) - g(x)g(y),$$
$$g(x+y) = f(x)g(y) + g(x)f(y).$$

If $f'(0) = 0$, prove that $(f(x))^2 + (g(x))^2 = 1$ for all x.

Motivation. We can use calculus to reach the intermediate goal of proving that the $H(x) = f(x)^2 + g(x)^2$ is constant. In other words, we try to prove $H'(x) = 0$ by differentiating the given functional equations. This motivates the first solution.

Alternatively, the hypotheses and conclusion in the problem are recognizable as identities satisfied by $f(x) = \cos x$ and $g(x) = \sin x$. If $f(x)$ *were* $\cos x$ and $g(x)$ *were* $\sin x$, then $h(x) = f(x) + ig(x)$ would satisfy $h(x+y) = h(x)h(y)$. This suggests the approach of the second solution.

Solution 1. Differentiate both sides with respect to y to obtain

$$f'(x+y) = f(x)f'(y) - g(x)g'(y),$$
$$g'(x+y) = f(x)g'(y) + g(x)f'(y).$$

Setting $y = 0$ yields

$$f'(x) = -g'(0)g(x) \quad \text{and} \quad g'(x) = g'(0)f(x).$$

Thus

$$2f(x)f'(x) + 2g(x)g'(x) = 0,$$

and therefore

$$f(x)^2 + g(x)^2 = C$$

for some constant C. Since f and g are nonconstant, $C \neq 0$. The identity

$$f(x+y)^2 + g(x+y)^2 = \left(f(x)^2 + g(x)^2\right)\left(f(y)^2 + g(y)^2\right),$$

implies $C = C^2$. But $C \neq 0$, so $C = 1$. ■

Solution 2. Define $h : \mathbb{R} \to \mathbb{C}$ by $h(x) = f(x) + ig(x)$. Then h is differentiable, and $h'(0) = bi$ for some $b \in \mathbb{R}$. The two given functional equations imply $h(x+y) = h(x)h(y)$. Differentiating with respect to y and substituting $y = 0$ yields $h'(x) = h(x)h'(0) = bi \cdot h(x)$, so $h(x) = Ce^{bix}$ for some $C \in \mathbb{C}$. From $h(0+0) = h(0)h(0)$ we obtain $C = C^2$. If $C = 0$, then h would be identically zero, and f and g would be constant, contradiction. Thus $C = 1$. Finally, for any $x \in \mathbb{R}$,

$$f(x)^2 + g(x)^2 = |h(x)|^2 = |e^{bix}|^2 = 1.$$ ■

Remark. It follows that $f(x) = \cos(bx)$ and $g(x) = \sin(bx)$ for some nonzero $b \in \mathbb{R}$.

Remark. Solution 1 is very close to what one gets if one writes out Solution 2 in terms of real and imaginary parts.

Remark. Suppose we change the problem by dropping the assumption that f and g are differentiable, and assume only that f and g are continuous, and that $f'(0)$ exists and is zero. Then we could still conclude $f(x)^2 + g(x)^2 = 1$ by following Solution 2, because it is true that any continuous function $h : \mathbb{R} \to \mathbb{C}$ satisfying $h(x + y) = h(x)h(y)$ is either identically zero, or of the form $h(x) = e^{bx}$ for some $b \in \mathbb{C}$. This is a consequence of the following lemma, sometimes attributed to Cauchy.

Lemma. *Suppose $f : \mathbb{R} \to \mathbb{R}$ is a continuous function such that*

$$f(x + y) = f(x) + f(y). \tag{1}$$

Then $f(x) = cx$ for some $c \in \mathbb{R}$.

Proof. By substituting $x = y = 0$ into (1), we find $f(0) = 0$. Let $c = f(1)$. It is not hard to show that $f(x) = cx$ for rational x (by using (1) repeatedly). Since $f(x) - cx$ is a continuous function that vanishes on a dense set (the rationals), it must be 0. □

Reinterpretation. Solution 2 can be understood in a more sophisticated context. If h is not identically zero, $h : \mathbb{R} \to \mathbb{C}^*$ is a continuous homomorphism between real Lie groups, so by [War, Theorem 3.38], it is a homomorphism of Lie groups. (In other words, it is automatically analytic.) Since homomorphisms of Lie groups are determined by their induced Lie algebra homomorphisms [War, Theorem 3.16], h must be identically zero, or of the form $h(x) = e^{bx}$ for some $b \in \mathbb{C}$.

B3. (38, 11, 4, 0, 0, 0, 0, 0, 0, 5, 7, 49, 99)

Does there exist a real number L such that, if m and n are integers greater than L, then an $m \times n$ rectangle may be expressed as a union of 4×6 and 5×7 rectangles, any two of which intersect at most along their boundaries?

Answer. Yes, such a real number L exists.

The solution will use a generalization of the following well-known result:

Theorem 1. *If a and b are positive integers, then there exists a number g such that every multiple of $\gcd(a, b)$ greater than g may be written in the form $ra + sb$, where r and s are nonnegative integers.*

This is sometimes called the "Postage Stamp Theorem" because if $\gcd(a, b) = 1$, then every amount of postage greater than g cents can be paid for with a-cent and b-cent stamps. In this case, g may be taken to be $ab - a - b$, but no smaller: $ab - a - b$ is not of the form $ra + sb$ with $r, s \geq 0$. For further discussion, see [NZM, Section 5.1]; the theorem appears as Problem 16.

Proof. Suppose first that $\gcd(a, b) = 1$. Then $0, a, 2a, \ldots, (b - 1)a$ is a complete set of residues modulo b. Thus, for any integer k greater than $(b-1)a-1$, $k - qb = ja$ for some $q \geq 0$, $j = 0, 1, 2, \ldots, b - 1$, hence the claim for this special case.

For general a and b, write $a = da_0$ and $b = db_0$, where $d = \gcd(a, b)$ and $\gcd(a_0, b_0) = 1$. We showed that all sufficiently positive integers are expressible as $ra_0 + sb_0$. Multiplying by d, we find that all sufficiently positive multiples of d are expressible as $ra + sb$. □

Solution. We begin by forming 20×6 and 20×7 rectangles. From Theorem 1, we may form $20 \times n$ rectangles for n sufficiently large. We may also form 35×5 and 35×7 rectangles, hence $35 \times n$ rectangles for n sufficiently large. We may further form 42×4 and 42×5 rectangles, hence $42 \times n$ rectangles for n sufficiently large.

Since $\gcd(20, 35) = 5$, there exists a multiple m_0 of 5, relatively prime to 42 and independent of n, for which we may form an $m_0 \times n$ rectangle. Finally, since $\gcd(m_0, 42) = 1$, we may use $m_0 \times n$ and $42 \times n$ rectangles to form $m \times n$ rectangles for all m and n sufficiently large. ∎

Related question. By using the explicit value of g in the discussion after the statement of the Theorem 1, this approach shows that one can construct an $m \times n$ rectangle if $m \geq 41 \cdot 54 = 2214$ and $n \geq 30$. Hence one can take $L = 2213$ in the original problem. Dave Savitt has shown that the conditions $m \geq 65$ and $n \geq 80$ suffice. Hence one can take $L = 79$. How much better can one do? We do not know the smallest possible L.

Related question. Problem 1971A5 [PutnamII, p. 15] is related. See also [Hon2, Ch. 13]:

> A game of solitaire is played as follows. After each play, according to the outcome, the player receives either a or b points (a and b are positive integers with a greater than b), and his score accumulates from play to play. It has been noticed that there are thirty-five non-attainable scores and that one of these is 58. Find a and b.

Remark. Let $a_1, \ldots, a_n$ be positive integers with greatest common divisor 1. The problem of determining the greatest integer $g = g(a_1, \ldots, a_n)$ not expressible as $\sum_{i=1}^{n} r_i a_i$ with r_i nonnegative integers is called the Frobenius Problem, because according to A. Brauer [Br], Frobenius mentioned it in his lectures. There is an extensive literature on the problem, dating back at least to the nineteenth century [Shar]. For each fixed n the problem can be solved in polynomial time, but the problem as a whole is NP-hard [Ka]. For the solution in some special cases, see [Sel]. The Frobenius Problem is relevant to the running time analysis of the sorting algorithm Shellsort [P1].

The reader may enjoy proving that g exists (say by induction on n).

Related question. Another generalization of Theorem 1 is Problem 3 from the 1983 International Mathematical Olympiad [IMO79–85, p. 6] In the language of the preceding remark, it asks for a proof that $g(bc, ca, ab) = 2abc - ab - bc - ca$, where a, b, and c be positive integers, no two of which have a common divisor greater than 1.

Remark. Suppose that we insist that the length 4 side of each 4×6 rectangle be parallel to the length m side of the big $m \times n$ rectangle, and suppose we similarly restrict the orientation of the 5×7 rectangles. Then the answer to the problem becomes NO!

Let us explain why. Label the 1×1 squares in the $m \times n$ rectangle in the obvious way with (i, j) where i, j are integers satisfying $0 \leq i \leq m-1$ and $0 \leq j \leq n-1$. Write the monomial $x^i y^j$ in the square (i, j). Then the sum of the monomials inside an $a \times b$ rectangle equals $x^i y^j (1+x+\cdots+x^{a-1})(1+y+\cdots+y^{b-1})$ for some nonnegative integers i, j. If the $m \times n$ rectangle is a disjoint union of 4×6 and 5×7 rectangles (disjoint except at boundaries), then the polynomial $f(x, y) = (1+x+\cdots+x^{m-1})(1+y+\cdots+y^{n-1})$

is in the ideal of $\mathbb{C}[x,y]$ generated by $(1+x+\cdots+x^{4-1})(1+y+\cdots+y^{6-1})$ and $(1+x+\cdots+x^{5-1})(1+y+\cdots+y^{7-1})$. In particular,

$$f(x,y)=\frac{(x^m-1)(y^n-1)}{(x-1)(y-1)}$$

must vanish at $(e^{2\pi i/4},e^{2\pi i/7})$. But if 4 does not divide m and 7 does not divide n, then $f(e^{2\pi i/4},e^{2\pi i/7})$ is *nonzero* and hence it is impossible to dissect the $m\times n$ rectangle in the specified way.

This argument is related to a beautiful proof [Wag, Proof 1] of the following remarkable fact.

Theorem 2. *If a rectangle is tiled by rectangles each of which has at least one integer side, then the tiled rectangle has at least one integer side.*

Remark. Theorem 1 also comes up in other advanced contexts. For example, it corresponds to the fact that two measures of the "nonsmoothness" of the complex singularity given by $y^p=x^q$ in $\mathbb{C}^2$ are the same [ACGH, p. 60].

B4. (21, 1, 7, 0, 0, 0, 0, 0, 23, 1, 37, 123)

Suppose p is an odd prime. Prove that

$$\sum_{j=0}^{p}\binom{p}{j}\binom{p+j}{j}\equiv 2^p+1 \pmod{p^2}.$$

Solution 1. The sum $\sum_{j=0}^{p}\binom{p}{j}\binom{p+j}{j}$ is the coefficient of x^p in

$$\begin{aligned}\sum_{j=0}^{p}\binom{p}{j}(1+x)^{p+j}&=\left(\sum_{j=0}^{p}\binom{p}{j}(1+x)^j\right)(1+x)^p\\&=((1+x)+1)^p(1+x)^p\\&=(2+x)^p(1+x)^p,\end{aligned}$$

so this coefficient equals

$$\sum_{k=0}^{p}\binom{p}{k}\binom{p}{p-k}2^k.$$

But p divides $\binom{p}{k}$ for $k\neq 0$. Thus,

$$\begin{aligned}\sum_{j=0}^{p}\binom{p}{j}\binom{p+j}{p}&\equiv\binom{p}{0}\binom{p}{p}2^0+\binom{p}{p}\binom{p}{0}2^p\\&\equiv 1+2^p \pmod{p^2}.\end{aligned}$$

■

Remark. If at each step of Solution 1, one writes the coefficient of x^p explicitly, one obtains a solution that does not use generating functions, but instead uses Vandermonde's identity (mentioned in 1987B2) at the first equality sign.

Solution 2. Modulo p, the sets $\{1, 2, \ldots, p-1\}$ and $\{1^{-1}, 2^{-1}, \ldots, (p-1)^{-1}\}$ are the same (where the inverses are modulo p). Thus

$$\begin{aligned}\binom{2p}{p} &= \frac{p+1}{1} \cdot \frac{p+2}{2} \cdots \frac{p+(p-1)}{p-1} \cdot \frac{2p}{p} \\ &\equiv (1^{-1}p+1)(2^{-1}p+1)\cdots((p-1)^{-1}p+1)\cdot 2 \\ &\equiv 2(1p+1)(2p+1)\cdots((p-1)p+1) \\ &\equiv 2\left(\left(\sum_{k=1}^{p-1} k\right)p+1\right) \\ &\equiv 2 \pmod{p^2}.\end{aligned}$$

Although $1^{-1}, \ldots, (p-1)^{-1}$ are only defined modulo p, they appear in the equations above multiplied by p, so the terms make sense modulo p^2.

$$\begin{aligned}\sum_{j=0}^{p}\binom{p}{j}\binom{p+j}{j} &= 1+\sum_{j=1}^{p-1}\binom{p}{j}\frac{p+1}{1}\frac{p+2}{2}\cdots\frac{p+j}{j}+\binom{2p}{p} \\ &\equiv 1+\sum_{j=1}^{p-1}\binom{p}{j}(1^{-1}p+1)(2^{-1}p+1)\cdots(j^{-1}p+1)+2 \\ &\equiv 1+\sum_{j=1}^{p-1}\binom{p}{j}+2 \qquad \left(\text{since } p \mid \tbinom{p}{j} \text{ for } 1 \le j \le p-1\right) \\ &\equiv \sum_{j=0}^{p}\binom{p}{j}+1 \\ &\equiv 2^p+1 \pmod{p^2}.\end{aligned}$$

■

Remark. Here is another way to prove $\binom{2p}{p} \equiv 2 \pmod{p^2}$ for odd primes p. We have

$$\binom{2p}{p} = 2\,\frac{(p+1)(p+2)\cdots(p+p-1)}{1\cdot 2\cdots(p-1)} \tag{1}$$

and

$$\begin{aligned}(x+1)(x+2)\cdots(x+p-1) &= \prod_{j=1}^{(p-1)/2}(x+j)(x+p-j) \\ &= \prod_{j=1}^{(p-1)/2}(x^2+px+j(p-j)) \\ &= (p-1)!+\lambda px+x^2f(x),\end{aligned}$$

for some $\lambda \in \mathbb{Z}$ and $f(x) \in \mathbb{Z}[x]$. Substituting $x = p$ yields

$$(p+1)(p+2)\cdots(p+p-1) \equiv (p-1)! \pmod{p^2},$$

which we substitute into (1). Since $(p-1)!$ is prime to p, we deduce $\binom{2p}{p} \equiv 2 \pmod{p^2}$.

Remark. In fact, if $p > 3$, then $\binom{2p}{p} \equiv 2 \pmod{p^3}$; this is *Wolstenholme's Theorem.* Another version of this theorem is that if p is a prime greater than 3, then the numerator of $1 + \frac{1}{2} + \cdots + \frac{1}{p-1}$ is divisible by p^2. (See [Bak1, p. 26] or the remarks following 1997B3.) It is not hard to derive one version from the other; see below.

Remark. Wolstenholme's Theorem can be generalized. For example, for $p \geq 5$,

$$\binom{2p}{p} \equiv \binom{2}{1} \pmod{p^3}$$

$$\binom{2p^2}{p^2} \equiv \binom{2p}{p} \pmod{p^6},$$

$$\binom{2p^3}{p^3} \equiv \binom{2p^2}{p^2} \pmod{p^9},$$

and so on. Like Wolstenholme's Theorem, these assertions can be recast in terms of a sum of reciprocals. Namely, write

$$\binom{2p^n}{p^n} = \prod_{i=1}^{p^n} \left(\frac{p^n + i}{i} \right)$$

and divide up the terms into those with i divisible by p and those with i not divisible by p. This gives

$$\binom{2p^n}{p^n} = \left(\prod_{j=1}^{p^{n-1}} \frac{p^n + pj}{pj} \right) \left(\prod_{j=1}^{p^{n-1}} \prod_{i=1}^{p-1} \frac{p^n + pj + i}{pj + i} \right).$$

The term in the first parentheses is precisely $\binom{2p^{n-1}}{p^{n-1}}$. Therefore

$$\begin{aligned} \frac{\binom{2p^n}{p^n}}{\binom{2p^{n-1}}{p^{n-1}}} &= \prod_{j=1}^{p^{n-1}} \prod_{i=1}^{p-1} \left(1 + \frac{p^n}{pj + i} \right) \\ &\equiv 1 + p^n \sum_{j=1}^{p^{n-1}} \sum_{i=1}^{p-1} \frac{1}{pj + i} + p^{2n} \sum_{pj+i \neq pj'+i'} \frac{1}{(pj + i)(pj' + i')} \pmod{p^{3n}}. \end{aligned}$$

The last sum is divisible by p^n. Thus the assertions follow from the congruence

$$\sum_{j=1}^{p^{n-1}} \sum_{i=1}^{p-1} \frac{1}{pj + i} \equiv 0 \pmod{p^{2n}},$$

which is discussed in remarks following 1997B3.

Remark. Congruences between binomial coefficients can be proved using analytic properties of Morita's p-adic gamma function $\Gamma_p(x)$. See [Robe, p. 380] for such a proof of the following theorem of Kazandzidis, which generalizes the congruences of the previous remark: for all primes $p \geq 5$ we have

$$\binom{pn}{pk} \equiv \binom{n}{k} \pmod{p^m},$$

where p^m is the largest power of p dividing $p^3nk(n-k)\binom{n}{k}$. The same congruence holds for $p=3$, but with p^3 replaced by 3^2. (In other words, one loses a factor of 3.)

B5. (38, 4, 3, 0, 3, 0, 1, 0, 9, 3, 50, 102)

Let p be an odd prime and let $\mathbb{Z}_p$ denote (the field of) integers modulo p.

How many elements are in the set

$$\{x^2 : x \in \mathbb{Z}_p\} \cap \{y^2+1 : y \in \mathbb{Z}_p\}?$$

Warning. In current mathematics, especially in number theory, the notation $\mathbb{Z}_p$ is usually reserved for the ring of p-adic integers. If one intends the additive group of integers modulo p, it is safer to write $\mathbb{Z}/p\mathbb{Z}$ or its abbreviation $\mathbb{Z}/p$, or $\mathbb{F}_p$ if, as in this problem, one wishes to consider the set as a field. More generally, for any prime power $q=p^m$ with p prime and $m \geq 1$, there is a unique finite field with q elements up to isomorphism [Se1, Theorem 1], and it is denoted $\mathbb{F}_q$, or GF(q) in older literature.

Answer. The number of elements in the intersection is $\lceil p/4 \rceil = \lfloor (p+3)/4 \rfloor$. In other words, the answer is

$$\frac{p+3}{4} \text{ if } p \equiv 1 \pmod 4, \qquad \text{and} \qquad \frac{p+1}{4} \text{ if } p \equiv 3 \pmod 4.$$

Solution 1. Let S be the set of solutions to $x^2=y^2+1$ over $\mathbb{F}_p$. The linear change of coordinates $(u,v)=(x+y,x-y)$ is invertible since $\det\begin{pmatrix}1 & 1\\ 1 & -1\end{pmatrix} = -2$ is nonzero in $\mathbb{F}_p$. Hence $|S|$ equals the number of solutions to $uv=1$ over $\mathbb{F}_p$. There is one possible v for each nonzero u, and no v for $u=0$, so $|S|=p-1$.

The problem asks for the size of the image of the map $\phi: S \to \mathbb{F}_p$ taking (x,y) to x^2. If $z=x^2$ for some $(x,y)\in S$, then $\phi^{-1}(z)=\{(\pm x, \pm y)\}$, which has size 4, except in the cases $z=1$ (in which case $x=\pm 1$ and $y=0$, making $\phi^{-1}(z)$ of size 2) and $z=0$ (in which case $x=0$ and $y^2=-1$, again making $\phi^{-1}(z)$ of size 2); the latter exception occurs if and only if -1 is a square in $\mathbb{F}_p$. Hence $|S|=4|\phi(S)|-2-2c$, where c is 1 or 0 according as -1 is a square in $\mathbb{F}_p$ or not. Thus the answer to the problem is

$$|\phi(S)| = \frac{(p-1)+2+2c}{4} = \frac{p+1+2c}{4}.$$

This should be an integer, so $c=1$ if $p \equiv 1 \pmod 4$ and $c=0$ if $p \equiv 3 \pmod 4$. In either case, $|\phi(S)| = \lceil p/4 \rceil$ as claimed. ∎

Remark. Alternatively, one can count the solutions to $x^2=y^2+1$ over $\mathbb{F}_p$ by explicit parametrization:

$$x = \frac{r+r^{-1}}{2}, \qquad y = \frac{r-r^{-1}}{2}$$

for $r \in \mathbb{F}_p^*$. (Given a solution (x,y), take $r=x+y$, so that $r^{-1}=x-y$.) This parametrization shows that the conic $x^2=y^2+1$ is birationally equivalent to the line over $\mathbb{F}_p$; this is not surprising in light of the remark after Solution 2 to 1987B3.

Solution 2. We use the *Legendre symbol*, defined by

$$\left(\frac{a}{p}\right) = \begin{cases} 1, & \text{if } a \equiv k^2 \pmod p \text{ for some } k \not\equiv 0 \pmod p \\ 0, & \text{if } a \equiv 0 \pmod p \\ -1, & \text{otherwise.} \end{cases}$$

The number of nonzero squares in $\mathbb{F}_p$ equals the number of nonsquares, so

$$\sum_{a=0}^{p-1} \left(\frac{a-k}{p}\right) = 0 \tag{1}$$

for any $k \in \mathbb{Z}$.

In this solution only, if P is a statement, let $[P]$ be 1 if P is true, and 0 if P is false. Let $\mathbb{F}_p^2$ denote the set of squares in $\mathbb{F}_p$, including 0. The problem asks us to compute

$$N = \sum_{a=0}^{p-1} [a \in \mathbb{F}_p^2] \cdot [a-1 \in \mathbb{F}_p^2].$$

Substituting the identity

$$[a \in \mathbb{F}_p^2] = \frac{1}{2}\left(1 + \left(\frac{a}{p}\right) + [a=0]\right)$$

we obtain

$$\begin{aligned} N &= \frac{1}{4}\left(1 + \left(\frac{-1}{p}\right) + 1 + \left(\frac{1}{p}\right) + \sum_{a=0}^{p-1}\left(1 + \left(\frac{a}{p}\right)\right)\left(1 + \left(\frac{a-1}{p}\right)\right)\right) \\ &= \frac{1}{2}\left[\left(\frac{-1}{p}\right) = 1\right] + \frac{1}{2} + \frac{1}{4}\sum_{a=0}^{p-1}\left(1 + \left(\frac{a}{p}\right) + \left(\frac{a-1}{p}\right) + \left(\frac{a}{p}\right)\left(\frac{a-1}{p}\right)\right) \\ &= \frac{1}{2}\left[\left(\frac{-1}{p}\right) = 1\right] + \frac{1}{2} + \frac{p}{4} + \frac{1}{4}\sum_{a=0}^{p-1}\left(\frac{a}{p}\right)\left(\frac{a-1}{p}\right), \end{aligned}$$

by (1) twice. For $k \in \mathbb{Z}$, let $S(k) = \sum_{a=0}^{p-1}\left(\frac{a}{p}\right)\left(\frac{a-k}{p}\right)$. We want $S(1)$. For k not divisible by p, the substitution $a = kb$, with b running over the residue classes modulo p, shows that

$$\begin{aligned} S(k) &= \sum_{b=0}^{p-1}\left(\frac{k}{p}\right)\left(\frac{b}{p}\right)\left(\frac{k}{p}\right)\left(\frac{b-1}{p}\right) \\ &= \sum_{b=0}^{p-1}\left(\frac{b}{p}\right)\left(\frac{b-1}{p}\right) \quad \left(\text{since } \left(\tfrac{k}{p}\right)^2 = 1\right) \\ &= S(1). \end{aligned}$$

Also,

$$\sum_{k=0}^{p-1} S(k) = \sum_{a=0}^{p-1}\left(\frac{a}{p}\right)\sum_{k=0}^{p-1}\left(\frac{a-k}{p}\right) = 0,$$

since each inner sum is zero by (1). Thus

$$S(1) = -\frac{S(0)}{p-1} = -\frac{p-1}{p-1} = -1.$$

Hence

$$N = \frac{1}{2}\left[\left(\frac{-1}{p}\right) = 1\right] + \frac{p+1}{4}.$$

But N is an integer. Thus if $p \equiv 1 \pmod 4$, then $\left(\frac{-1}{p}\right) = 1$ and $N = (p+3)/4$; if $p \equiv 3 \pmod 4$, then $\left(\frac{-1}{p}\right) = -1$ and $N = (p+1)/4$. ■

Remark. In each solution, we proved:

Corollary. *The number -1 is a square modulo an odd prime p if and only if $p \equiv 1$* (mod 4).

From this and unique factorization in the ring of Gaussian integers

$$\mathbb{Z}[i] = \{\, a + bi : a, b \in \mathbb{Z}\,\},$$

one can prove:

Theorem. *Every prime $p \equiv 1 \pmod 4$ is a sum of two squares.*

Hint: There is some x such that $x^2 \equiv -1 \pmod p$, i.e., there is some integer k such that $kp = (x-i)(x+i)$; consider prime factorizations of both sides, and note that p is not a factor of $x \pm i$.

A remarkable one-sentence proof of the theorem, due to D. Zagier [Z], is the following:

The involution on the finite set $S = \{\,(x,y,z) \in \mathbb{Z}^{\geq 0} : x^2 + 4yz = p\,\}$ defined by

$$(x,y,z) \mapsto \begin{cases} (x+2z, z, y-x-z) & \text{if } x < y - z \\ (2y-x, y, x-y+z) & \text{if } y - z < x < 2y \\ (x-2y, x-y+z, y) & \text{if } x > 2y \end{cases}$$

has exactly one fixed point $(1, 1, (p-1)/4)$, so $\#S$ is odd and the involution defined by $(x,y,z) \mapsto (x,z,y)$ also has a fixed point.

Using the corollary and theorem above, one can prove the well-known theorem that a positive integer can be expressed as a sum of two squares if and only if every prime congruent to 3 modulo 4 appears with even power in the factorization of n. One can even count the number of ways to express a positive integer as a sum of two squares [IR, pp. 278–280].

Remark (Gauss and Jacobi sums). In Solution 1 we counted the number of solutions to $x^2 = y^2 + 1$ in $\mathbb{F}_p$. Much more generally, given $a_1, \dots, a_r \in \mathbb{F}_p^*$, $\ell_1, \dots, \ell_r \geq 1$, and $b \in \mathbb{F}_p^*$ one can express the number of solutions over $\mathbb{F}_p$ to

$$a_1 x_1^{\ell_1} + a_2 x_2^{\ell_2} + \cdots + a_r x_r^{\ell_r} = b \tag{2}$$

in terms of Jacobi sums, which we will define here. First, a *multiplicative character* on $\mathbb{F}_p$ is a homomorphism $\chi : \mathbb{F}_p^* \to \mathbb{C}^*$, extended to a function on $\mathbb{F}_p$ by defining $\chi(0) = 0$. If $\chi_1, \dots, \chi_\ell$ are multiplicative characters, the *Jacobi sum* is defined by

$$J(\chi_1, \dots, \chi_\ell) = \sum_{t_1 + \cdots + t_\ell = 1} \chi_1(t_1)\, \chi_2(t_2) \cdots \chi_\ell(t_\ell),$$

where the sum is taken over all ℓ-tuples $(t_1, \ldots, t_\ell)$ with the t_i in $\mathbb{F}_p$ summing to 1. Jacobi sums are related to *Gauss sums*, which are sums of the form $g_a(\chi) = \sum_{t=0}^{p-1} \chi(t)\zeta^{at}$ where χ is a multiplicative character on $\mathbb{F}_p$, $\zeta = e^{2\pi i/p}$, and $a \in \mathbb{Z}$. See Chapters 6 and 8 of [IR] for more details. See also [Lan2, Chapter IV,§3] for more general Gauss sums, over $\mathbb{Z}/n\mathbb{Z}$ for n not necessarily prime, and over finite fields not necessarily of prime order. Theorem 5 in Chapter 8 of [IR] gives the formula for the number of solutions to (2) in terms of Jacobi sums.

Remark (the Weil Conjectures). The formula just mentioned led Weil [Weil2] to formulate his famous conjectures (now proven) about the number of points on varieties over finite fields. He proved the conjectures himself in the case of curves [Weil1]. Their proof for arbitrary algebraic varieties uses ideas of Grothendieck and Deligne, among others. For a survey, see [Har, Appendix C]. See Proof 3 of the stronger result following 1998B6 for one application of the conjectures.

B6. (2, 0, 0, 0, 1, 0, 1, 2, 0, 4, 30, 173)

Let a and b be positive numbers. Find the largest number c, in terms of a and b, such that

$$a^x b^{1-x} \le a\frac{\sinh ux}{\sinh u} + b\frac{\sinh u(1-x)}{\sinh u}$$

for all u with $0 < |u| \le c$ and for all x, $0 < x < 1$. (Note: $\sinh u = (e^u - e^{-u})/2$.)

Answer. The largest c for which the inequality holds for $0 < |u| \le c$ is $c = |\ln(a/b)|$.

Motivation. The key idea here is to strip away the inessential details. First, since the inequality is even in u, we may assume $u > 0$. Also, we may assume $a \ge b$ without loss of generality. A second step is to realize that the substitution $v = e^u$ will remove the exponentials, leaving

$$a^x b^{1-x} \le a\frac{v^x - v^{-x}}{v - v^{-1}} + b\frac{v^{1-x} - v^{-(1-x)}}{v - v^{-1}}.$$

Multiplying by $v - v^{-1}$ (which is positive) yields

$$a^x b^{1-x} v - a^x b^{1-x} v^{-1} \le av^x - av^{-x} + bv^{1-x} - bv^{-1+x}.$$

Since this equation is homogeneous in $\{a, b\}$, we can divide by b and set $r = a/b$, yielding

$$r^x v - r^x v^{-1} \le rv^x - rv^{-x} + v^{1-x} - v^{-1+x}.$$

At this point, some inspiration is necessary. Notice that the terms involving x are of the form $r^{\pm x}$ or $v^{\pm x}$. This suggests the substitution $v = r^{\pm 1}$; since $v, r \ge 1$ we substitute $v = r$. Then after this substitution, the inequality becomes $0 \ge 0$, i.e., equality holds for all x! This strongly suggests that $c = \ln r$, and that a good strategy would be to check that the inequality only "gets better" as $|u|$ decreases from c, and "gets worse" as $|u|$ increases from c.

This intuition then leads directly to the following argument.

Solution 1. We will show that $c = |\ln(a/b)|$ by proving that the inequality is satisfied if and only if $0 < |u| \le |\ln(a/b)|$. The right-hand side is an even function of

u; hence it suffices to consider $u > 0$. Replacing x by $1-x$ and interchanging a and b preserves the inequality, so we may assume $a \geq b$.

We will show that for each $x \in (0,1)$, the function

$$F(u) = a\frac{\sinh ux}{\sinh u} + b\frac{\sinh u(1-x)}{\sinh u} - a^x b^{1-x}$$

of u is decreasing for $u > 0$. For this, it suffices to show that for each $\alpha \in (0,1)$, the function

$$f(u) = \frac{\sinh \alpha u}{\sinh u}$$

is decreasing for $u > 0$, since $F(u)$ is a constant plus a sum of positive multiples of two such functions f. Differentiating with respect to u yields

$$f'(u) = \frac{\alpha \cosh(\alpha u)\sinh u - \sinh(\alpha u)\cosh u}{\sinh^2 u},$$

which for $u > 0$ is negative if and only if the value of the function

$$g(u) = \alpha \tanh u - \tanh(\alpha u)$$

is negative. The negativity of $g(u)$ for $u > 0$ follows from $g(0) = 0$ and the negativity of

$$g'(u) = \frac{\alpha}{\cosh^2 u} - \frac{\alpha}{\cosh^2 \alpha u},$$

which in turn follows since cosh is increasing on $\mathbb{R}^+$. Thus $F(u)$ is decreasing for $u > 0$.

If $a > b$ then $F(u)$ is zero at $u = \ln(a/b)$. If $a = b$ then $\lim_{u\to 0^+} F(u) = 0$ by L'Hôpital's Rule [Spv, Ch. 11, Theorem 9]. Since $F(u)$ is decreasing for $u > 0$, in both cases we have $F(u) \geq 0$ for $0 < u \leq \ln(a/b)$ and $F(u) < 0$ for $u > \ln(a/b)$, as desired. ■

Solution 2. As before, we may assume $u > 0$ and $a \geq b$. Taking ln of each side and rearranging, we find that the inequality is equivalent to

$$x(\ln(a\sinh u)) + (1-x)(\ln(b\sinh u)) \leq \ln(a\sinh ux + b\sinh u(1-x)),$$

which, if we define

$$f(x) = a\sinh ux + b\sinh u(1-x),$$

can be written as

$$x\ln f(1) + (1-x)\ln f(0) \leq \ln f(x).$$

This will hold if u is such that $\ln f(x)$ is concave for $x \in (0,1)$, and will fail if $\ln f(x)$ is strictly convex on $(0,1)$.

We compute

$$\begin{aligned}
\frac{d}{dx}\ln f &= \frac{f'}{f} \\
\frac{d^2}{dx^2}\ln f &= \frac{ff''-(f')^2}{f^2} \\
f' &= au\cosh ux - bu\cosh u(1-x) \\
f'' &= au^2\sinh ux + bu^2\sinh u(1-x) \\
ff''-(f')^2 &= (a^2+b^2)u^2(\sinh^2 ux - \cosh^2 ux) \\
&\quad + 2abu^2\big(\sinh ux\sinh u(1-x) + \cosh ux\cosh u(1-x)\big) \\
&= (a^2+b^2)u^2(-1) + 2abu^2\cosh\big(ux+u(1-x)\big) \\
&= u^2(-a^2-b^2+2ab\cosh u)
\end{aligned}$$

so $\frac{d^2}{dx^2}\ln f$ has constant sign for $x \in (0,1)$, and the following are equivalent:

$$\begin{aligned}
\frac{d^2}{dx^2}\ln f &\le 0 \qquad \text{for } x\in(0,1) \\
-a^2-b^2+2ab\cosh u &\le 0 \\
-a^2-b^2+abe^u+abe^{-u} &\le 0 \\
-(a-e^ub)(a-e^{-u}b) &\le 0 \\
-(a-e^ub) &\le 0 \qquad (\text{since } u>0 \text{ and } a\ge b>0) \\
u &\le \ln(a/b).
\end{aligned}$$

■

Remark. Taking the limit as $u \to 0$ in the inequality of the problem yields

$$ax + b(1-x) \ge a^x b^{1-x},$$

a weighted version of the AM-GM Inequality mentioned at the end of 1985A2.

The Fifty-Third William Lowell Putnam Mathematical Competition December 5, 1992

A1. (31, 82, 42, 10, 0, 0, 0, 7, 23, 6, 2, 0)

Prove that $f(n) = 1 - n$ is the only integer-valued function defined on the integers that satisfies the following conditions:

(i) $f(f(n)) = n$, **for all integers** n;

(ii) $f(f(n+2)+2) = n$ **for all integers** n;

(iii) $f(0) = 1$.

Solution. If $f(n) = 1-n$, then $f(f(n)) = f(1-n) = 1-(1-n) = n$, so (i) holds. Similarly, $f(f(n+2)+2) = f((-n-1)+2) = f(1-n) = n$, so (ii) holds. Clearly (iii) holds, and so $f(n) = 1-n$ satisfies the conditions.

Conversely, suppose f satisfies the three given conditions. By (i), $f(0) = 1$ and $f(1) = 0$. From condition (ii), $f(f(f(n+2)+2)) = f(n)$, and applying (i) yields $f(n+2)+2 = f(n)$, or equivalently $f(n+2) = f(n) - 2$. Easy inductions in both directions yields $f(n) = 1-n$. ∎

A2. (157, 1, 0, 0, 0, 0, 0, 0, 2, 14, 14, 15)

Define $C(\alpha)$ to be the coefficient of x^{1992} in the power series expansion about $x = 0$ of $(1+x)^\alpha$. Evaluate

$$\int_0^1 C(-y-1)\left(\frac{1}{y+1}+\frac{1}{y+2}+\frac{1}{y+3}+\cdots+\frac{1}{y+1992}\right)dy.$$

Answer. The value of the integral is 1992.

Solution. From the binomial theorem, we see that

$$C(\alpha) = \alpha(\alpha-1)\cdots\frac{\alpha-1991}{1992!},$$

so $C(-y-1) = (y+1)\cdots(y+1992)/1992!$. Therefore

$$C(-y-1)\left(\frac{1}{y+1}+\cdots+\frac{1}{y+1992}\right) = \frac{d}{dy}\left(\frac{(y+1)\cdots(y+1992)}{1992!}\right).$$

(The same formula for the derivative of a factored polynomial came up in 1991A3.)

Hence the integral in question is

$$\begin{aligned}\int_0^1 \frac{d}{dy}\left(\frac{(y+1)\cdots(y+1992)}{1992!}\right)dy &= \left.\frac{(y+1)\cdots(y+1992)}{1992!}\right|_0^1 \\ &= \frac{1993!-1992!}{1992!} = 1992.\end{aligned}$$

∎

A3. (55, 20, 7, 0, 0, 0, 0, 0, 16, 7, 45, 53)

For a given positive integer m, find all triples (n, x, y) of positive integers, with n relatively prime to m, which satisfy $(x^2+y^2)^m = (xy)^n$.

Answer. There are no solutions if m is odd. If m is even, the only solution is $(n, x, y) = (m+1, 2^{m/2}, 2^{m/2})$.

Solution. Note that $x^2 + y^2 > xy$, so $n > m$. Let $d = \gcd(x,y)$, so $x = ad$ and $y = bd$ where $\gcd(a,b) = 1$. Then $(a^2+b^2)^m = d^{2(n-m)}(ab)^n$. If p is a prime factor of a, then p divides the right side of this equation, but not the left. Hence $a = 1$, and similarly $b = 1$. Thus $2^m = d^{2(n-m)}$. Now m must be even, so say $m = 2k$, from which $2^k = d^{n-2k}$, so $n - 2k$ divides k. Since $\gcd(m,n) = 1$, we have $n - 2k = \gcd(n-2k, k) = \gcd(n,k) = 1$, so $n = 2k+1$. Thus $d = x = y = 2^k$, and the solutions are as claimed. ■

A4. (17, 6, 7, 0, 0, 0, 2, 0, 73, 18, 47, 33)

Let f be an infinitely differentiable real-valued function defined on the real numbers. If

$$f\left(\frac{1}{n}\right) = \frac{n^2}{n^2+1}, \qquad n = 1,2,3,\ldots,$$

compute the values of the derivatives $f^{(k)}(0)$, $k = 1,2,3,\ldots$.

Answer. We have

$$f^{(k)}(0) = \begin{cases} (-1)^{k/2}k! & \text{if } k \text{ is even;} \\ 0 & \text{if } k \text{ is odd.} \end{cases}$$

Solution. Let $g(x) = 1/(1+x^2)$ and $h(x) = f(x) - g(x)$. The value of $g^{(k)}(0)$ is $k!$ times the coefficient of x^k in the Taylor series $1/(1+x^2) = \sum_{m=0}^{\infty}(-1)^m x^{2m}$, and the value of $h^{(k)}(0)$ is zero by the lemma below (which arguably is the main content of this problem). Thus

$$f^{(k)}(0) = g^{(k)}(0) + h^{(k)}(0) = \begin{cases} (-1)^{k/2}k! & \text{if } k \text{ is even;} \\ 0 & \text{if } k \text{ is odd.} \end{cases}$$ ■

Lemma. *Suppose h is an infinitely differentiable real-valued function defined on the real numbers such that $h(1/n) = 0$ for $n = 1,2,3,\ldots$. Then $h^{(k)}(0) = 0$ for all nonnegative integers k.*

This lemma can be proved in many ways (and is a special case of a more general result stating that if $h : \mathbb{R} \to \mathbb{R}$ is infinitely differentiable, if $k \geq 0$, and if $a \in \mathbb{R}$, then $h^{(k)}(a)$ is determined by the values of h on any sequence of distinct real numbers tending to a).

Proof 1 (Rolle's Theorem). Since $h(x) = 0$ for a sequence of values of x strictly decreasing to 0, $h(0) = 0$. By Rolle's Theorem, $h'(x)$ has zeros between the zeros of $h(x)$; hence $h'(x) = 0$ for a sequence strictly decreasing to 0, so $h'(0) = 0$. Repeating this argument inductively, with $h^{(n)}(x)$ playing the role of $h(x)$, proves the lemma. □

Proof 2 (Taylor's Formula). We prove that $h^{(n)}(0) = 0$ by induction. The $n = 0$ case follows as in the previous proof, by continuity. Now assume that $n > 0$, and $h^{(k)}(0) = 0$ is known for $k < n$. Recall Taylor's Formula (Lagrange's form, see e.g. [Ap1, Section 7.7]) which states that for any $x > 0$ and integer $n > 0$, there exists $\theta_n \in [0,x]$ such that

$$h(x) = h(0) + h'(0)x + \cdots + h^{(n-1)}(0)x^{n-1}/(n-1)! + h^{(n)}(\theta_n)x^n/n!.$$

By our inductive hypothesis,

$$h(0) = \cdots = h^{(n-1)}(0) = 0.$$

Hence by taking $x = 1, 1/2, 1/3, \ldots$, we get $h^{(n)}(\theta_m) = 0$, where $0 \leq \theta_m \leq 1/m$. But $\lim_{m\to\infty} \theta_m = 0$, so by continuity $h^{(n)}(0) = 0$. □

Proof 3 (J. P. Grossman). By continuity, $h(0) = 0$. Let k be the smallest nonnegative integer such that $h^{(k)}(0) \neq 0$. We assume $h^{(k)}(0) > 0$; the same argument applies if $h^{(k)}(0) < 0$. Then there exists ϵ such that $h^{(k)}(x) > 0$ on $(0, \epsilon]$. Repeated integration shows that $h(x) > 0$ on $(0, \epsilon]$, a contradiction. □

Proof 4 sketch (explicit computation). By definition,

$$h'(0) = \lim_{\epsilon\to 0} \frac{f(\epsilon) - f(0)}{\epsilon}.$$

More generally, if h is infinitely differentiable in a neighborhood of 0, then

$$h^{(k)}(0) = \lim_{\epsilon\to 0} \frac{\sum_{j=0}^{k} \binom{k}{j}(-1)^{k-j} h(j\epsilon)}{\epsilon^k}. \tag{1}$$

(This can be proved by applying L'Hôpital's Rule k times to the expression on the right.) Then choose $\epsilon = 1/n$ where n runs over the multiples of $\operatorname{lcm}(1, \ldots, k)$, to obtain $h^{(k)}(x) = 0$. □

Remark. The formula (1) holds under the weaker assumption that $h^{(k)}(0)$ exists. To prove this, apply L'Hôpital's Rule $k - 1$ times, and then write the resulting expression as a combination of limits of the form

$$\lim_{\epsilon\to 0} \frac{h^{(k-1)}(j\epsilon) - h^{(k-1)}(0)}{j\epsilon},$$

each of which equals $h^{(k)}(0)$, by definition.

Remark. Note that $h(x)$ need not be the zero function! An infinitely differentiable function need not be represented by its Taylor series at a point, i.e., it need not be analytic. For example, consider

$$h(x) = \begin{cases} e^{-1/x^2} \sin(\pi/x) & \text{for } x \neq 0, \\ 0 & \text{for } x = 0. \end{cases}$$

It is infinitely differentiable for all x, and all of its derivatives are 0 at $x = 0$. (It satisfies the hypotheses of the lemma!)

A5. (1, 9, 1, 0, 0, 0, 0, 0, 5, 3, 72, 112)

For each positive integer n, let

$$a_n = \begin{cases} 0 & \textbf{if the number of 1's in the binary representation of } n \textbf{ is even,} \\ 1 & \textbf{if the number of 1's in the binary representation of } n \textbf{ is odd.} \end{cases}$$

Show that there do not exist positive integers k and m such that

$$a_{k+j} = a_{k+m+j} = a_{k+2m+j},$$

for $0 \leq j \leq m - 1$.

Solution 1. The sequence begins

$$0,1,1,0,1,0,0,1,1,0,0,1,0,1,1,0,1,0,0,1,0,1,1,0,0,1,1,0,1,0,0,1,\ldots.$$

The problem is to show that there are not "three identical blocks in a row".

The definition of a_n implies that $a_{2n} = a_n = 1 - a_{2n+1}$.

Suppose that there exist k, m as in the problem; we may assume that m is minimal for such examples.

Suppose first that m is odd. We'll suppose $a_k = a_{k+m} = a_{k+2m} = 0$; the case $a_k = 1$ can be treated similarly. Since either k or $k+m$ is even, $a_{k+1} = a_{k+m+1} = a_{k+2m+1} = 1$. Again, since either $k+1$ or $k+m+1$ is even, $a_{k+2} = a_{k+m+2} = a_{k+m+2} = 0$. By this means, we see that the terms $a_k, a_{k+1}, \ldots, a_{k+m-1}$ alternate between 0 and 1. Then since $m-1$ is even, $a_{k+m-1} = a_{k+2m-1} = a_{k+3m-1} = 0$. But, since either $k+m-1$ or $k+2m-1$ is even, that would imply that $a_{k+m} = a_{k+2m} = 1$, a contradiction.

Thus m must be even. Extracting the terms with even indices in

$$a_{k+j} = a_{k+m+j} = a_{k+2m+j}, \quad \text{for } 0 \le j \le m-1,$$

and using the fact that $a_r = a_{r/2}$ for even r, we get

$$a_{\lceil k/2\rceil+i} = a_{\lceil k/2\rceil+(m/2)+i} = a_{\lceil k/2\rceil+m+i}, \quad \text{for } 0 \le i \le m/2-1.$$

(The even numbers $\ge k$ are $2\lceil k/2\rceil$, $2\lceil k/2\rceil+2, \ldots$.) This contradicts the minimality of m.

Hence there are no such k and m. ■

Solution 2 (J.P. Grossman).

Lemma. *Let $c(i,j)$ be the number of carries when i is added to j in binary. Then $a_{i+j} \equiv a_i + a_j + c(i,j) \pmod 2$.*

Proof. Perform the addition column by column as taught in grade school, writing a "1" above the next column every time there is a carry. Then modulo 2, the total number of 1's appearing "above the line" is $a_i + a_j + c(i,j)$, and the total number of 1's in the sum is a_{i+j}. But in each column the digit of the sum is congruent modulo 2 to the sum of the 1's that appear above it. Summing over all columns yields $a_{i+j} \equiv a_i + a_j + c(i,j) \pmod 2$. □

We are asked to show that there is no block of $2m$ consecutive integers such that $a_k = a_{k+m}$ for all k in the block. Suppose such a block exists.

If $2^n \le m < 2^{n+1}$, then m has $n+1$ bits (binary digits), and since $2m \ge 2^{n+1}$, any $2m$ consecutive integers will exhibit every possible pattern of $n+1$ low bits. Let κ be the integer in this block with the $n+1$ low bits zero. Then $c(\kappa, m) = 0$, so the lemma gives $a_{\kappa+m} \equiv a_\kappa + a_m \pmod 2$. Since $a_{\kappa+m} = a_\kappa$, we obtain $a_m = 0$. Running this argument in reverse, we have $a_{k+m} \equiv a_k + a_m \pmod 2$ and $c(k,m) \equiv 0 \pmod 2$ for *all* k in the block.

In particular, $c(k,m) = 0$ for the k with the $n+1$ low bits equal to 0 except for one 1, so m cannot contain the pattern 01. Thus the binary expansion of m has the form $1\cdots10\cdots0$; that is, $m = 2^{n+1} - 2^s$ for some $s \ge 0$.

Suppose we can find k_1 and k_2 in the block such that $\lfloor k_1/2^{n+1}\rfloor = \lfloor k_2/2^{n+1}\rfloor$, $k_1 \equiv 2^s \pmod{2^{n+1}}$, and $k_2 \equiv 2^{s+1} \pmod{2^{n+1}}$. Then $c(k_1,m) = c(k_2,m)+1$, contradicting the fact that $c(k, m)$ is even for each k in the block. Thus no such pair (k_1, k_2) exists.

On the other hand, since $2m \geq 2^{n+1}$, there does exist k in the block such that $k \equiv 2^s \pmod{2^{n+1}}$. The smallest element of the block must be at least $k - 2^{n+1} + 1$, or else we could take $k_1 = k - 2^{n+1}$ and $k_2 = k_1 + 2^s$. The largest element must be at most $k + 2^s - 1$, or else we could take $k_1 = k$ and $k_2 = k_1 + 2^s$. Thus the block has length at most

$$2^{n+1} + 2^s - 1 < 2^{n+1} + 2^{s+2} - 2^{s+1} - 1 \leq 2^{n+2} - 2^{s+1} - 1 = 2m - 1,$$

which is a contradiction.

Related question. Show that if two adjacent blocks are identical, then the length of each is a power of 2. Show that any power of 2 is achievable.

Remark. This sequence is sometimes called the Thue-Morse sequence; see [AS] for examples of its ubiquity in mathematics, including its use by Machgielis Euwe (chess world champion 1935–37) to show that infinite games of chess may occur despite the so-called "German rule" which states that a draw occurs if the same sequence of moves occurs three times in succession.

Related question. Another fact discussed in [AS] is the following (see there for further references). The result of Christol mentioned in the remark following 1989A6 implies that the generating function $F(x) = \sum_{n=0}^{\infty} a_n x^n$ considered as a power series with coefficients in $\mathbb{F}_2$, is algebraic over the field $\mathbb{F}_2(x)$ of rational functions with coefficients in $\mathbb{F}_2$. We leave it to the reader to verify that

$$(x+1)^3 F(x)^2 + (x+1)^2 F(x) + x = 0.$$

On the other hand, if the coefficients of $F(x)$ are considered to be rational, then one can show that $F(x)$ is transcendental over $\mathbb{Q}(x)$, and even that $F(1/2)$ is a transcendental number.

Related questions. The Thue-Morse sequence has a self-similarity property: if one replaces each "0" with "0, 1", and each "1" with "1,0", one recovers the original sequence. The "extraction of terms with even indices" in Solution 1 is just reversing this self-similarity.

Here are a few more problems and results relating to "arithmetic self-similarity".

(a) Show that the sequence $s(0), s(1), s(2), \ldots$, defined by $s(3m) = 0$, $s(3m+1) = s(m)$, $s(3m+2) = 1$, also has the property that there are not "three identical blocks in a row". This problem is given in [Hal, p. 156], along with a discussion of its motivation in chess.

(b) 1993A6.

(c) Problem 3 on the 1988 International Mathematical Olympiad [IMO88, p. 37] is:

A function f is defined on the positive integers by

$$f(1) = 1, \quad f(3) = 3, \quad f(2n) = f(n),$$

$$f(4n+1) = 2f(2n+1) - f(n), \quad f(4n+3) = 3f(2n+1) - 2f(n),$$

for all positive integers n. Determine the number of positive integers n, less than or equal to 1988, for which $f(n) = n$.

(d) Also, [YY, p. 12] has a series of similar problems. For example, this Putnam problem answers the question Yaglom and Yaglom pose in Problem 124(b). The subsequent problem is harder:

> Show that there exist arbitrarily long sequences consisting of the digits 0, 1, 2, 3, such that no digit or sequence of digits occurs twice in succession. Show that there are solutions in which the digit 0 does not occur; thus three digits is the minimum we need to construct sequences of the desired type.

For a link to a discrete version of dynamical systems, [YY] suggests [MH].

A6. (9, 3, 4, 0, 0, 0, 0, 0, 0, 10, 32, 22, 123)
Four points are chosen at random on the surface of a sphere. What is the probability that the center of the sphere lies inside the tetrahedron whose vertices are at the four points? (It is understood that each point is independently chosen relative to a uniform distribution on the sphere.)

Answer. The probability is 1/8.

Solution 1. Set up a coordinate system so that the sphere is centered at the origin of $\mathbb{R}^3$. Identify points with vectors in $\mathbb{R}^3$.

Let P_1, P_2, P_3, P_4 be the four random points on the sphere. We may suppose that the choice of each P_i is made in two steps: first a random point Q_i is chosen, then a random sign $\epsilon_i \in \{-1, 1\}$ is chosen and we set $P_i = \epsilon_i Q_i$.

The probability that Q_3 is in the linear subspace spanned by Q_1 and Q_2 is zero. Similar statements hold for any three of the Q_i, so we may assume that every three of the Q_i are linearly independent. The probability that Q_4 is in the plane through Q_1, Q_2, Q_3 is zero, so we may assume that $Q_1Q_2Q_3Q_4$ is a nondegenerate tetrahedron T_Q. We may also assume that for any choices of the ϵ_i, the tetrahedron T_P with the P_i as vertices is nondegenerate.

The linear map $L_Q : \mathbb{R}^4 \to \mathbb{R}^3$ sending (w_1, w_2, w_3, w_4) to $\sum w_i Q_i$ is surjective, so $\ker L_Q$ is 1-dimensional, generated by some (w_1, w_2, w_3, w_4). Since every three Q_i are linearly independent, every w_i is nonzero.

We claim that 0 lies in the interior of T_Q if and only if all the w_i have the same sign. If 0 lies in the interior of T_Q, then 0 is a convex combination of the vertices in which each occurs with nonzero coefficient; i.e., $0 = \sum_{i=1}^4 r_i Q_i$ for some $r_1, r_2, r_3, r_4 > 0$ with sum 1. Then (w_1, w_2, w_3, w_4) is a multiple of (r_1, r_2, r_3, r_4), so the w_i have the same sign. Conversely if the w_i have the same sign, then $w_1 + w_2 + w_3 + w_4 \neq 0$ and $0 = \sum_{i=1}^4 r_i Q_i$ where $(r_1, r_2, r_3, r_4) = \frac{1}{w_1+w_2+w_3+w_4}(w_1, w_2, w_3, w_4)$ is such that $r_1, r_2, r_3, r_4 > 0$ sum to 1, so 0 lies in the interior of T_Q.

Fix Q_1, Q_2, Q_3, Q_4 and (w_1, w_2, w_3, w_4). Then $(\epsilon_1 w_1, \epsilon_2 w_2, \epsilon_3 w_3, \epsilon_4 w_4)$ is a generator of $\ker L_P$, where L_P is defined using the P_i instead of the Q_i. The numbers $\epsilon_1 w_1$, $\epsilon_2 w_2$, $\epsilon_3 w_3$, $\epsilon_4 w_4$ have the same sign (all plus or all minus) for exactly 2 of the $2^4 = 16$ choices of $(\epsilon_1, \epsilon_2, \epsilon_3, \epsilon_4)$. Thus, conditioned on a choice of the Q_i, the probability that T_P contains the origin in its interior is $2/16 = 1/8$. This is the same for any $(Q_1, \dots, Q_4)$, so the overall probability that 0 lies inside T_P also is 1/8. ■

Remark. As noted in [HoS], the same argument proves the following generalization.

Let $\mathbb{R}^n$ be endowed with a probability measure μ that is symmetric with respect to the map $x \mapsto -x$, and such that, when $n+1$ points are chosen independently with respect to μ, with probability one their convex hull is a simplex. Then the probability that the origin is contained in the simplex generated by $n+1$ such random points is $1/2^n$.

The original problem is the case $n = 3$ with μ equal to the uniform distribution. The case $n = 2$ with the uniform distribution appears, along with a strategy for approaching it, in [Hon1, Essay 1].

Remark. The article [HoS] mentions also the following result of J. G. Wendel [Wen], which generalizes the original result in a different direction: if N points are chosen at random from the surface of the unit sphere in $\mathbb{R}^n$, the probability that there exists a hemisphere containing all N points equals

$$2^{-N+1} \sum_{k=0}^{n-1} \binom{N-1}{k}.$$

Remark. Here is a bizarre related fact: if five points are chosen at random from a ball in $\mathbb{R}^3$, the probability that one of them is contained in the tetrahedron generated by the other four is $9/143$ [So].

We end with a variant of the solution to the original problem.

Solution 2. Pick the first three points first; call them A, B, C. Consider the spherical triangle T defined by those points. The center of the sphere lies in the convex hull of A, B, C, and another point P if and only if it lies in the convex hull of T and P. This happens if and only if P is antipodal to some point of T. So the desired probability is the expected fraction of the sphere's surface area which is covered by T.

Denote the antipode to a point P by P'. We consider the eight spherical triangles ABC, $A'BC$, $A'B'C$, ABC', $AB'C'$, $A'B'C'$. Denote these by $T_0, T_1, \dots, T_7$; we regard T_i as a function of the random variables A, B, C. There is an automorphism of our probability space defined by $(A, B, C) \mapsto (A', B, C)$, so T_0 and T_1 have the same distribution. By choosing similar automorphisms, T_0 and T_i have the same distribution for all i. In particular, the expected fraction of the sphere covered by T_i is independent of i. On the other hand, the triangles $T_0, \dots, T_7$ cover the sphere (with overlap of measure zero), since they are the eight regions formed by the three great circles obtained by extending the sides of spherical triangle ABC, so the probability we seek is $1/8$. ∎

B1. (145, 15, 4, 0, 0, 0, 0, 0, 6, 14, 11, 8)

Let S be a set of n distinct real numbers. Let A_S be the set of numbers that occur as averages of two distinct elements of S. For a given $n \geq 2$, what is the smallest possible number of elements in A_S?

Answer. The smallest possible number of elements of A_S is $2n - 3$.

Solution. Let $x_1 < x_2 < \cdots < x_n$ represent the elements of S. Then

$$\frac{x_1 + x_2}{2} < \frac{x_1 + x_3}{2} < \cdots < \frac{x_1 + x_n}{2} < \frac{x_2 + x_n}{2} < \frac{x_3 + x_n}{2} < \cdots < \frac{x_{n-1} + x_n}{2}$$

represent $2n-3$ distinct elements of A_S, so A_S has at least $2n-3$ distinct elements.

On the other hand, if we take $S=\{1,2,\ldots,n\}$, the elements of A_S are $\frac{3}{2}$, $\frac{4}{2}$, $\frac{5}{2}$, $\ldots$, $\frac{2n-1}{2}$. There are only $2n-3$ such numbers; thus there is a set A_S with at most $2n-3$ distinct elements. ■

Related question. This is a generalization of Problem 2 of the 1991 Asian-Pacific Mathematical Olympiad [APMO],

> Suppose there are 997 points given on a plane. If every two points are joined by a line segment with its midpoint coloured in red, show that there are at least 1991 red points on the plane. Can you find a special case with exactly 1991 red points?

B2. (159, 10, 7, 0, 0, 0, 0, 0, 1, 4, 13, 9)

For nonnegative integers n and k, define $Q(n,k)$ to be the coefficient of x^k in the expansion of $(1+x+x^2+x^3)^n$. Prove that

$$Q(n,k)=\sum_{j=0}^{k}\binom{n}{j}\binom{n}{k-2j},$$

where $\binom{a}{b}$ is the standard binomial coefficient. (Reminder: For integers a and b with $a\geq 0$, $\binom{a}{b}=\frac{a!}{b!\,(a-b)!}$ for $0\leq b\leq a$, with $\binom{a}{b}=0$ otherwise.)

Solution. Write $(1+x+x^2+x^3)^n$ as $(1+x^2)^n(1+x)^n$. The coefficient of x^k gets contributions from the x^{2j} term in the first factor (with coefficient $\binom{n}{j}$) times the x^{k-2j} term in the second factor (with coefficient $\binom{n}{k-2j}$). ■

Remark. This solution can also be stated in the language of generating functions:

$$\begin{aligned}\sum_{k\geq 0}Q(n,k)x^k &= (1+x+x^2+x^3)^n\\ &= (1+x^2)^n(1+x)^n\\ &= \sum_{j\geq 0}\binom{n}{j}x^{2j}\sum_{i\geq 0}\binom{n}{i}x^i\\ &= \sum_{j\geq 0}\sum_{i\geq 0}x^{2j+i}\binom{n}{j}\binom{n}{i}\\ &= \sum_{k\geq 0}x^k\sum_{j\geq 0}\binom{n}{j}\binom{n}{k-2j}.\end{aligned}$$

B3. (23, 11, 10, 0, 0, 0, 0, 0, 27, 24, 71, 37)

For any pair (x,y) of real numbers, a sequence $(a_n(x,y))_{n\geq 0}$ is defined as follows:

$$\begin{aligned}a_0(x,y)&=x,\\ a_{n+1}(x,y)&=\frac{(a_n(x,y))^2+y^2}{2},\qquad \text{for } n\geq 0.\end{aligned}$$

Find the area of the region $\{(x,y)|(a_n(x,y))_{n\geq 0}$ converges$\}$.

Answer. The area is $4+\pi$.

Solution. The region of convergence is shown in Figure 22; it is a (closed) square $\{(x, y) : -1 \le x, y \le 1\}$ of side 2 with (closed) semicircles of radius 1 centered at $(\pm 1, 0)$ described on two opposite sides.

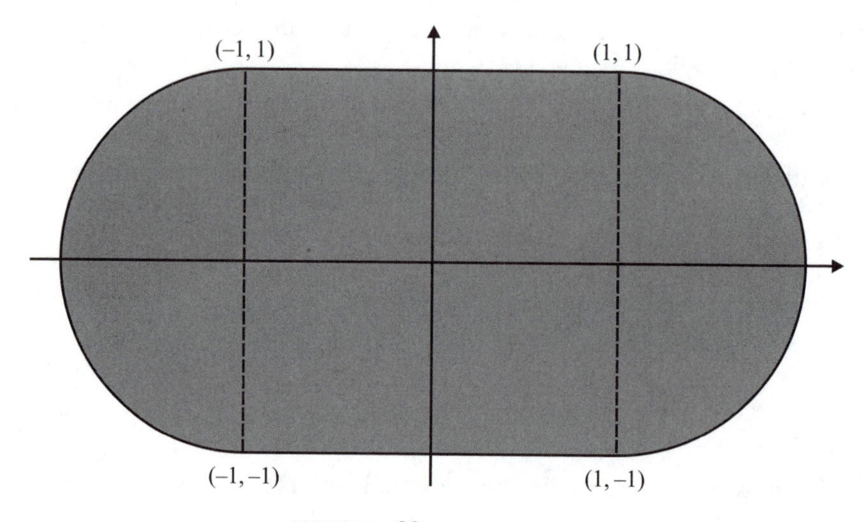

FIGURE 22.
The region of convergence.

Note that (x, y), $(-x, y)$, $(x, -y)$, and $(-x, -y)$ produce the same sequence after the first step, so will restrict attention to the first quadrant $(x, y \ge 0)$, and use symmetry to deal with the other three.

Fix y; we will determine for which x (x, y) is in the region. Let $f(w) = (w^2 + y^2)/2$, so $a_n(x, y) = f^n(x)$. If the limit exists and is L, then $f(L) = L$, so $L = 1 \pm \sqrt{1 - y^2}$ (call these values L_+ and L_-). It will be useful to observe that

$$f(w) - L = (w - L)(w + L)/2. \tag{1}$$

In particular, if $y > 1$, the limit cannot exist; we now assume $y \le 1$.

If $w > L_+$, then $f(w) > w$ (from (1) for $L = L_+$, and $w + L_+ > 2$). Hence if $x > 1 + \sqrt{1 - y^2}$, the sequence x, $f(x)$, $f(f(x))$, $\ldots$, cannot converge.

If $x = L_+$, then x, $f(x)$, $f(f(x))$, $\ldots$ is the constant sequence (L_+), and hence converges.

If $0 \le w < L_+$, then

(i) $f(w)$ satisfies the same inequality, i.e., $0 \le f(w) < L_+$ (the left inequality is immediate, and the right follows from (1) for $L = L_+$),

(ii) $|f(w) - L_-| < |w - L_-|$ (from (1) for L_- and $w + L_- < L_+ + L_- = 2$) , and

(iii) $w - L_-$ and $f(w) - L_-$ have the same sign (from (1) for $L = L_-$).

Then (i)–(iii) imply that if $0 \le x < L_+$, then

$$x,\ f(x),\ f(f(x)),\ \ldots$$

converges to L_-. ■

Remark. Problem 1996A6 involves a similar dynamical system.

Related question. For another problem involving a dynamical system, observe what happens when you repeatedly hit the cosine button on a calculator, and explain this phenomenon. In essence, this is Problem 1952/7 [PutnamI, p. 37]:

> Given any real number N_0, if $N_{j+1} = \cos N_j$, prove that $\lim_{j\to\infty} N_j$ exists and is independent of N_0.

Literature note. For introductory reading on dynamical systems, see [De] or [Mi].

B4. (35, 11, 13, 0, 0, 0, 0, 0, 12, 5, 48, 79)

Let $p(x)$ be a nonzero polynomial of degree less than 1992 having no nonconstant factor in common with $x^3 - x$. Let

$$\frac{d^{1992}}{dx^{1992}}\left(\frac{p(x)}{x^3 - x}\right) = \frac{f(x)}{g(x)}$$

for polynomials $f(x)$ and $g(x)$. Find the smallest possible degree of $f(x)$.

Answer. The smallest possible degree of $f(x)$ is 3984.

Solution. By the Division Algorithm, we can write $p(x) = (x^3 - x)q(x) + r(x)$, where $q(x)$ and $r(x)$ are polynomials, the degree of $r(x)$ is at most 2, and the degree of $q(x)$ is less than 1989. Then

$$\frac{d^{1992}}{dx^{1992}}\left(\frac{p(x)}{x^3 - x}\right) = \frac{d^{1992}}{dx^{1992}}\left(\frac{r(x)}{x^3 - x}\right).$$

Write $r(x)/(x^3 - x)$ in the form

$$\frac{A}{x-1} + \frac{B}{x} + \frac{C}{x+1}.$$

Because $p(x)$ and $x^3 - x$ have no common factor, neither do $r(x)$ and $x^3 - x$, and thus A, B, and C are nonzero. Thus

$$\begin{aligned} &\frac{d^{1992}}{dx^{1992}}\left(\frac{r(x)}{x^3 - x}\right) \\ &\quad = 1992!\left(\frac{A}{(x-1)^{1993}} + \frac{B}{x^{1993}} + \frac{C}{(x+1)^{1993}}\right) \\ &\quad = 1992!\left(\frac{Ax^{1993}(x+1)^{1993} + B(x-1)^{1993}(x+1)^{1993} + C(x-1)^{1993}x^{1993}}{(x^3 - x)^{1993}}\right). \end{aligned}$$

Since A, B, C are nonzero, the numerator and denominator have no common factor. Expanding the numerator yields

$$(A + B + C)x^{3986} + 1993(A - C)x^{3985} + 1993(996A - B + 996C)x^{3984} + \cdots.$$

From $A = C = 1$, $B = -2$ we see that the degree can be as low as 3984. A lower degree would imply $A + B + C = 0$, $A - C = 0$, $996A - B + 996C = 0$, implying that $A = B = C = 0$, a contradiction.

Expressing $\frac{d^{1992}}{dx^{1992}}\left(\frac{p(x)}{x^3 - x}\right)$ in other than lowest terms can only increase the degree of the numerator. (Nowhere in the problem does it say that $f(x)/g(x)$ is to be in lowest terms.) ∎

Remark (Noam Elkies). Although the proposer of this problem presumably had polynomials with real or complex coefficients in mind, the same solution works for polynomials over a field of characteristic greater than 1992. When the characteristic is less than 1992, one can show that $\frac{d^{1992}}{dx^{1992}}\left(\frac{p(x)}{x^3-x}\right)$ is always zero, so the problem does not make much sense.

B5. (62, 4, 4, 0, 0, 0, 0, 3, 6, 2, 49, 73)
Let D_n denote the value of the $(n-1)\times(n-1)$ determinant

$$\begin{vmatrix} 3 & 1 & 1 & 1 & \cdots & 1 \\ 1 & 4 & 1 & 1 & \cdots & 1 \\ 1 & 1 & 5 & 1 & \cdots & 1 \\ 1 & 1 & 1 & 6 & \cdots & 1 \\ \vdots & \vdots & \vdots & \vdots & \ddots & \vdots \\ 1 & 1 & 1 & 1 & \cdots & n+1 \end{vmatrix}.$$

Is the set $\{D_n/n!\}_{n\geq 2}$ bounded?

Answer. No. Each of the solutions below shows that

$$D_n = n!\left(1+\frac{1}{2}+\cdots+\frac{1}{n}\right).$$

Thus the sequence $\{D_n/n!\}_{n\geq 2}$ is the nth partial sum of the harmonic series, which is unbounded as $n\to\infty$.

Solution 1. Subtract the first row from each of the other rows, to get

$$D_n = \det\begin{pmatrix} 3 & 1 & 1 & 1 & \cdots & 1 \\ -2 & 3 & 0 & 0 & \cdots & 0 \\ -2 & 0 & 4 & 0 & \cdots & 0 \\ -2 & 0 & 0 & 5 & \cdots & 0 \\ \vdots & \vdots & \vdots & \vdots & \ddots & \vdots \\ -2 & 0 & 0 & 0 & \cdots & n \end{pmatrix}. \tag{1}$$

Then for $2\leq i\leq n-1$, add $2/(i+1)$ times the ith column to the first column to obtain

$$D_n = \det\begin{pmatrix} 3+\frac{2}{3}+\frac{2}{4}+\cdots+\frac{2}{n} & 1 & 1 & 1 & \cdots & 1 \\ 0 & 3 & 0 & 0 & \cdots & 0 \\ 0 & 0 & 4 & 0 & \cdots & 0 \\ 0 & 0 & 0 & 5 & \cdots & 0 \\ \vdots & \vdots & \vdots & \vdots & \ddots & \vdots \\ 0 & 0 & 0 & 0 & \cdots & n \end{pmatrix}.$$

The resulting matrix is upper triangular, so the determinant is the product of the diagonal elements, which is

$$n!\left(1+\frac{1}{2}+\cdots+\frac{1}{n}\right).$$

■

Solution 2 (J.P. Grossman). We derive a recursion for D_{n+1}. Expand (1) along the last row to get $D_n = nD_{n-1} + (n-1)!$. Divide by $n!$ to obtain

$$\frac{D_n}{n!} = \frac{D_{n-1}}{(n-1)!} + \frac{1}{n},$$

from which the result follows. ∎

Solution 3. Define

$$D_{n+1}(a_1, \ldots, a_n) = \det \begin{pmatrix} 1+a_1 & 1 & 1 & 1 & \cdots & 1 \\ 1 & 1+a_2 & 1 & 1 & \cdots & 1 \\ 1 & 1 & 1+a_3 & 1 & \cdots & 1 \\ 1 & 1 & 1 & 1+a_4 & \cdots & 1 \\ \vdots & \vdots & \vdots & \vdots & \ddots & \vdots \\ 1 & 1 & 1 & 1 & \cdots & 1+a_n \end{pmatrix}.$$

The problem asks for $D_n(2, 3, \ldots, n)$. We will prove the identity

$$D_{n+1}(a_1, \ldots, a_n) = \prod_{i=1}^{n} a_i + \sum_{i=1}^{n} \prod_{\substack{j=1 \\ j \neq i}}^{n} a_j. \tag{2}$$

This formula follows immediately from the recursion

$$D_{n+1}(a_1, \ldots, a_n) = a_n D_n(a_1, \ldots, a_{n-1}) + a_{n-1} D_n(a_1, \ldots, a_{n-2}, 0).$$

To prove this recursion, subtract the $(n-1)$st column from the nth column, and then expand along the nth column.

If all the a_i are nonzero, we can write the polynomial $D_n(a_1, \ldots, a_{n-1})$ in the form

$$D_n(a_1, \ldots, a_{n-1}) = a_1 a_2 \cdots a_{n-1} \left(1 + \frac{1}{a_1} + \frac{1}{a_2} + \cdots + \frac{1}{a_{n-1}}\right).$$

This problem is the special case $a_i = i + 1$. ∎

Remark. This formula can also be used for a generalization of Problem 1993B5; see page 188.

Solution 4. The following result is Problem 7 of Part VII of [PS], and appeared as Problem 1978A2 [PutnamII, p. 31].

Let a, b, p_1, p_2, ... , p_n be real numbers with $a \neq b$. Define

$$f(x) = (p_1 - x)(p_2 - x)(p_3 - x) \cdots (p_n - x).$$

Then

$$\det \begin{pmatrix} p_1 & a & a & a & \cdots & a & a \\ b & p_2 & a & a & \cdots & a & a \\ b & b & p_3 & a & \cdots & a & a \\ b & b & b & p_4 & \cdots & a & a \\ \vdots & \vdots & \vdots & \vdots & \ddots & \vdots & \vdots \\ b & b & b & b & \cdots & p_{n-1} & a \\ b & b & b & b & \cdots & b & p_n \end{pmatrix} = \frac{bf(a) - af(b)}{b-a}. \tag{3}$$

The left side of (3) is a polynomial in b, and hence continuous in b. Letting $a = 1$, $b \to 1$, and using L'Hôpital's Rule, we find

$$\det \begin{pmatrix} p_1 & 1 & 1 & 1 & \cdots & 1 & 1 \\ 1 & p_2 & 1 & 1 & \cdots & 1 & 1 \\ 1 & 1 & p_3 & 1 & \cdots & 1 & 1 \\ 1 & 1 & 1 & p_4 & \cdots & 1 & 1 \\ \vdots & \vdots & \vdots & \vdots & \ddots & \vdots & \vdots \\ 1 & 1 & 1 & 1 & \cdots & p_{n-1} & 1 \\ 1 & 1 & 1 & 1 & \cdots & 1 & p_n \end{pmatrix} = f(1) - f'(1).$$

Letting $p_i = a_i + 1$, we recover (2), using for example equation (1) on page 136. ■

Remark. There are many other approaches that proceed by applying row and column operations to find a recursion for D_n. Here we sketch one more.

Subtract the next-to-last column from the last column, and then subtract the next-to-last row from the last row, and expand along the last row to obtain

$$D_{n+1} = (2n+1)D_n - n^2 D_{n-1}.$$

If we set $r_n = D_n/n!$, this becomes $r_{n+1} - r_n = \frac{n}{n+1}(r_n - r_{n-1})$. One proves $r_{n+1} - r_n = 1/(n+1)$ by induction, and finally

$$r_n = 1 + \frac{1}{2} + \frac{1}{3} + \cdots + \frac{1}{n}.$$

B6. (0, 0, 0, 0, 0, 0, 0, 0, 5, 4, 39, 155)

Let $\mathcal{M}$ be a set of real $n \times n$ matrices such that

(i) $I \in \mathcal{M}$, where I is the $n \times n$ identity matrix;

(ii) if $A \in \mathcal{M}$ and $B \in \mathcal{M}$, then either $AB \in \mathcal{M}$ or $-AB \in \mathcal{M}$, but not both;

(iii) if $A \in \mathcal{M}$ and $B \in \mathcal{M}$, then either $AB = BA$ or $AB = -BA$;

(iv) if $A \in \mathcal{M}$ and $A \neq I$, there is at least one $B \in \mathcal{M}$ such that $AB = -BA$.

Prove that $\mathcal{M}$ contains at most n^2 matrices.

Solution 1 (Noam Elkies). Suppose A, B are in $\mathcal{M}$. By (iii), $AB = \epsilon BA$, where $\epsilon = \pm 1$, for any B in $\mathcal{M}$. Then

$$AAB = A\epsilon BA = \epsilon^2 BAA = BAA,$$

so A^2 commutes with any B in $\mathcal{M}$; of course the same is true of $-A^2$. On the other hand, by (ii), A^2 or $-A^2$ is in $\mathcal{M}$. Let C be the one that is in $\mathcal{M}$.

If C is not I, then by (iv) we can find a B in $\mathcal{M}$ such that $CB = -BC$. But we know $CB = BC$ for any B in $\mathcal{M}$. Thus $CB = 0$, which is impossible by (ii).

We conclude that $C = I$. In other words, for any A in $\mathcal{M}$, $A^2 = \pm I$.

Now suppose $\mathcal{M}$ has more than n^2 matrices. The space of real $n \times n$ matrices has dimension n^2, so we can find a nontrivial linear relation $\sum_{D \in \mathcal{M}} x_D D = 0$. Pick such a relation with the smallest possible number of nonzero x_D. We will construct a smaller relation, obtaining a contradiction and finishing the proof.

Pick an A with x_A nonzero, and multiply by it on the right: $\sum_{D \in \mathcal{M}} x_D DA = 0$. In light of (ii) the matrices DA run over $\mathcal{M}$ modulo sign; so we have a new relation

$\sum_{E\in\mathcal{M}} y_E E = 0$. The point of this transformation is that now the coefficient y_I of I is $\pm x_A$, which is nonzero.

Pick any D other than I such that y_D is nonzero. By (iv), we can pick B in $\mathcal{M}$ such that $DB = -BD$. Multiply $\sum_{E\in\mathcal{M}} y_E E = 0$ by B on both the left and the right, and add:

$$\sum_{E\in\mathcal{M}} y_E(BE + EB) = 0.$$

Now by (iii) we have $BE + EB = (1 + \epsilon_{BE})BE$, where $\epsilon_{BE} = \pm 1$. In particular, $\epsilon_{BI} = 1$ (clear) and $\epsilon_{BD} = -1$ (by construction of B). So we get

$$\sum_{E\in\mathcal{M}} y_E(1 + \epsilon_{BE})BE = 0,$$

where at least one term does not disappear and at least one term does disappear. As before, the matrices BE run over $\mathcal{M}$ modulo sign. So we have a relation with fewer terms, as desired. ∎

Remark. This proof did not need (i) in an essential way. The only modification to the proof needed to avoid using (i), is that instead of arranging for y_I to be nonzero, we might have y_{-I} nonzero instead.

In fact, if $\mathcal{M}$ is nonempty, (i) follows from (ii), (iii), and (iv), as we now show. Choose $A \in \mathcal{M}$. Then $\pm A^2 \in \mathcal{M}$, so $\pm I \in \mathcal{M}$ by the first half of Solution 1. But $-I \in \mathcal{M}$ contradicts (iv).

Remark. The result holds not only for real matrices, but also for matrices with entries in any field k: if the characteristic of k is not 2, the proof proceeds as before; if the characteristic is 2, then (ii) implies that $\mathcal{M}$ is empty.

Remark. An argument similar to that in the latter half of the solution proves the following result of field theory ("independence of characters"): if L is a finite extension of a field K, then the automorphisms of L over K are linearly independent in the K-vector space of K-linear maps $L \to L$ [Ar, Theorem 12].

Solution 2. We prove the result more generally for complex matrices, by induction on n.

If $n = 1$, then the elements of $\mathcal{M}$ commute so (iv) cannot be satisfied unless $\mathcal{M} = \{I\}$. Suppose that $n > 1$ and that the result holds for sets of complex matrices of smaller dimension.

We may assume $|\mathcal{M}| > 1$, so by (iv), there exist $C, D \in \mathcal{M}$ with $CD = -DC$. Fix such C, D. As in Solution 1, $C^2 = \pm I$, so the eigenvalues of C are $\pm\lambda$ where $\lambda = 1$ or i. Furthermore, $\mathbb{C}^n = V_\lambda \oplus V_{-\lambda}$, where V_λ, $V_{-\lambda}$ are the eigenspaces corresponding to λ and $-\lambda$ (i.e., the nullspaces of $(C - \lambda I)$, $(C + \lambda I)$) respectively. (Here we follow the convention that the matrices are acting on $\mathbb{C}^n$ by right-multiplication.) Observe that if $X \in \mathcal{M}$ then

$$CX = XC \implies (C \pm \lambda I)X = X(C \pm \lambda I) \implies V_{\pm\lambda}X = V_{\pm\lambda};$$

$$CX = -XC \implies (C \pm \lambda I)X = (-1)X(C \mp \lambda I) \implies V_{\pm\lambda}X = V_{\mp\lambda}.$$

In particular, since $V_\lambda D = V_{-\lambda}$, $\dim(V_\lambda) = \dim(V_{-\lambda}) = n/2$.

Let $\mathcal{N} = \{ X \in \mathcal{M} : CX = XC, DX = XD \}$. If $Y \in \mathcal{M}$ then exactly one of Y, YC, YD, YCD is in $\mathcal{N}$. It follows that $|\mathcal{N}| = |\mathcal{M}|/4$.

For $X \in \mathcal{N}$, let $\phi(X)$ be the $(n/2) \times (n/2)$ matrix representing, with respect to a basis of V_λ, the linear transformation given by $v \mapsto vX$ for $v \in V_\lambda$. Then ϕ is injective. To see this, assume $\phi(X) = \phi(Y)$, so $vX = vY$ for $v \in V_\lambda$; but if $v \in V_{-\lambda}$ then $vD \in V_\lambda$, so $vXD = vDX = vDY = vYD$, which again implies $vX = vY$; since X and Y induce the same transformations of both V_λ and $V_{-\lambda}$, it follows that $X = Y$.

It suffices finally to show that $\phi(\mathcal{N})$, a set of $(n/2)\times(n/2)$ complex matrices, satisfies (i), (ii), (iii), (iv), for then, by induction, $|\phi(\mathcal{N})| \leq (n/2)^2$, whence $|\mathcal{M}| = 4|\mathcal{N}| = 4|\phi(\mathcal{N})| \leq n^2$.

Conditions (i), (ii), (iii) for $\phi(\mathcal{N})$ are inherited from those of $\mathcal{M}$. To show (iv), let $\phi(A) \in \phi(\mathcal{N})$, with $\phi(A)$ not the $(n/2) \times (n/2)$ identity matrix. Then $A \neq I$ (since ϕ is injective) and $AB = -BA$ for some $B \in \mathcal{M}$. Let B' be the element of $\{B, BC, BD, BCD\}$ belong to $\mathcal{N}$. Since $AB' = -B'A$, $\phi(A)\phi(B') = -\phi(B')\phi(A)$. ∎

Solution 3. Again we prove the result more generally for complex matrices. We will use the following facts about the set S of irreducible complex representations of a finite group G up to equivalence:

1. The number of conjugacy classes of G is $|S|$. [Se2, Theorem 7]
2. The number of one-dimensional representations in S is $|G/G'|$, where G' is the commutator subgroup of G. (This is a consequence of the previous fact applied to G/G', since all one-dimensional representations must be trivial on G'.)
3. The sum of the squares of the dimensions of the representations in S equals $|G|$. [Se2, Corollary 2]

As in Solution 1, we have $A^2 = \pm I$ for any $A \in \mathcal{M}$. Thus any finite subset $\{A_1, \dots, A_k\} \subseteq \mathcal{M}$ generates a finite group G_0, whose elements are of the form

$$\pm A_{i_1} A_{i_2} \cdots A_{i_m}$$

where $i_1 < i_2 < \cdots < i_m$. If $A \neq \pm I$ is in the center of G_0, then A or $-A$ belongs to $\mathcal{M}$, so by (iv) some B in $\mathcal{M}$ does not commute with A. Let G_1 be the group generated by $A_1, \dots, A_k, B$. If there were some A' in G_0 such that $A'B$ is central in G_1, then

$$A'BA = AA'B = A'AB = -A'BA,$$

giving a contradiction. Hence G_1 has a strictly smaller center than G_0. By repeating this enlargement process, we can find a finite set $A_1, \dots, A_{k'}$ of elements of $\mathcal{M}$ ($k' > k$) generating a finite group G with center $Z = \{\pm I\}$. Note that $|G| \geq 2k$.

If $X \in G - Z$, then X has precisely two conjugates, namely itself and $-X$. Thus G has $1 + |G|/2$ conjugacy classes, and therefore G has $1 + |G|/2$ inequivalent irreducible representations over $\mathbb{C}$. The number of inequivalent representations of dimension 1 is $|G/G'|$. Since $G' = \{\pm I\} = Z$, this number is $|G|/2$. The remaining irreducible representation η has dimension $\sqrt{|G|/2}$, since the sum of the squares of the dimensions of the irreducible representations equals $|G|$. Then η must occur in the representation $G \hookrightarrow \mathrm{GL}_n(\mathbb{C})$, since Z is in the kernel of all the 1-dimensional representations. Hence $n \geq \sqrt{|G|/2}$, or equivalently $2n^2 \geq |G|$.

Thus $k \leq |G|/2 \leq n^2$. Since all finite subsets of $\mathcal{M}$ have cardinality at most n^2, we have $|\mathcal{M}| \leq n^2$. ∎

Remark. If we knew in advance that $\mathcal{M}$ were finite, the proof would be cleaner: we could let G be the group $\{\pm A : A \in \mathcal{M}\}$.

Group theory interlude. Here we collect some facts from group theory that will be used in the next remark to classify all possible $\mathcal{M}$. First recall some definitions, from pages 4, 8, and 183 of [Gor]. Let Z denote the center of a group G. If p is a prime, then G is called a *p-group* if G is finite of order a power of p. An *elementary abelian p-group* is a finite abelian group killed by p, hence isomorphic to $(\mathbb{Z}/p\mathbb{Z})^k$ for some integer $k \geq 0$. A *special p-group* is a p-group G such that either G is elementary abelian, or $\{1\} \subsetneqq G' = Z \subsetneqq G$ with Z and G/Z elementary abelian. Finally an *extraspecial p-group* is a nonabelian special p-group G with $|Z| = p$.

There is a relatively simple classification of extraspecial p-groups [Gor, p. 204]. We reproduce the classification for $p = 2$ here. First one has the dihedral group D of order 8 and the quaternion group Q of order 8. Identify the center of each of these with $\{\pm 1\}$. For integers $a, b \geq 0$ not both zero, we let $G_{a,b}$ denote the quotient of $D^a \times Q^b$ by $H \subset \{\pm 1\}^{a+b}$, where H is the kernel of the multiplication map $\{\pm 1\}^{a+b} \to \{\pm 1\}$. (This quotient is called a *central product.*) A group G is an extraspecial 2-group if and only if $G \cong G_{a,b}$ for some a, b as above. Moreover, $G_{a,b} \cong G_{a',b'}$ if and only if $a + b = a' + b'$ and $b \equiv b' \pmod 2$. For convenience, we also define $G_{0,0} = \mathbb{Z}/2$, although it is not extraspecial. Then $|G_{a,b}| = 2^{2a+2b+1}$ for all $a, b \geq 0$.

The argument of Solution 3 shows that over $\mathbb{C}$, $G_{a,b}$ has 2^{2a+2b} representations of dimension 1 that are trivial on the center, and one other irreducible representation η, of dimension 2^{a+b}. We can construct η explicitly as follows. Let $\mathbb{H} = \mathbb{R} + \mathbb{R}i + \mathbb{R}j + \mathbb{R}k$ denote Hamilton's ring of quaternions, in which $i^2 = j^2 = k^2 = -1$, $ij = -ji = k$, $jk = -kj = i$, and $ki = -ik = j$. The 2-dimensional representations η_D and η_Q of D and Q are obtained by complexifying the real representation $D \to \mathrm{GL}_2(\mathbb{R})$ given by symmetries of the square and the inclusion $Q = \{\pm 1, \pm i, \pm j, \pm k\} \to \mathbb{H}^*$, respectively, noting that $\mathbb{H} \otimes_{\mathbb{R}} \mathbb{C} \cong \mathrm{M}_2(\mathbb{C})$. The tensor product of these irreducible representations gives an irreducible representation of $D^a \times Q^b$ whose kernel is exactly H; this induces η. The corresponding simple factor of the group algebra $\mathbb{C}G_{a,b}$ is the complexification of a simple factor of $\mathbb{R}G_{a,b}$ isomorphic to $\mathrm{M}_2(\mathbb{R})^{\otimes a} \otimes_{\mathbb{R}} \mathbb{H}^{\otimes b}$. The latter is isomorphic to $\mathrm{M}_{2^{a+b}}(\mathbb{R})$ or $M_{2^{a+b-1}}(\mathbb{H})$ according as b is even or odd, since the Brauer group $\mathrm{Br}(\mathbb{R})$ is of order 2, generated by the class of $\mathbb{H}$. (For the definition of the Brauer group, see Section 4.7 of [J].) Hence η is the complexification of a representation η' over $\mathbb{R}$ if and only if b is even, i.e., if and only if $G_{a,b} \cong G_{\alpha,0}$ for some $\alpha \geq 0$. If b is odd, we instead let η' be the unique faithful $\mathbb{R}$-irreducible representation; it is of dimension $2 \dim \eta$, and is obtained by identifying the $\mathbb{C}^n$ for η with $\mathbb{R}^{2n}$.

Remark. We are now ready to classify all possible $\mathcal{M}$. The argument of Solution 3 shows that if $\mathcal{M}$ satisfies (i), (ii), (iii), (iv), then $G = \{\pm A : A \in \mathcal{M}\}$ is a finite group such that

(1) the center Z has order 2 and contains the commutator subgroup G'; and

(2) there is a faithful representation $\rho : G \to \mathrm{GL}_n(\mathbb{R})$ identifying Z with $\{\pm I\}$.

Conversely, given a group G with $\rho : G \to \mathrm{GL}_n(\mathbb{R})$ satisfying (1) and (2), any set $\mathcal{M}$ of coset representatives for $\{\pm I\}$ in $\rho(G)$ with $I \in \mathcal{M}$ satisfies (i), (ii), (iii), (iv).

Suppose that G and ρ satisfy (1) and (2). The argument at the beginning of Solution 1 shows that every element of G/Z has order dividing 2; hence inversion on G/Z is a homomorphism, and G/Z is an elementary abelian 2-group. Thus G is an extraspecial 2-group or $G \cong \mathbb{Z}/2\mathbb{Z}$. (In particular $|\mathcal{M}| = |G|/2$ is always a power of 4.) Since ρ is nontrivial on Z, η' must occur in ρ. Conversely, if $G = G_{a,b}$ for some $a, b \geq 0$, and $\rho : G_{a,b} \to \mathrm{M}_n(\mathbb{R})$ is a real representation containing η', then G and ρ satisfy (1) and (2), and hence we obtain all possibilities for $\mathcal{M}$ by choosing a set of coset representatives for $\{\pm I\}$ in $\rho(G)$ with $I \in \mathcal{M}$, for such G and ρ.

Remark. We can also determine all situations in the original problem in which equality holds, i.e., in which $\mathcal{M}$ is a set of n^2 matrices in $\mathrm{M}_n(\mathbb{R})$ satisfying (i), (ii), (iii), (iv). In this case, $G \cong G_{a,b}$ for some $a \geq 0$ and $b \in \{0, 1\}$, $|G| = 2n^2 = 2^{2a+2b+1}$, and ρ is n-dimensional. Then $\dim \eta = 2^{a+b} = n$, and $\dim \eta'$ equals n or $2n$ according as $b = 0$ or $b = 1$. But ρ contains η', so $\rho \cong \eta'$ and $b = 0$. It follows that equality is possible if and only if $n = 2^a$ for some $a \geq 0$, and in that case $\mathcal{M}$ is uniquely determined up to a change of signs of its elements not equal to I and up to an overall conjugation.

Similarly, if we relax the conditions of the problem to allow $\mathcal{M}$ to contain complex matrices, then the equality cases arise from $n = 2^a$, $G \cong G_{a,0}$ or $G \cong G_{a-1,1}$ (the latter being possible only if $a \geq 1$), and $\rho \cong \eta$.

The Fifty-Fourth William Lowell Putnam Mathematical Competition
December 4, 1993

A1. (185, 2, 0, 0, 0, 0, 0, 0, 1, 0, 14, 5)

The horizontal line $y = c$ intersects the curve $y = 2x - 3x^3$ in the first quadrant as in the figure.† Find c so that the areas of the two shaded regions are equal.

Answer. The area of the two regions are equal when $c = 4/9$.

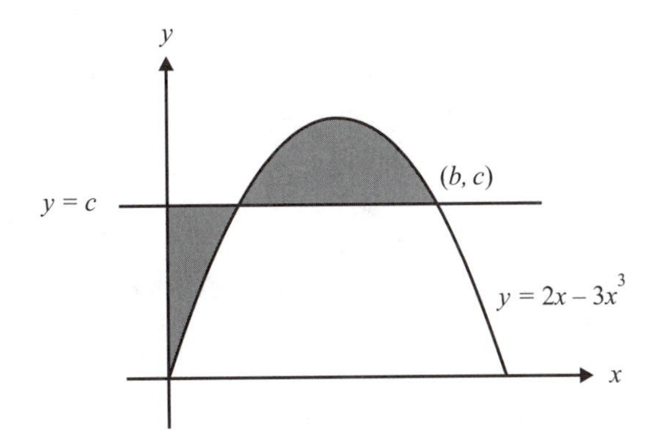

FIGURE 23.

Solution. Let (b, c) denote the rightmost intersection point. (See Figure 23.) We wish to find c such that

$$\int_0^b \left((2x - 3x^3) - c\right) dx = 0.$$

This leads to $b^2 - (3/4)b^4 - bc = 0$. After substituting $c = 2b - 3b^3$ and solving, we find a unique positive solution, namely $b = 2/3$. Thus $c = 4/9$. To validate the solution, we check that $(2/3, 4/9)$ is rightmost among the intersection points of $y = 4/9$ and $y = 2x - 3x^3$: the zeros of $2x - 3x^3 - 4/9 = (2/3 - x)(3x^2 + 2x - 2/3)$ other than $2/3$ are $(-1 \pm \sqrt{3})/3$, which are less than $2/3$. ■

A2. (146, 21, 6, 0, 0, 0, 0, 0, 8, 1, 17, 8)

Let $(x_n)_{n \geq 0}$ be a sequence of nonzero real numbers such that

$$x_n^2 - x_{n-1}x_{n+1} = 1 \textbf{ for } n = 1, 2, 3, \ldots.$$

Prove there exists a real number a such that $x_{n+1} = ax_n - x_{n-1}$ for all $n \geq 1$.

Solution 1. Since the terms are nonzero, we can define $a_n = (x_{n+1} + x_{n-1})/x_n$ for $n \geq 1$. It suffices to show $a_1 = a_2 = \cdots$. But $x_{n+1}^2 - x_n x_{n+2} = 1 = x_n^2 - x_{n-1}x_{n+1}$,

---

† The figure is omitted here, since it is almost identical to Figure 23 used in the solution.

so

$$\begin{aligned} x_{n+1}(x_{n+1}+x_{n-1}) &= x_n(x_{n+2}+x_n) \\ \frac{x_{n+1}+x_{n-1}}{x_n} &= \frac{x_{n+2}+x_n}{x_{n+1}} \\ a_n &= a_{n+1} \end{aligned}$$

since $x_n, x_{n+1} \neq 0$. The result follows by induction. ■

Solution 2. Given x_0, x_1, x_2, let $a = (x_0 + x_2)/x_1$. From $x_1^2 - x_0x_2 = 1$, we get $x_1^2 - ax_1x_0 + x_0^2 = 1$.

Now

$$\begin{aligned} x_2^2 - x_1(ax_2 - x_1) &= (ax_1 - x_0)^2 - x_1(a(ax_1 - x_0) - x_1) \\ &= x_1^2 - ax_1x_0 + x_0^2 = 1, \end{aligned}$$

so $x_3 = ax_2 - x_1$.

By essentially the same algebra (and a simple induction), $x_{n+1} = ax_n - x_{n-1}$ for all $n \geq 1$. ■

Solution 3. Let $a = (x_0 + x_2)/x_1$. Consider a sequence defined by $y_{n+1} - ay_n + y_{n-1} = 0$ where $y_i = x_i$ for $i = 0, 1, 2$; we wish to show that $y_i = x_i$ for all nonnegative integers i.

If $a \neq \pm 2$, then by the theory of linear recursive sequences (discussed after the solution to 1988A5), the general solution is

$$y_n = Ar^n + Bs^n,$$

where A and B are constants, and r and s are the roots of the characteristic equation $t^2 - at + 1 = 0$. Then $rs = 1$, and $(r - s)^2 = a^2 - 4$ by the quadratic formula or by the identity $(r - s)^2 = (r + s)^2 - 4rs$. Now

$$\begin{aligned} y_n^2 - y_{n+1}y_{n-1} &= (Ar^n + Bs^n)^2 - (Ar^{n-1} + Bs^{n-1})(Ar^{n+1} + Bs^{n+1}) \\ &= -AB(rs)^{n-1}(r - s)^2 \\ &= -AB(a^2 - 4), \end{aligned}$$

which is independent of n. Thus $y_n^2 - y_{n+1}y_{n-1} = y_1^2 - y_2y_0 = 1$ for all n, so $y_n = x_n$ by induction on n.

If $a = \pm 2$, then the general solution is

$$y_n = (A + Bn)(\pm 1)^n.$$

Then

$$y_n^2 - y_{n+1}y_{n-1} = B^2,$$

which is independent of n. Hence as before, $y_n^2 - y_{n+1}y_{n-1} = y_1^2 - y_2y_0 = 1$ for all n, so $y_n = x_n$ for all n as well. ■

Solution 4. For all n,

$$\det\begin{pmatrix} x_{n-1}+x_{n+1} & x_n+x_{n+2} \\ x_n & x_{n+1}\end{pmatrix} = \det\begin{pmatrix} x_{n-1} & x_n \\ x_n & x_{n+1}\end{pmatrix} + \det\begin{pmatrix} x_{n+1} & x_{n+2} \\ x_n & x_{n+1}\end{pmatrix}$$
$$= -1+1=0.$$

Thus $(x_{n-1}+x_{n+1}, x_n+x_{n+2}) = a_n(x_n, x_{n+1})$ for some scalar a_n. Hence

$$\frac{x_{n+1}+x_{n-1}}{x_n} = \frac{x_{n+2}+x_n}{x_{n+1}}$$

for all $n \geq 1$, since $x_n,\ x_{n+1} \neq 0$, so we are done by induction. ■

Remark. In a similar manner, one can prove that if $(x_n)_{n\geq 0}$ is a sequence of nonzero real numbers such that

$$\det\begin{pmatrix} x_n & x_{n+1} & \cdots & x_{n+k} \\ x_{n+1} & x_{n+2} & \cdots & x_{n+k+1} \\ \vdots & \vdots & \ddots & \vdots \\ x_{n+k} & x_{n+k+1} & \cdots & x_{n+2k}\end{pmatrix} = cr^n$$

for $n = 1, 2, 3, \ldots$, then there exist real numbers $a_1, \ldots, a_k$ such that

$$x_{n+k+1} = a_1 x_{n+k} + a_2 x_{n+k-1} + \cdots + a_k x_{n+1} + (-1)^k r x_n.$$

Related question. The Fibonacci numbers have a similar structure: $F_{n+1} - F_n - F_{n-1} = 0$, and $F_n^2 - F_{n-1}F_{n+1} = (-1)^{n+1}$. How do the solutions generalize to this case? Can they be generalized even further?

A3. (2, 11, 24, 0, 0, 0, 0, 0, 27, 8, 54, 81)

Let $\mathcal{P}_n$ be the set of subsets of $\{1, 2, \ldots, n\}$. Let $c(n,m)$ be the number of functions $f : \mathcal{P}_n \to \{1, 2, \ldots, m\}$ such that $f(A \cap B) = \min\{f(A), f(B)\}$. Prove that

$$c(n,m) = \sum_{j=1}^{m} j^n.$$

Solution 1. Let $S = \{1, 2, \ldots, n\}$ and $S_i = S - \{i\}$. First we show that $f \mapsto (f(S), f(S_1), f(S_2), \ldots, f(S_n))$ defines a bijection between the set of allowable functions f and the set of $(n+1)$-tuples $(j, a_1, a_2, \ldots, a_n)$ of elements of $\{1, 2, \ldots, m\}$ such that $j \geq a_i$ for $i = 1, 2, \ldots, n$. This map is well defined, since for any allowable f, we have $f(S_i) = f(S_i \cap S) = \min\{f(S_i), f(S)\} \leq f(S)$. The inverse map takes the tuple $(j, a_1, a_2, \ldots, a_n)$ to the function f such that $f(S) = j$ and $f(T) = \min_{i \notin T} a_i$ for $T \subsetneq S$. This function satisfies $f(A \cap B) = \min\{f(A), f(B)\}$, and these two constructions are inverse to each other.

Now we count the $(n+1)$-tuples. For fixed j, the possibilities for each a_i are $1, 2, \ldots, j$, so there are j^n possibilities for $(a_1, a_2, \ldots, a_n)$. Summing over j, which can be anywhere from 1 to m, yields $c(n,m) = \sum_{j=1}^{m} j^n$. ■

Solution 2. We will use induction on m. The base case, $m = 1$, states that $c(n,1) = 1$ for all n, which is obvious.

Suppose $m \geq 2$. Given $f : \mathcal{P}_n \to \{1, 2, \ldots, m\}$ *not identically* 1 such that $f(A \cap B) = \min\{f(A), f(B)\}$, let S_f be the intersection of all $A \in \mathcal{P}_n$ such that $f(A) \geq 2$. The property implies that $f(S_f) \geq 2$, and that if $T \supseteq S_f$ then $f(T) \geq 2$. Hence the sets T for which $f(T) \geq 2$ are exactly those that contain S_f.

Given $S \in \mathcal{P}_n$, how many f's have $S_f = S$? To give such an f is the same as specifying a function

$$g : \{T \in \mathcal{P}_n : T \supseteq S\} \to \{2, 3, \ldots, m\}$$

satisfying $g(A \cap B) = \min\{g(A), g(B)\}$. There is an intersection-preserving bijection from $\{T \in \mathcal{P}_n : T \supseteq S\}$ to the set of subsets of $\{1, 2, \ldots, n\} - S$ (the bijection maps T to $T - S$), and an order-preserving bijection from $\{2, 3, \ldots, m\}$ to $\{1, 2, \ldots, m-1\}$, so the number of such functions g equals $c(n - \#S, m-1)$.

Thus, remembering to count the identically 1 function, we have

$$\begin{aligned}
c(n, m) &= 1 + \sum_S c(n - \#S, m-1) \\
&= 1 + \sum_{k=0}^{n} \binom{n}{k} c(n-k, m-1) \\
&= 1 + \sum_{k=0}^{n} \binom{n}{k} \sum_{j=1}^{m-1} j^{n-k} \qquad \text{(inductive hypothesis)} \\
&= 1 + \sum_{j=1}^{m-1} \sum_{k=0}^{n} \binom{n}{k} j^{n-k} \\
&= 1 + \sum_{j=1}^{m-1} (1+j)^n \\
&= \sum_{j=1}^{m} j^n,
\end{aligned}$$

completing the inductive step. ∎

A4. (3, 2, 0, 0, 0, 0, 0, 0, 0, 0, 44, 158)

Let $x_1, x_2, \ldots, x_{19}$ be positive integers each of which is less than or equal to 93. Let $y_1, y_2, \ldots, y_{93}$ be positive integers each of which is less than or equal to 19. Prove that there exists a (nonempty) sum of some x_i's equal to a sum of some y_j's.

Solution 1. We move a pebble among positions numbered $-18, -17, \ldots, 0, 1, 2, \ldots, 93$, until it revisits a location. The pebble starts at position 0. Thereafter, if the pebble is at position t, we move it as follows. If $t \leq 0$, choose some unused x_i, move the pebble to $t + x_i$, and then discard that x_i. If $t > 0$, choose some unused y_j, move the pebble to $t - y_j$, and then discard that y_j. Since $x_i \leq 93$ and $y_j \leq 19$, the pebble's position stays between -18 and 93.

In order to continue this process until a location is revisited, we must show that there is always an unused x_i or y_j as needed. If $t \leq 0$ and a revisit has not yet occurred, then one x_i has been used after visiting each nonpositive position *except the current one*, so the total number of x_i's used so far is at most $19 - 1 = 18$, and at least one x_i remains. Similarly, if $t > 0$ and a revisit has not yet occurred, then one y_j has been used after visiting each positive position except the current one, so the total number of y_j's used so far is at most $93 - 1 = 92$, and at least one y_j remains.

Since there are only finitely many x_i's and y_j's to be used, the algorithm must eventually terminate with a revisit. The steps between the two visits of the same position constitute a sum of some x_i's equal to a sum of some y_j's. ∎

Our next solution is similar to Solution 1, but we dispense with the algorithmic interpretation.

Solution 2. For the sake of generality, replace 19 and 93 in the problem statement by m and n respectively. Define $X_k = \sum_{i=1}^{k} x_i$ and $Y_\ell = \sum_{j=1}^{\ell} y_j$. Without loss of generality, assume $X_m \geq Y_n$. For $1 \leq \ell \leq n$, define $f(\ell)$ by

$$X_{f(\ell)} \leq Y_\ell < X_{f(\ell)+1},$$

so $0 \leq f(\ell) \leq m$. Let $g(\ell) = Y_\ell - X_{f(\ell)}$. If $g(\ell) = 0$ for some ℓ, we are done. Otherwise,

$$g(\ell) = Y_\ell - X_{f(\ell)} < x_{f(\ell)+1} \leq n,$$

so $0 < g(\ell) \leq n - 1$ whenever $1 \leq \ell \leq n$. Hence by the Pigeonhole Principle, there exist $\ell_0 < \ell_1$ such that $g(\ell_0) = g(\ell_1)$. Then

$$\sum_{i=f(\ell_0)+1}^{f(\ell_1)} x_i = X_{f(\ell_1)} - X_{f(\ell_0)} = Y_{\ell_1} - Y_{\ell_0} = \sum_{j=\ell_0+1}^{\ell_1} y_j. \qquad ∎$$

Solution 3 (based on an idea of Noam Elkies). With the same notation as in the previous solution, without loss of generality $X_m \geq Y_n$. If equality holds, we are done, so assume $X_m > Y_n$. By the Pigeonhole Principle, two of the $(m+1)(n+1)$ sums $X_i + Y_j$ $(0 \leq i \leq m, 0 \leq j \leq n)$ are congruent modulo X_m, say

$$X_{i_1} + Y_{j_1} \equiv X_{i_2} + Y_{j_2} \pmod{X_m}.$$

But the difference

$$(X_{i_1} - X_{i_2}) + (Y_{j_1} - Y_{j_2}) \tag{1}$$

lies strictly between $-2X_m$ and $+2X_m$, so it equals 0 or $\pm X_m$. Clearly $i_1 \neq i_2$; without loss of generality $i_1 > i_2$, so (1) equals 0 or X_m. If $j_1 < j_2$, then (1) must be 0, so $X_{i_1} - X_{i_2} = Y_{j_2} - Y_{j_1}$ are two equal subsums. If $j_1 > j_2$, then (1) must be X_m, and $Y_{j_1} - Y_{j_2} = X_m - (X_{j_1} - X_{j_2})$ are two equal subsums. ∎

A5. (3, 3, 0, 0, 0, 0, 0, 0, 1, 16, 37, 147)

Show that

$$\int_{-100}^{-10} \left(\frac{x^2 - x}{x^3 - 3x + 1}\right)^2 dx + \int_{\frac{1}{101}}^{\frac{1}{11}} \left(\frac{x^2 - x}{x^3 - 3x + 1}\right)^2 dx + \int_{\frac{101}{100}}^{\frac{11}{10}} \left(\frac{x^2 - x}{x^3 - 3x + 1}\right)^2 dx$$

is a rational number.

Solution 1. The polynomial $x^3 - 3x + 1$ changes sign in each of the intervals $[-2, -1]$, $[1/3, 1/2]$, $[3/2, 2]$, so it has no zeros outside these intervals. Hence in the problem, the integrand is continuous on the three ranges of integration.

By the substitutions $x = 1/(1-t)$ and $x = 1 - 1/t$, the integrals over $[1/101, 1/11]$ and $[101/100, 11/10]$ are respectively converted into integrals over $[-100, -10]$. The integrand

$$Q(x) = \left(\frac{x^2 - x}{x^3 - 3x + 1}\right)^2,$$

is invariant under each of the substitutions $x \to 1/(1-x)$ and $x \to 1 - 1/x$. Hence the sum of the three given integrals is expressible as

$$\int_{-100}^{-10} \left(\frac{x^2 - x}{x^3 - 3x + 1}\right)^2 \left(1 + \frac{1}{x^2} + \frac{1}{(1-x)^2}\right) dx. \tag{1}$$

But

$$\frac{1}{Q(x)} = \left(x + 1 - \frac{1}{x} - \frac{1}{x-1}\right)^2,$$

so the last integral is of the form $\int u^{-2} du$. Hence its value is

$$-\frac{x^2 - x}{x^3 - 3x + 1}\bigg|_{-100}^{-10},$$

which is rational. ∎

Solution 2. Set

$$f(t) = \int_{-100}^{t} \left(\frac{x^2 - x}{x^3 - 3x + 1}\right)^2 dx + \int_{\frac{1}{101}}^{1/(1-t)} \left(\frac{x^2 - x}{x^3 - 3x + 1}\right)^2 dx$$
$$+ \int_{\frac{101}{100}}^{1-1/t} \left(\frac{x^2 - x}{x^3 - 3x + 1}\right)^2 dx$$

for $-100 \le t \le -10$. We want $f(-10)$. By the Fundamental Theorem of Calculus,

$$f'(t) = Q(t) + Q\left(\frac{1}{1-t}\right)\frac{1}{(1-t)^2} + Q\left(1 - \frac{1}{t}\right)\frac{1}{t^2}.$$

We find that $Q(1/(1-x)) = Q(1-1/x) = Q(x)$ (in fact $x \mapsto 1/(1-x)$ and $x \mapsto 1-1/x$ are inverses of each other), so

$$f(-10) = \int_{-100}^{-10} \left(\frac{x^2 - x}{x^3 - 3x + 1}\right)^2 \left(1 + \frac{1}{x^2} + \frac{1}{(1-x)^2}\right) dx.$$

The integrand equals

$$\frac{x^4 - 2x^3 + 3x^2 - 2x + 1}{(x^3 - 3x + 1)^2}.$$

We guess that this is the derivative of a quotient

$$\frac{Ax^2 + Bx + C}{x^3 - 3x + 1}$$

and solve for the undetermined coefficients, obtaining $A = -1$, $B = 1$, $C = 0$. Thus

$$f(-10) = \left.\frac{-x^2+x}{x^3-3x+1}\right|_{-100}^{-10},$$

which is rational. ■

Remark. Both solutions reached the formula (1) in the middle, so we could construct two more solutions by matching the first half of Solution 1 with the second half of Solution 2, or vice versa.

Remark. The actual value of the sum of the three integrals is $\dfrac{11131110}{107634259}$.

Related question. Let k be a positive integer, and suppose $f(x) \in \mathbb{Q}[x]$ has degree at most $3k-2$. Then

$$\int_{-100}^{-10} \frac{f(x)}{(x^3-3x+1)^k}\,dx + \int_{\frac{1}{101}}^{\frac{1}{11}} \frac{f(x)}{(x^3-3x+1)^k}\,dx + \int_{\frac{101}{100}}^{\frac{11}{10}} \frac{f(x)}{(x^3-3x+1)^k}\,dx \qquad (1)$$

is rational.

Proof. We will use the theory of differentials on the projective line $\mathbb{P}^1$. Let $\tau(x) = 1 - 1/x$. Then τ, as an automorphism of $\mathbb{P}^1$, permutes the three zeros of $p(x) = x^3 - 3x + 1$ cyclically, since a calculation shows $p(1-1/x) = -x^{-3}p(x)$. In particular, τ^3 is the identity, since an automorphism of $\mathbb{P}^1$ is determined its action on three distinct points, or by direct calculation.

Let ω be the differential $\frac{f(x)}{(x^3-3x+1)^k}\,dx$ on $\mathbb{P}^1$. Because $\deg f \le 3k-2$, the substitution $x = 1/t$, $dx = -t^{-2}\,dt$ shows that ω has no pole at ∞, so ω is regular on all of $\mathbb{P}^1$ except possibly at the zeros of $x^3 - 3x + 1$.

The substitutions used in Solution 1 transform (1) into $\int_{-100}^{-10} \eta$ where $\eta = \omega + \tau^*\omega + (\tau^2)^*\omega$. Then η is regular outside the zeros of $x^3 - 3x + 1$. Since $\tau^*\eta = \eta$, the residues of η at the three zeros are equal, but the sum of the residues of any differential is zero, so all the residues are zero. Hence $\eta = dg$ for some rational function g on $\mathbb{P}^1$ having poles at most at the zeros of $x^3 - 3x + 1$. Adding a constant to g, we may assume $g(0) = 0$. We want to show that g has rational coefficients. Write g in lowest terms, with monic denominator. Any automorphism σ of $\mathbb{C}$ over $\mathbb{Q}$ applied to the coefficients of g results in another rational function g_1 with $dg_1 = \eta$ and $g_1(0) = 0$. Then $d(g - g_1) = 0$, so $g - g_1$ is constant, and evaluating at 0 shows that $g = g_1$. This holds for all σ, so $g \in \mathbb{Q}(x)$. Finally, (1) equals $\int_{-100}^{-10} dg = g(-10) - g(-100)$, which is rational. □

Remark. We prove that if instead $f(x) \in \mathbb{Q}[x]$ has degree exactly $3k-1$, then (1) is irrational. Such $f(x)$ can be expressed as a nonzero rational multiple of $f_1(x) = (3x^2-3)(x^3-3x+1)^{k-1}$ plus a polynomial in $\mathbb{Q}[x]$ of degree at most $3k-2$, so it suffices to consider $f(x) = f_1(x)$. Then (1) becomes

$$\ln p(x)\big|_{-100}^{-10} + \ln p(x)\big|_{1/101}^{1/11} + \ln p(x)\big|_{101/100}^{11/10} = \ln\left(\frac{347193581883670000}{49251795138277608547}\right),$$

which is irrational, since the only rational number $q > 0$ with $\ln q \in \mathbb{Q}$ is $q = 1$.

Remark. One can strengthen "irrational" to "transcendental" in the previous remark, since Lindemann proved in 1882 that all nonzero numbers that are logarithms of algebraic numbers are transcendental [FN, Corollary 2.2]. (Recall that a complex number is *algebraic* if it is the zero of some nonzero polynomial with rational coefficients, and *transcendental* otherwise.)

Remark. We can explain why for the polynomial $p(x) = x^3 - 3x + 1$ there was a fractional linear transformation with rational coefficients permuting its zeros cyclically. An automorphism of $\mathbb{P}^1$ is uniquely determined by its action on any three distinct points, so given any cubic polynomial $p(x) \in \mathbb{Q}[x]$ with distinct zeros, there is a unique fractional linear transformation $\tau \in L(x)$ permuting the zeros, where L is the splitting field of $p(x)$. If $\sigma \in \mathrm{Gal}(L/\mathbb{Q})$, then σ acts on τ, and the action of ${}^{\sigma}\tau$ on the three zeros is obtained by *conjugating* the action of τ by the permutation of the zeros given by σ. In particular, ${}^{\sigma}\tau = \tau$ for all $\sigma \in \mathrm{Gal}(L/\mathbb{Q})$ if and only if the $\mathrm{Gal}(L/\mathbb{Q})$ is contained in the cyclic subgroup of order 3 of the permutations of the three zeros. Hence $\tau \in \mathbb{Q}(x)$ if and only if the discriminant of $p(x)$ is a square. In the problem, the discriminant of $p(x) = x^3 - 3x + 1$ is 81.

Remark. This action of the group of order three on the projective line appeared in Problem 1971B2 [PutnamII, p. 15]:

> Let $F(x)$ be a real valued function defined for all real x except for $x = 0$ and $x = 1$ and satisfying the functional equation $F(x) + F((x-1)/x) = 1 + x$. Find all functions $F(x)$ satisfying these conditions.

A6. (0, 1, 0, 0, 0, 0, 1, 3, 3, 11, 36, 152)
The infinite sequence of 2's and 3's

$$2, 3, 3, 2, 3, 3, 3, 2, 3, 3, 3, 2, 3, 3, 2, 3, 3, 3, 2, 3, 3, 3, 2, 3, 3, 3, 2, 3, 3, 2, 3, 3, 3, 2, \ldots$$

has the property that, if one forms a second sequence that records the number of 3's between successive 2's, the result is identical to the given sequence. Show that there exists a real number r such that, for any n, the nth term of the sequence is 2 if and only if $n = 1 + \lfloor rm \rfloor$ for some nonnegative integer m. (Note: $\lfloor x \rfloor$ denotes the largest integer less than or equal to x.)

Motivation. Assuming the result, we derive the value of r. Fix a large integer m. The result implies that the mth 2 in the sequence occurs at the nth term, where $n \approx rm$. Interleaved among these m 2's are $n - m$ 3's. By the self-generation property of the sequence, these 2's and 3's describe an initial segment with n 2's separated by blocks containing a total of $2m + 3(n - m)$ 3's. But the length of this segment should also approximate r times the number of 2's in the segment, so $n + 2m + 3(n - m) \approx rn$. Substituting $n \approx rm$, dividing by m, and taking the limit as $m \to \infty$ yields $r + 2 + 3(r - 1) = r \cdot r$, so $r^2 - 4r + 1 = 0$. Clearly $r \geq 1$, so $r = 2 + \sqrt{3}$. Note that $3 < r < 4$.

Solution. Let $r = 2 + \sqrt{3}$. The sequence is uniquely determined by the self-generation property and its first few terms. Therefore it suffices to show that the sequence $a_0, a_1, a_2, \ldots$ has the self-generation property when we define $a_n = 2$ if $n = \lfloor rm \rfloor$ for some m, and $a_n = 3$ otherwise. (Note that the first term is a_0.)

The self-generation property for (a_n) is equivalent to

$$\lfloor r(n+1)\rfloor - \lfloor rn\rfloor = \begin{cases} 3 & \text{if } n = \lfloor rm\rfloor \text{ for some } m, \\ 4 & \text{otherwise.} \end{cases}$$

Since $\lfloor r(n+1)\rfloor - \lfloor rn\rfloor$ is always 3 or 4, we seek to prove:

$$n \text{ is of the form } \lfloor rm\rfloor \qquad \Longleftrightarrow \qquad \lfloor r(n+1)\rfloor - \lfloor rn\rfloor = 3.$$

This is a consequence of the following series of equivalences, in which we use the identity

$$\lceil z/r\rceil = \lceil 4z - rz\rceil = 4z - \lfloor rz\rfloor \tag{1}$$

for integer z:

n is of the form $\lfloor rm\rfloor$.

$\Longleftrightarrow$ There is an m for which $n+1 > rm > n$.

$\Longleftrightarrow$ There is an m for which $(n+1)/r > m > n/r$.

$\Longleftrightarrow$ $\lceil (n+1)/r\rceil - \lceil n/r\rceil = 1$.

$\Longleftrightarrow$ $4(n+1) - \lfloor (n+1)r\rfloor - 4n + \lfloor nr\rfloor = 1$ (by (1) with $z = n$ and $z = n+1$).

$\Longleftrightarrow$ $\lfloor (n+1)r\rfloor - \lfloor nr\rfloor = 3$. ∎

Remark. A similar argument shows that $a_n = 3$ if and only if n has the form $\lfloor sm\rfloor$ where $s = (1+\sqrt{3})/2$ and m is a positive integer. Hence the positive integers are disjointly partitioned into the two sequences $(\lfloor rm\rfloor)_{m>0}$ and $(\lfloor sm\rfloor)_{m>0}$. Two sequences forming a partition of the positive integers are called *complementary sequences.* These are discussed in [Hon1, Essay 12].

Complementary sequences of the form $(\lfloor rm\rfloor)$ and $(\lfloor sm\rfloor)$ are called *Beatty sequences.* A short argument shows that if $(\lfloor rm\rfloor)$ and $(\lfloor sm\rfloor)$ are complementary, then r and s are irrational and $1/r + 1/s = 1$. Conversely, Problem 1959A6 [PutnamI, p. 57] asks:

> Prove that, if x and y are positive irrationals such that $1/x + 1/y = 1$, then the sequences $\lfloor x\rfloor, \lfloor 2x\rfloor, \dots, \lfloor nx\rfloor, \dots$ and $\lfloor y\rfloor, \lfloor 2y\rfloor, \dots, \lfloor ny\rfloor, \dots$ together include every positive integer exactly once.

This result is known as *Beatty's Theorem.* For further discussion, see [PutnamI, p. 514]. For another Beatty-related problem, see Problem 1995B6.

Related question. How far can you generalize this? For example, the infinite sequence of 1's and 2's starting $1, 2, 1, 2, 2, 1, \dots$ (where the leading term is considered the 0th term) has the property that, if one forms a second sequence that records the number of 2's between successive 1's, the result is identical to the identical to the given sequence. Then the nth term of the sequence is 1 if and only if $n = \lfloor m\tau^2\rfloor$, where $\tau = (1+\sqrt{5})/2$ is the golden mean.

Related questions. This problem is reminiscent of Problem 3 on the 1978 International Mathematical Olympiad [IMO79–85, p. 1]:

> The set of all positive integers is the union of two disjoint subsets
>
> $$\{f(1), f(2), \dots, f(n), \dots\}, \ \{g(1), g(2), \dots, g(n), \dots\},$$

where

$$f(1) < f(2) < \cdots < f(n) < \cdots ,$$

$$g(1) < g(2) < \cdots < g(n) < \cdots ,$$

and $g(n) = f(f(n)) + 1$ for all $n \geq 1$. Determine $f(240)$.

The two subsets again turn out to form a Beatty sequence.

Another related result, due to E. Dijkstra [Hon3, p. 13], is the following.

> Suppose $f(n)$, $n = 0, 1, \ldots$ is a nondecreasing sequence of nonnegative integers which is unbounded above. Let the "converse function" $g(n)$ $(n \geq 0)$ be the number of values m for which $f(m) \leq n$. Then the converse function of g is f, and the sequences $F(n) = f(n) + n$ and $G(n) = g(n) + n$ are complementary.

Yet another problem, related to the previous two, is the following [New, Problem 47]:

> Suppose we "sieve" the integers as follows: take $a_1 = 1$ and then delete $a_1 + 1 = 2$. The smallest untouched integer is 3, which we call a_2, and then we delete $a_2 + 2 = 5$. The next untouched integer is $4 = a_3$, and we delete $a_3 + 3 = 7$, etc. The sequence $a_1, a_2, \ldots$ is then
>
> $$1, 3, 4, 6, 8, 9, 11, 12, 14, 16, 17, \ldots.$$
>
> Find a formula for a_n.

The sieve just described arises in the analysis of Wythoff's game with two heaps [Wy], [BCG, pp. 62, 76], [Hon1].

B1. (83, 29, 9, 0, 0, 0, 0, 0, 6, 10, 38, 32)

Find the smallest positive integer n such that for every integer m, with $0 < m < 1993$, there exists an integer k for which

$$\frac{m}{1993} < \frac{k}{n} < \frac{m+1}{1994}.$$

Answer. The smallest positive integer n satisfying the condition is $n = 3987$.

Lemma 1. *Suppose a, b, c, and d are positive numbers, and $\frac{a}{b} < \frac{c}{d}$. Then*

$$\frac{a}{b} < \frac{a+c}{b+d} < \frac{c}{d}.$$

This lemma is sometimes called the *mediant property* or the *règle des nombres moyens* [Nel, pp. 60–61]. It appears without proof in the *Triparty en la Science des Nombres*, which was written by the French physician Nicholas Chuquet in 1484 (although his work went unpublished until 1880). It is not hard to verify this inequality directly, but there is an even easier way of remembering it: If a sports team wins a of b games in the first half of a season, and c of d games in the second half, then its overall record $((a + c)/(b + d))$ is between its records in its two halves (a/b and c/d). Alternatively, Figure 24 gives a "proof without words." For three more proofs without words of this result, see [Nel, pp. 60–61].

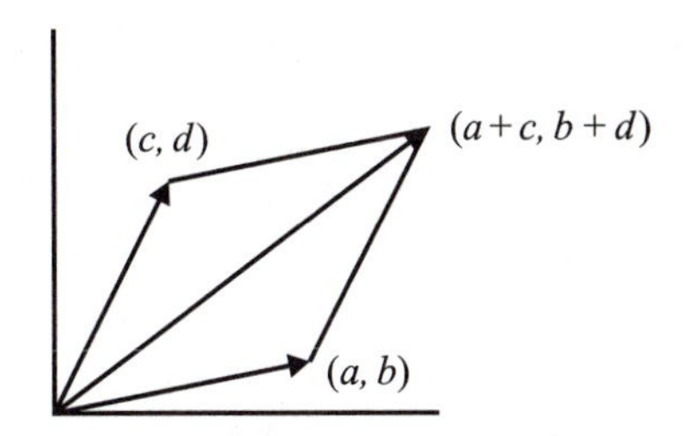

FIGURE 24.
A proof without words of Lemma 1.

Solution 1. By Lemma 1,

$$\frac{m}{1993} < \frac{2m+1}{3987} < \frac{m+1}{1994}.$$

We will show that 3987 is best possible. If

$$\frac{1992}{1993} < \frac{k}{n} < \frac{1993}{1994} \tag{1}$$

then

$$\frac{1}{1993} > \frac{n-k}{n} > \frac{1}{1994},$$

so

$$1993 < \frac{n}{n-k} < 1994.$$

Clearly $n - k \neq 1$, so $n - k \geq 2$. Thus $n > 1993(n - k) \geq 3986$, and $n \geq 3987$. ■

Solution 2. Subtracting everything in the desired inequality from 1, and using the change of variables $M = 1993 - m$, $K = n - k$, the problem becomes: determine the smallest positive integer n such that for every integer M with $1993 > M > 0$, there exists an integer K for which

$$\frac{M}{1993} > \frac{K}{n} > \frac{M}{1994}, \quad \text{or equivalently,} \quad 1993K < nM < 1994K.$$

For $M = 1$, K cannot be 1 and hence is at least 2, so $n > 1993 \cdot 2 = 3986$. Thus $n \geq 3987$. On the other hand $n = 3987$ works, since then for each M, $K = 2M$ satisfies the inequalities. ■

Solution 3 (Naoki Sato).

Lemma 2. *Let a, b, c, d, p, and q be positive integers such that $a/b < p/q < c/d$, and $bc - ad = 1$. Then $p \geq a + c$ and $q \geq b + d$.*

Proof. Since $bp - aq > 0$, $bp - aq \geq 1$. Also, $cq - dp > 0$, so $cq - dp \geq 1$. Hence $d(bp - aq) + b(cq - dp) \geq b + d$, which simplifies to $(bc - ad)q \geq b + d$. But $bc - ad = 1$, so $q \geq b + d$. The proof of $p \geq a + c$ is similar. □

Now

$$\frac{1992}{1993} < \frac{k}{n} < \frac{1993}{1994}$$

for some k, and $1993 \cdot 1993 - 1992 \cdot 1994 = 1$, so $n \geq 1993 + 1994 = 3987$. And $n = 3987$ works, by Lemma 1. ■

Remark. Looking at (1) was key. What clues suggest that it would be helpful to look at large m?

Remark. This problem and Lemma 2 especially are related to the beautiful topic of Farey series; see [Hon1, Essay 5].

B2. (89, 0, 1, 0, 0, 0, 0, 0, 7, 3, 42, 65)

Consider the following game played with a deck of $2n$ cards numbered from 1 to $2n$. The deck is randomly shuffled and n cards are dealt to each of two players, A and B. Beginning with A, the players take turns discarding one of their remaining cards and announcing its number. The game ends as soon as the sum of the numbers on the discarded cards is divisible by $2n+1$. The last person to discard wins the game. Assuming optimal strategy by both A and B, what is the probability that A wins?

Answer. The probability that A wins is 0, i.e., B can always win.

Solution. Player B can always win, because B can always guarantee that A will not win on the next move: B holds one more card than A, and each of A's cards causes at most one of B's cards to be a fatal play. Hence B has at least one safe play. Player B wins on the last move if not earlier, since the sum of the numbers on all the cards is $n(2n+1)$. ∎

Remark. David Savitt gives the following 'alternate solution': "I maintain that A, playing optimally, will decline to participate!"

B3. (111, 16, 13, 0, 0, 0, 0, 0, 4, 20, 10, 33)

Two real numbers x and y are chosen at random in the interval $(0,1)$ with respect to the uniform distribution. What is the probability that the closest integer to x/y is even? Express the answer in the form $r+s\pi$, where r and s are rational numbers.

Answer. The limit is $(5-\pi)/4$. That is, $r=5/4$ and $s=-1/4$.

Solution. The probability that x/y is exactly half an odd integer is 0, so we may safely ignore this possibility.

The closest integer to $\frac{x}{y}$ is even if and only if $0<\frac{x}{y}<\frac{1}{2}$ or $\frac{4n-1}{2}<\frac{x}{y}<\frac{4n+1}{2}$ for some integer $n \geq 1$. The former occurs inside the triangle with vertices $(0,0)$, $(0,1)$, $(\frac{1}{2},1)$, whose area is $\frac{1}{4}$. The latter occurs inside the triangle $(0,0)$, $(1,\frac{2}{4n-1})$, $(1,\frac{2}{4n+1})$, whose area is $\frac{1}{4n-1}-\frac{1}{4n+1}$. These regions are shown in Figure 25.

Hence the total area is

$$\frac{1}{4}+\frac{1}{3}-\frac{1}{5}+\frac{1}{7}+\cdots.$$

Comparing this with Leibniz's formula

$$\frac{\pi}{4}=1-\frac{1}{3}+\frac{1}{5}-\frac{1}{7}+\cdots$$

shows that the total area is $(5-\pi)/4$. ∎

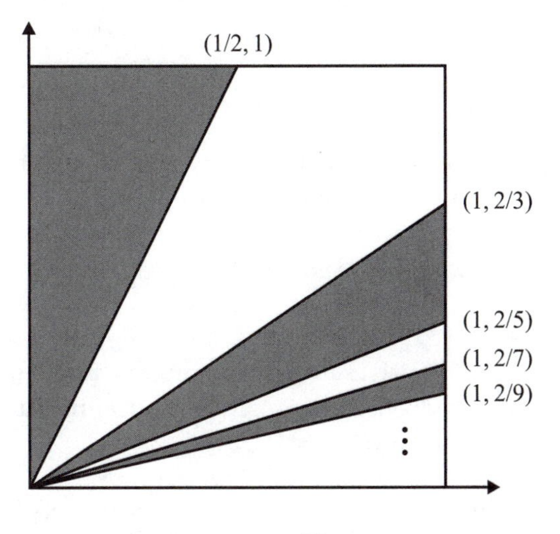

FIGURE 25.

Remark. Leibniz's formula can be derived as follows:

$$\begin{aligned}\frac{\pi}{4} &= \arctan 1 \\ &= \int_0^1 \frac{dx}{1+x^2} \\ &= \int_0^1 \left((1-x^2)+(x^4-x^6)+(x^8-x^{10})+\cdots\right) dx \\ &= \int_0^1 (1-x^2)\,dx + \int_0^1 (x^4-x^6)\,dx + \int_0^1 (x^8-x^{10})\,dx + \cdots \\ &= \left(1-\frac{1}{3}\right)+\left(\frac{1}{5}-\frac{1}{7}\right)+\left(\frac{1}{9}-\frac{1}{11}\right)+\cdots.\end{aligned}$$

The interchange of integral and summation in the penultimate step is justified by the Monotone Convergence Theorem [Ru, p. 319], since each integrand $x^{4k} - x^{4k+2}$ is nonnegative on $[0, 1]$.

Remark. This gives an interesting connection between π and number theory (and gives an "experimental" way to estimate π). Here are more links. In fact, $6/\pi^2$ is both the probability that a pair of positive integers are relatively prime, and the probability that a positive integer is squarefree (where these probabilities are appropriately defined as limits). More generally, for any $k \geq 2$, the probability that a set of k positive integers has greatest common divisor 1, and the probability that a positive integer is kth-power-free, is $1/\zeta(k)$, where

$$\zeta(k) = \frac{1}{1^k} + \frac{1}{2^k} + \frac{1}{3^k} + \cdots$$

is the Riemann zeta function. (See [NZM, Theorem 8.25] for a proof in the case $n = 2$; the general case is similar.) It can be shown that $\zeta(2n)$ is a rational multiple of π^{2n}

for all positive integers n, and in fact

$$\zeta(2n) = \frac{2^{2n-1}B_n}{(2n)!}\pi^{2n}$$

where B_n is the nth Bernoulli number [HW, Section 17.2]. (See 1996B2 for another appearance of Bernoulli numbers.) The special case $\zeta(2) = \pi^2/6$ can be computed using Fourier analysis [Fo, Exercise 8.17 or 8.21], trigonometry [NZM, Appendix A.3], or multivariable change of variables [Sim, p. 748].

B4. (0, 0, 0, 0, 0, 0, 0, 0, 1, 1, 61, 144)

The function $K(x,y)$ is positive and continuous for $0 \le x \le 1, 0 \le y \le 1$, and the functions $f(x)$ and $g(x)$ are positive and continuous for $0 \le x \le 1$. Suppose that for all x, $0 \le x \le 1$,

$$\int_0^1 f(y)K(x,y)\,dy = g(x) \qquad \textbf{and} \qquad \int_0^1 g(y)K(x,y)\,dy = f(x).$$

Show that $f(x) = g(x)$ for $0 \le x \le 1$.

Solution. For convenience of notation, define the linear operator T by

$$(Th)(x) = \int_0^1 h(y)K(x,y)\,dy$$

for any continuous function h on $[0,1]$. Then $Tf = g$ and $Tg = f$.

By the Extreme Value Theorem [Ap1, Section 3.16], the continuous function $f(x)/g(x)$ attains a minimum value on $[0,1]$, say r. Thus $f(x) - rg(x) \ge 0$ on $[0,1]$, with equality somewhere. Suppose that $f - rg$ is not identically zero on $[0,1]$. Then by continuity $f - rg$ is positive on some interval, so $T(f - rg)$ is positive on $[0,1]$, and so is $T^2(f - rg)$. But by linearity, $T^2(f - rg) = T(g - rf) = f - rg$, which is zero somewhere. This contradiction shows that $f - rg = 0$. Thus $Tg = rg$, and $g = T^2g = r^2g$, so $r^2 = 1$. Also $r \ge 0$, so $r = 1$. Hence $f = g$. ■

Remark. This result also follows from a continuous analogue of the Perron-Frobenius Theorem, that every square matrix with positive entries has a unique positive eigenvector. Namely, if $F(x,y)$ is a continuous function from $[0,1] \times [0,1]$ to $\mathbb{R}^+$, then there exists a continuous function $f(x)$ from $[0,1]$ to $\mathbb{R}^+$ such that $cf(x) = \int_{0,1} F(x,y)f(y)\,dy$ for some $c > 0$, and f is unique up to a scalar multiple. The article of Birkhoff [Bi] cited in the remarks following 1997A6 proves this continuous analogue as well as the original theorem.

To apply this result to the given problem, note that T^2 has eigenvectors f and g which are both positive of eigenvalue 1. The result implies then that f and g are scalar multiples of each other, so f is an eigenvector of T of eigenvalue ± 1. Clearly the plus sign is correct, so $f = Tf = g$.

Stronger result. For further discussion along these lines, see [Hol], where a short proof of the following generalization is given.

> Let X be a normed linear space, and let C be a strict cone in X. Let T be a linear operator on X mapping $C - \{0\}$ into the interior of C. If $x \in C$ and $T^kx = x$ for some integer $k \ge 1$, then $Tx = x$.

B5. (6, 1, 0, 0, 0, 0, 0, 0, 1, 7, 65, 127)

Show there do not exist four points in the Euclidean plane such that the pairwise distances between the points are all odd integers.

All of the following solutions involve finding a contradiction modulo some power of 2. The first solution is long, but uses little more than coordinate geometry.

Solution 1. For real numbers x and y, and for a positive integer n, let "$x \equiv y \pmod{n}$" mean that $x - y$ is an integer divisible by n. Choose a coordinate system in which the four points are $(0,0)$, $(a,0)$, (r,s), (x,y). Here a is an odd integer, and we may assume $a > 0$. The square of an odd integer is congruent to 1 modulo 8, so if all the pairwise distances are odd integers, we have

$$\begin{aligned} r^2 + s^2 &\equiv 1 \pmod{8} \\ (r-a)^2 + s^2 &\equiv 1 \pmod{8} \\ x^2 + y^2 &\equiv 1 \pmod{8} \\ (x-a)^2 + y^2 &\equiv 1 \pmod{8} \\ (x-r)^2 + (y-s)^2 &\equiv 1 \pmod{8}. \end{aligned}$$

Subtracting the first two yields $2ar \equiv a^2 \pmod{8}$. Thus r is a rational number whose denominator is a multiple of 2 and a divisor of $2a$. The same is true of x. Therefore we can multiply all coordinates by the odd integer a, to reduce to the case where the denominators of r and x are both 2. Then the congruence $2ar \equiv a^2 \pmod{8}$ between integers implies $r \equiv a/2 \pmod{4}$. If $r = a/2 + 4b$, then

$$r^2 = a^2/4 + 4ab + 16b^2 \equiv a^2/4 \pmod{4},$$

so the first congruence implies

$$s^2 \equiv 1 - r^2 \equiv 1 - a^2/4 \pmod{4}.$$

Similarly $x \equiv a/2 \pmod{4}$ and $y^2 \equiv 1 - a^2/4 \pmod{4}$. Also

$$x - r \equiv a/2 - a/2 \equiv 0 \pmod{4},$$

so $(x-r)^2 \in 16\mathbb{Z}$, and the last of the five congruences yields

$$(y-s)^2 \equiv 1 - (x-r)^2 \equiv 1 \pmod{8}.$$

We will derive a contradiction from the congruences $s^2 \equiv y^2 \equiv 1 - a^2/4 \pmod{4}$ and $(y-s)^2 \equiv 1 \pmod{8}$ obtained above. First,

$$\begin{aligned} (y+s)^2 &\equiv 2y^2 + 2s^2 - (y-s)^2 \\ &\equiv 2(1 - a^2/4) + 2(1 - a^2/4) - 1 \\ &\equiv 3 - a^2 \\ &\equiv 2 \pmod{4}. \end{aligned}$$

Multiplying this integer congruence by $(y-s)^2 \equiv 1 \pmod{8}$ yields $(y^2 - s^2)^2 \equiv 2 \pmod{4}$. But y^2 and s^2 are rational by the beginning of this paragraph, so $y^2 - s^2$ is a rational number with square congruent to 2 modulo 4. This is impossible. ∎

Solution 2.

Lemma. *If* $r = \cos\alpha$, $s = \cos\beta$, *and* $t = \cos(\alpha+\beta)$, *then* $1 - r^2 - s^2 - t^2 + 2rst = 0$.

Proof. We have

$$\begin{aligned} \cos(\alpha+\beta) &= \cos\alpha\cos\beta - \sin\alpha\sin\beta, \\ (\cos(\alpha+\beta) - \cos\alpha\cos\beta)^2 &= \sin^2\alpha\sin^2\beta, \\ (t - rs)^2 &= (1-r^2)(1-s^2), \end{aligned}$$

from which the result follows. □

Suppose that O, A, B, C are four points in the plane such that $a = OA$, $b = OB$, $c = OC$, $x = BC$, $y = CA$, $z = AB$ are all odd integers. Using the Law of Cosines, let

$$\begin{aligned} r = \cos\angle AOB &= \frac{a^2+b^2-z^2}{2ab} \\ s = \cos\angle BOC &= \frac{b^2+c^2-x^2}{2bc} \\ t = \cos\angle AOC &= \frac{c^2+a^2-y^2}{2ca}. \end{aligned}$$

But $\angle AOB + \angle BOC = \angle AOC$ as directed angles, so the lemma implies $1 - r^2 - s^2 - t^2 + 2rst = 0$. Substituting the values of r, s, t, and multiplying by $4a^2b^2c^2$ yields

$$\begin{aligned} 4a^2b^2c^2 - c^2(a^2+b^2-z^2)^2 - a^2(b^2+c^2-x^2)^2 - b^2(c^2+a^2-y^2)^2 \\ + (a^2+b^2-z^2)(b^2+c^2-x^2)(c^2+a^2-y^2) = 0. \end{aligned}$$

The square of an odd integer is 1 modulo 4, so we obtain

$$4 - 1 - 1 - 1 + 1 \equiv 0 \pmod 4,$$

a contradiction. ■

Solution 3 (Manjul Bhargava). Suppose $\vec{v}_1$, $\vec{v}_2$, $\vec{v}_3$ are vectors in 3-space. Then the volume V of the parallelepiped spanned by the vectors is given by $|\vec{v}_1 \cdot (\vec{v}_2 \times \vec{v}_3)| = |\det M|$, where $M = (\vec{v}_1\vec{v}_2\vec{v}_3)$ is the 3×3 matrix with entries given by the vectors $\vec{v}_1$, $\vec{v}_2$, $\vec{v}_3$. Since $\det M = \det M^T$,

$$V^2 = M \cdot M^T = \det\begin{pmatrix} \vec{v}_1\cdot\vec{v}_1 & \vec{v}_1\cdot\vec{v}_2 & \vec{v}_1\cdot\vec{v}_3 \\ \vec{v}_2\cdot\vec{v}_1 & \vec{v}_2\cdot\vec{v}_2 & \vec{v}_2\cdot\vec{v}_3 \\ \vec{v}_3\cdot\vec{v}_1 & \vec{v}_3\cdot\vec{v}_2 & \vec{v}_3\cdot\vec{v}_3 \end{pmatrix}. \tag{1}$$

The volume of the tetrahedron spanned by the vectors is $V/6$. If the edges of the tetrahedron are $a = |\vec{v}_1|$, $b = |\vec{v}_2|$, $c = |\vec{v}_3|$, $x = |\vec{v}_2 - \vec{v}_3|$, $y = |\vec{v}_3 - \vec{v}_1|$, $z = |\vec{v}_1 - \vec{v}_2|$, then by the Law of Cosines, written as

$$2\vec{v}_i \cdot \vec{v}_j = |\vec{v}_i|^2 + |\vec{v}_j|^2 - |\vec{v}_i - \vec{v}_j|^2, \tag{2}$$

$$8V^2 = \det\begin{pmatrix} 2a^2 & a^2+b^2-z^2 & a^2+c^2-y^2 \\ a^2+b^2-z^2 & 2b^2 & b^2+c^2-x^2 \\ a^2+c^2-y^2 & b^2+c^2-x^2 & 2c^2 \end{pmatrix}.$$

Suppose now that there were 4 points as described in the problem. Locate one point at the origin, and let $\vec{v}_1$, $\vec{v}_2$, $\vec{v}_3$ be the vectors from the origin to the other three; they are coplanar, so $V = 0$. Since squares of odd integers are congruent to 1 modulo 8,

$$8V^2 \equiv \det \begin{pmatrix} 2 & 1 & 1 \\ 1 & 2 & 1 \\ 1 & 1 & 2 \end{pmatrix} \equiv 4 \pmod{8},$$

which gives a contradiction. ■

Remark. Suppose $\vec{v}_1, \vec{v}_2, \ldots, \vec{v}_n$ are vectors in $\mathbb{R}^m$. Then the n-dimensional volume V of the parallelepiped spanned by the vectors is given by

$$V^2 = \det \begin{pmatrix} \vec{v}_1 \cdot \vec{v}_1 & \cdots & \vec{v}_1 \cdot \vec{v}_n \\ \vdots & \ddots & \vdots \\ \vec{v}_n \cdot \vec{v}_1 & \cdots & \vec{v}_n \cdot \vec{v}_n \end{pmatrix}. \tag{3}$$

(Note that m need not equal n.) This generalizes (1).

Solution 4. Suppose as before there were 4 such points; define $\vec{v}_1$, $\vec{v}_2$, $\vec{v}_3$ as in the previous solution. Then $\vec{v}_i \cdot \vec{v}_i \equiv 1 \pmod{8}$, and from (2), $2\vec{v}_i \cdot \vec{v}_j \equiv 1 \pmod{8}$ for $i \neq j$ as well.

No three of the points can be collinear. Hence $\vec{v}_3 = x\vec{v}_1 + y\vec{v}_2$ for some scalars x and y. Then

$$\begin{aligned} 2\vec{v}_1 \cdot \vec{v}_3 &= 2x\vec{v}_1 \cdot \vec{v}_1 + 2y\vec{v}_1 \cdot \vec{v}_2 \\ 2\vec{v}_2 \cdot \vec{v}_3 &= 2x\vec{v}_2 \cdot \vec{v}_1 + 2y\vec{v}_2 \cdot \vec{v}_2 \\ 2\vec{v}_3 \cdot \vec{v}_3 &= 2x\vec{v}_3 \cdot \vec{v}_1 + 2y\vec{v}_3 \cdot \vec{v}_2. \end{aligned} \tag{4}$$

Since $\vec{v}_1$ is not a scalar multiple of $\vec{v}_2$,

$$\det \begin{pmatrix} \vec{v}_1 \cdot \vec{v}_1 & \vec{v}_1 \cdot \vec{v}_2 \\ \vec{v}_2 \cdot \vec{v}_1 & \vec{v}_2 \cdot \vec{v}_2 \end{pmatrix} > 0$$

(by the two-dimensional version of (3)) so the first two equations in (4) have a unique rational solution for x, y, say $x = X/D$, $y = Y/D$, where X, Y, and D are integers. We may assume $\gcd(X, Y, D) = 1$. Then multiplying (4) through by D we have

$$\begin{aligned} D &\equiv 2X + Y \pmod{8} \\ D &\equiv X + 2Y \pmod{8} \\ 2D &\equiv X + Y \pmod{8}. \end{aligned}$$

Adding the first two congruences and subtracting the third gives $2X + 2Y \equiv 0 \pmod{8}$, so, by the third congruence, D is even. But then the first two congruences force X and Y to be even, giving a contradiction. ■

Solution 5. If $P_0, \dots, P_n$ are the vertices of a simplex in $\mathbb{R}^n$, and $d_{ij} = (P_iP_j)^2$, then its volume V satisfies

$$V^2 = \frac{(-1)^{n+1}}{2^n(n!)^2} \det \begin{pmatrix} 0 & 1 & 1 & \cdots & 1 \\ 1 & d_{00} & d_{01} & \cdots & d_{0n} \\ 1 & d_{10} & d_{11} & \cdots & d_{1n} \\ \vdots & \vdots & \vdots & \ddots & \vdots \\ 1 & d_{n0} & d_{n1} & \cdots & d_{nn} \end{pmatrix}.$$

This is proved on page 98 of [Bl], where this determinant is called a *Cayley-Menger determinant.* In particular, the formula for $n = 3$ (which goes back to Euler) implies that the edge lengths a, b, c, x, y, z (labelled as in Solution 2) of a degenerate tetrahedron satisfy

$$\det \begin{pmatrix} 0 & 1 & 1 & 1 & 1 \\ 1 & 0 & z^2 & y^2 & a^2 \\ 1 & z^2 & 0 & x^2 & b^2 \\ 1 & y^2 & x^2 & 0 & c^2 \\ 1 & a^2 & b^2 & c^2 & 0 \end{pmatrix} = 0.$$

If a, b, c, x, y, z are odd integers, we get a contradiction modulo 8. ■

Remark. See 1988B5 for another example of showing a matrix is invertible by reducing modulo some number.

Remark. Either Solution 3 or 5, along with a formula from Solution 3 of 1992B5, can be used to prove the following generalization:

> Suppose n is an integer not divisible by 8. Show that there do not exist n points in $\mathbb{R}^{n-2}$ such that the pairwise distances between the points are all odd integers.

On the other hand, we do not know whether or not there exist 8 points in $\mathbb{R}^6$ such that the pairwise distances between the points are all odd integers.

B6. (2, 0, 0, 0, 0, 0, 0, 0, 1, 2, 59, 143)

Let S be a set of three, not necessarily distinct, positive integers. Show that one can transform S into a set containing 0 by a finite number of applications of the following rule: Select two of the three integers, say x and y, where $x \le y$ and replace them with $2x$ and $y - x$.

The following three solutions seem related only by the central role played by powers of 2.

Solution 1 (attributed to Garth Payne). It suffices to show that (a, b, c) with $0 < a \le b \le c$ can be transformed into (b', a', c'), where a' is the remainder when b is divided by a. (Hence any set with no zeros can be transformed into a set with a smaller element.)

Let $b = qa + r$ with $r < a$; let $q = q_0 + 2q_1 + 4q_2 + \cdots + 2^k q_k$ be the binary representation of q (so $q_i = 0$ or 1, and $q_k = 1$). Define $g_0(d, e, f) = (2d, e, f - d)$ and $g_1(d, e, f) = (2d, e - d, f)$. Then

$$g_{q_k}(\cdots(g_{q_1}(g_{q_0}(a, b, c)))\cdots) = (b', r, c')$$

describes a sequence of legal moves. ■

Solution 2. Let a, b, c be the elements of S. We prove the result by strong induction on $a+b+c$, a quantity that is preserved by applications of the rule. For the sake of obtaining a contradiction, assume that S cannot be transformed to a triple containing 0.

First we reduce to the case that exactly one of a, b, c is odd. If two are odd, apply the rule with those two, and then none are odd. If none are odd, divide all numbers by 2 and apply the inductive hypothesis. If three are odd, apply the rule once, and exactly one is odd. Once exactly one is odd, this will remain so.

Without loss of generality, a is odd and b and c are even. Let 2^n be the highest power of 2 dividing $b+c$. We will describe a series of moves leaving us with a', b', c', with a' odd and 2^{n+1} dividing $b'+c'$. Repeating such a series of moves results in a triple a'', b'', c'' with $b''+c''$ divisible by 2^m where $2^m > a+b+c$. Then since $b'', c'' > 0$, we have $b''+c'' \geq 2^m > a+b+c = a''+b''+c''$, contradicting $a'' > 0$.

We first apply a series of moves so that b and c are divisible by different powers of 2, and the one divisible by the smaller power of 2 (b, say) is also smaller. If b and c have the same number of factors of 2, then applying the rule to those two will yield both divisible by a higher power of 2, or one will have fewer factors of 2 than the other. Since $b+c$ is constant here, after a finite number of applications of the rule, b and c will not have the same number of factors of 2. Also, possibly after some additional moves (on b and c), the one of b and c divisible by the smaller power of 2 is also smaller.

Now if $a > b$, then apply the rule to (a, b); a remains odd, b is doubled, and $b+c$ is divisible by a higher power of 2 as desired.

On the other hand, if $a < b$, apply the rule first to (a, b), and then to $(b-a, c)$. (Note that $c > b > b-a$.) The result is the triple $(2a, 2b-2a, c-b+a)$. Now the odd number is $c-b+a$, and the sum of the even numbers is $2b$, which has one more factor of 2 than $b+c$. ■

Solution 3 (Dylan Thurston). As in Solution 2, we assume there is some triple a, b, c that cannot be transformed to a triple containing 0, and that exactly one of a, b, c is odd. We give a recipe showing that if two of the triples are divisible by 2^n, then we can reach a state where two are divisible by 2^{n+1}; since the sum of the triple is constant, this eventually gives a contradiction.

The result is trivial if both of the multiples of 2^n are already divisible by 2^{n+1}, or if neither are (just apply the rule once to that pair). So assume one is a multiple of 2^{n+1}, and the other is not, so the triple is (after renaming) $(a, b, c) \equiv (0, 2^n, x) \pmod{2^{n+1}}$ (where x is odd).

We can assume $a > c$. (Otherwise, apply the rule to (a, c) repeatedly until this is so.) Then apply the rule to (a, c), giving the triple $(a', b', c') \equiv (-x, 2^n, 2x) \pmod{2^{n+1}}$.

If $a' < b'$, then apply the rule to (a', b'), giving $(-2x, 2^n + x, 2x) \pmod{2^{n+1}}$. Repeated applications of the rule to $(2x, -2x)$ eventually produce $(0, 0) \pmod{2^{n+1}}$ (regardless of which is bigger at each stage).

If on the other hand $a' > b'$, apply the rule to (a', b') giving $(2^n - x, 0, 2x)$ $\pmod{2^{n+1}}$. Apply the rule repeatedly to the 0 and $2^n - x$ terms until the 0 is bigger, and then apply it once more to get $(-2x, 2^n + x, 2x) \pmod{2^{n+1}}$. Again apply the rule repeatedly to $(2x, -2x)$ to eventually produce $(0, 0) \pmod{2^{n+1}}$. ■

Related question. Problem 3 on the 1986 International Mathematical Olympiad [IMO86] is similar, although the solution is quite different:

> To each vertex of a regular pentagon an integer is assigned in such a way that the sum of all five numbers is positive. If three consecutive vertices are assigned the numbers x, y, z respectively and $y < 0$ then the following operation is allowed: the numbers x, y, z are replaced by $x + y$, $-y$, $z + y$ respectively. Such an operation is performed repeatedly as long as at least one of the five numbers is negative. Determine whether this procedure necessarily comes to an end after a finite number of steps.

The Fifty-Fifth William Lowell Putnam Mathematical Competition December 3, 1994

A1. (59, 59, 54, 21, 0, 0, 0, 0, 8, 0, 3, 2)

Suppose that a sequence a_1, a_2, a_3, ... satisfies $0 < a_n \le a_{2n} + a_{2n+1}$ for all $n \ge 1$. Prove that the series $\sum_{n=1}^{\infty} a_n$ diverges.

Solution. For $m \ge 1$, let $b_m = \sum_{i=2^{m-1}}^{2^m-1} a_i$. Summing $a_n \le a_{2n} + a_{2n+1}$ from $n = 2^{m-1}$ to $n = 2^m - 1$ yields $b_m \le b_{m+1}$ for all $m \ge 1$. For any $t \ge 1$,

$$\sum_{n=1}^{2^t-1} a_n = \sum_{m=1}^{t} b_m \ge tb_1 = ta_1,$$

which is unbounded as $t \to \infty$ since $a_1 > 0$, so $\sum_{n=1}^{\infty} a_n$ diverges.

Alternatively, assuming the series converges to a finite value L, we obtain the contradiction

$$\begin{aligned} L &= b_1 + (b_2 + b_3 + \cdots) \\ &\ge b_1 + (b_1 + b_2 + \cdots) \\ &= b_1 + L. \end{aligned}$$

■

Related question. An interesting variant is the following [New, Problem 87]:

> $(A_n)_{n\in\mathbb{Z}^+}$ is a sequence of positive numbers satisfying $A_n < A_{n+1} + A_{n^2}$ for all $n \in \mathbb{Z}^+$. Prove that $\sum_{n=1}^{\infty} A_n$ diverges.

A2. (169, 3, 2, 0, 0, 0, 0, 0, 1, 3, 22, 6)

Let A be the area of the region in the first quadrant bounded by the line $y = \frac{1}{2}x$, the x-axis, and the ellipse $\frac{1}{9}x^2 + y^2 = 1$. Find the positive number m such that A is equal to the area of the region in the first quadrant bounded by the line $y = mx$, the y-axis, and the ellipse $\frac{1}{9}x^2 + y^2 = 1$.

Answer. To make the areas equal, m must be $2/9$.

Solution 1. The linear transformation given by $x_1 = x/3$, $y_1 = y$ transforms the region R bounded by $y = x/2$, the x-axis, and the ellipse $x^2/9 + y^2 = 1$ into the region R' bounded by $y_1 = 3x_1/2$, the x_1-axis, and the circle $x_1^2 + y_1^2 = 1$; it also transforms

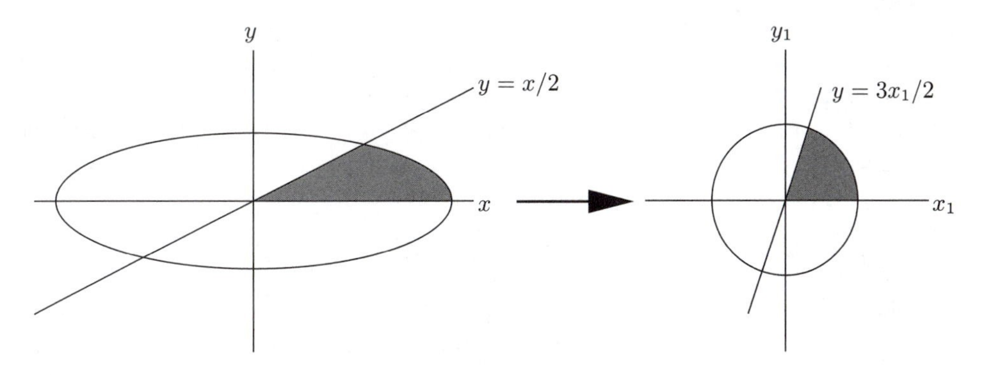

FIGURE 26.

the region S bounded by $y = mx$, the y-axis, and $x^2/9 + y^2 = 1$ into the region S'' bounded by $y_1 = 3mx_1$, the y_1-axis, and the circle. Since all areas are multiplied by the same (nonzero) factor under the linear transformation, R and S have the same area if and only if R' and S' have the same area. However, we can see by symmetry about the line $y_1 = x_1$ that this happens if and only if $3m = 2/3$, that is, $m = 2/9$. ■

Solution 2 (Noam Elkies). Apply the linear transformation $(x, y) \to (3y, x/3)$. This preserves area, and the ellipse $x^2/9 + y^2 = 1$. It switches the x and y axes, and takes $y = x/2$ to the desired line, $x/3 = (3y/2)$, i.e. $y = (2/9)x$. Thus $m = 2/9$. ■

Remark. There are, of course, less enlightened solutions. Setting up the integrals for the two areas yields the equation

$$\int_0^{3/\sqrt{13}} \left(\sqrt{9 - 9y^2} - 2y\right) dy = \int_0^{3/\sqrt{1+9m^2}} \left(\sqrt{1 - x^2/9} - mx\right) dx.$$

At this point, one might guess that a substitution $y = cx$ will transform one integral into the other, if c and m satisfy

$$\frac{3}{\sqrt{13}} = c\frac{3}{\sqrt{1 + 9m^2}}, \quad 3c = 1, \quad 2c^2 = m,$$

and in fact, $c = 1/3$ and $m = 2/9$ work. If this shortcut is overlooked, then as a last resort one could use trigonometric substitution to evaluate both sides: this yields

$$\frac{3}{2} \operatorname{Arcsin}\left(\frac{3}{\sqrt{13}}\right) = \frac{3}{2} \operatorname{Arcsin}\left(\frac{1}{\sqrt{1 + 9m^2}}\right).$$

Solving yields $m = 2/9$.

A3. (0, 10, 67, 0, 0, 0, 0, 0, 30, 31, 40, 28)

Show that if the points of an isosceles right triangle of side length 1 are each colored with one of four colors, then there must be two points of the same color which are at least a distance $2 - \sqrt{2}$ apart.

Motivation. The standard approach to such a problem would be to find five points in the triangle, no two closer than $2 - \sqrt{2}$; then two of them would have the same color. Unfortunately, such a set of five points does not exist, so a subtler argument is required.

Solution. Suppose that it is possible to color the points of the triangle with four colors so that any two points at least $2 - \sqrt{2}$ apart receive different colors. Suppose the vertices of the triangle are $A = (0, 0)$, $B = (1, 0)$, $C = (0, 1)$. Define also the points $D = (\sqrt{2} - 1, 0)$ and $E = (0, \sqrt{2} - 1)$ on the two sides, and $F = (2 - \sqrt{2}, \sqrt{2} - 1)$ and $G = (\sqrt{2} - 1, 2 - \sqrt{2})$ on the diagonal; see Figure 27. Note that

$$BD = BF = CE = CG = DE = DG = EF = 2 - \sqrt{2}.$$

The color of B must be different from the colors of the other named points. The same holds for C. Thus the vertices of the pentagram $AGDEF$ must each be painted one of the two remaining colors. Then two adjacent vertices of the pentagram must have the same color. But they are separated by at least $2 - \sqrt{2}$, so this is a contradiction. ■

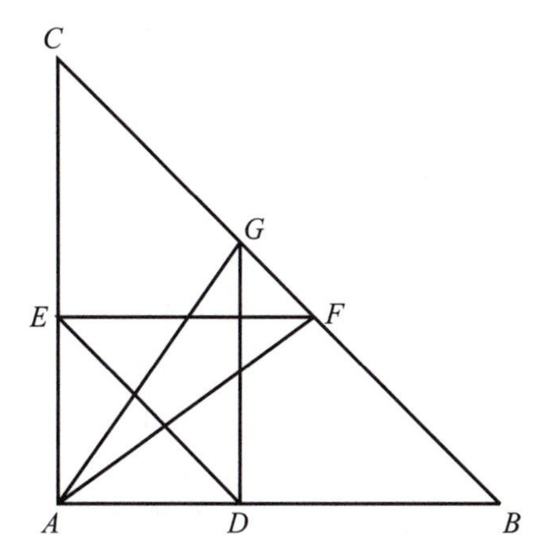

FIGURE 27.

Remark. The $2-\sqrt{2}$ in the problem is best possible. If we give triangles ADE, BDF, CEG, and quadrilateral $DEGF$ each their own color, coloring common edges arbitrarily, then no two points at distance *strictly* greater than $2-\sqrt{2}$ receive the same color.

Related questions. There are many problems of this sort. Other typical examples are Problems 1954M2 and 1960M2 [PutnamI, pp. 41, 58]:

> Consider any five points P_1, P_2, P_3, P_4, P_5 in the interior of a square S of side-length 1. Denote by d_{ij} the distance between the points P_i and P_j. Prove that at least one of the distances d_{ij} is less than $\sqrt{2}/2$. Can $\sqrt{2}/2$ be replaced by a smaller number in this statement?
>
> Show that if three points are inside a closed square of unit side, then some pair of them are within $\sqrt{6}-\sqrt{2}$ units apart.

Problem 1988A4 also is similar.

A4. (12, 17, 20, 0, 0, 0, 0, 0, 15, 3, 43, 96)

Let A and B be 2×2 matrices with integer entries such that A, $A+B$, $A+2B$, $A+3B$, and $A+4B$ are all invertible matrices whose inverses have integer entries. Show that $A+5B$ is invertible and that its inverse has integer entries.

Solution. A square matrix M with integer entries has an inverse with integer entries if and only if $\det M = \pm 1$: if N is such an inverse, $(\det M)(\det N) = \det(MN) = 1$ so $\det M = \pm 1$; conversely, if $\det M = \pm 1$, then $\pm M'$ is an inverse with integer entries, where M' is the classical adjoint of M. Let $f(x) = \det(A + xB)$. Then $f(x)$ is a polynomial of degree at most 2, such that $f(x) = \pm 1$ for $x = 0$, 1, 2, 3, and 4. Thus by the Pigeonhole Principle f takes one of these values three or more times. But the only polynomials of degree at most 2 that take the same value three times are constant polynomials. In particular, $\det(A + 5B) = \pm 1$, so $A + 5B$ has an inverse with integer entries. ■

A5. (20, 13, 4, 0, 0, 0, 0, 0, 6, 2, 57, 104)

Let $(r_n)_{n\geq 0}$ be a sequence of positive real numbers such that $\lim_{n\to\infty} r_n = 0$. Let S be the set of numbers representable as a sum

$$r_{i_1} + r_{i_2} + \cdots + r_{i_{1994}},$$

with $i_1 < i_2 < \cdots < i_{1994}$. Show that every nonempty interval (a, b) contains a nonempty subinterval (c, d) that does not intersect S.

Solution 1. We may permute the r_i to assume $r_0 \geq r_1 \geq \cdots$. This does not change S or the convergence to 0. If $b \leq 0$, the result is clear, so we assume $b > 0$.

Since $r_n \to 0$, only finitely many r_n exceed $b/2$. Thus we may choose a positive number a_1 so that $a < a_1 < b$ and $r_n \notin [a_1, b)$ for all n. Then for an element of $S \cap (a_1, b)$, there are only a finite number of possibilities for i_1 (since $0 < a_1/1994 \leq r_{i_1} < a_1$); let I_1 be the set of such i_1.

Choose a_2 so that $a_1 < a_2 < b$ and $r_{i_1} + r_n \notin [a_2, b)$ for all $i_1 \in I_1$ and $n \geq 0$. Then for each i_1, there are only a finite number of possibilities for i_2 (since $0 < (a_2 - r_{i_1})/1993 \leq r_{i_2} < a_2 - r_{i_1}$); let I_2 be the set of ordered pairs (i_1, i_2) of possibilities.

Similarly, inductively choose a_m so that $a_{m-1} < a_m < b$ and $r_{i_1} + \cdots + r_{i_{m-1}} + r_n \notin [a_m, b)$ for all $(i_1, \ldots, i_{m-1}) \in I_{m-1}$ and $n \geq 0$. The set I_m of ordered m-tuples $(i_1, \ldots, i_m)$ of possibilities is finite.

Then $(c, d) = (a_{1994}, b)$ does not intersect S. ■

Remark. A cleaner variation is to show that if A is a nowhere dense subset of $\mathbb{R}$, and $(r_n)_{n\geq 0}$ converges to 0, then $A + \{r_n\}$ is also nowhere dense (where $A + \{r_n\}$ denotes the set of numbers of the form $a + r_n$, where $a \in A$). Then the result follows by induction.

Solution 2. It suffices to show that any sequence in S contains a monotonically nonincreasing subsequence. For then, letting $(t_n)_{n\geq 0}$ be any strictly increasing sequence within (a, b), some (in fact, all but a finite number) of the intersections $S \cap (t_n, t_{n+1})$ would have to be empty. Otherwise, one could form a strictly increasing sequence $(s_n)_{n\geq 0}$ by taking $S_n \in S \cap (t_n, t_{n+1})$.

Let $(s_n)_{n\geq 0}$ be a sequence in S. For $n = 0, 1, 2, \ldots$, write

$$s_n = r_{f(n,1)} + r_{f(n,2)} + \cdots + r_{f(n,1994)} \quad \text{with} \quad f(n,1) < f(n,2) < \cdots < f(n,1994).$$

The sequence $(r_{f(n,1)})_{n\geq 0}$ has a monotonically nonincreasing subsequence (since $(r_n)_{n\geq 0}$ is a positive sequence converging to 0). Thus we may replace $(s_n)_{n\geq 0}$ by a subsequence for which $(r_{f(n,1)})_{n\geq 0}$ is monotonically nonincreasing. In a similar fashion, we pass to subsequences so that, successively, each of $(r_{f(n,2)})_{n\geq 0}$, $(r_{f(n,3)})_{n\geq 0}$, $\ldots$, $(r_{f(n,1994)})_{n\geq 0}$ may be assumed to be monotonically nonincreasing. The resulting $(s_n)_{n\geq 0}$ is monotonically nonincreasing. ■

Remark. A totally ordered set T is well-ordered if and only if every nonempty subset has a least element. This condition is equivalent to the condition that every sequence in T contain a nondecreasing subsequence. Hence $-S$ is well-ordered under the ordering induced from the reals. Solution 2 can be interpreted as proving that a finite sum of well-ordered subsets of the reals is well-ordered. See the remark on Zorn's Lemma in 1989B4 for more on well orderings.

Solution 3. Let C be the set $\{r_n : n \geq 0\} \cup \{0\}$. Then C is closed and bounded, so C is compact. Hence C^{1994} is compact, and its image S' under the "sum the coordinates" map $\mathbb{R}^{1994} \to \mathbb{R}$ is compact. Clearly $S \subset S'$.

Let (a, b) be a nonempty open interval. Since S' is countable, $(a, b) - S'$ is nonempty; it is open since S' is closed. Hence $(a, b) - S'$ includes a nonempty open interval. ■

Remark. The proofs above generalize to give the same conclusion for any convergent sequence $(r_n)_{n \geq 0}$.

A6. (5, 8, 10, 0, 0, 0, 0, 0, 7, 4, 34, 138)

Let f_1, f_2, ..., f_{10} be bijections of the set of integers such that for each integer n, there is some composition $f_{i_1} \circ f_{i_2} \circ \cdots \circ f_{i_m}$ of these functions (allowing repetitions) which maps 0 to n. Consider the set of 1024 functions

$$\mathcal{F} = \{f_1^{e_1} \circ f_2^{e_2} \circ \cdots \circ f_{10}^{e_{10}}\},$$

$e_i = 0$ or 1 for $1 \leq i \leq 10$. (f_i^0 is the identity function and $f_i^1 = f_i$.) Show that if A is any nonempty finite set of integers, then at most 512 of the functions in $\mathcal{F}$ map A to itself.

Solution. We say that a bijection of the integers to itself preserves a subset A if it restricts to a bijection of A.

Let G be the group of permutations of $\mathbb{Z}$ generated by the f_i. Then G has elements mapping any integer m to 0, and elements mapping 0 to any integer n, so G acts transitively on $\mathbb{Z}$. Hence no nonempty proper subset A can be preserved by all the f_i. It remains to prove the following lemma.

Lemma. *Let $f_1, \ldots, f_n$ be bijections $\mathbb{Z} \to \mathbb{Z}$, and let A be a subset of $\mathbb{Z}$. Suppose that some f_i does not preserve A. Then at most 2^{n-1} elements of*

$$\mathcal{F}_n = \{ f_1^{e_1} \circ f_2^{e_2} \circ \cdots \circ f_n^{e_n} : e_i = 0 \text{ or } 1 \}$$

preserve A.

Proof 1 of Lemma (inductive). The lemma is true for $n = 1$, so assume it it is true for all $n < k$ (for some $k > 1$), and false for $n = k$. Then by the Pigeonhole Principle, there are $e_1, \ldots, e_{k-1} \in \{0, 1\}$ such that both

$$f_1^{e_1} \circ f_2^{e_2} \circ \cdots \circ f_{k-1}^{e_{k-1}} \quad \text{and} \quad f_1^{e_1} \circ f_2^{e_2} \circ \cdots \circ f_{k-1}^{e_{k-1}} \circ f_k$$

fix A. Hence f_k preserves A as well. By the inductive hypothesis, at most 2^{k-2} elements of $\mathcal{F}_{k-1}$ fix A; thus at most 2^{k-1} elements of $\mathcal{F}_k$ fix A, giving a contradiction. □

Proof 2 of Lemma (noninductive). Let k be the largest integer such that f_k does *not* map A to itself, and suppose that more than 2^{n-1} of the functions $\mathcal{F}_n$ map A to itself. By the Pigeonhole Principle, there are

$$e_1, \ldots, e_{k-1}, e_{k+1}, \ldots, e_n \in \{0, 1\}$$

such that both

$$f_1^{e_1} \circ \cdots \circ f_{k-1}^{e_{k-1}} \circ f_{k+1}^{e_{k+1}} \circ \cdots \circ f_n^{e_n} \quad \text{and} \quad f_1^{e_1} \circ \cdots \circ f_{k-1}^{e_{k-1}} \circ f_k \circ f_{k+1}^{e_{k+1}} \circ \cdots \circ f_n^{e_n}$$

both fix A. Hence both $F_1 = f_1^{e_1} \circ \cdots \circ f_{k-1}^{e_{k-1}}$ and $F_1 = f_1^{e_1} \circ \cdots \circ f_{k-1}^{e_{k-1}} \circ f_k$ both map A to itself. But then $F_1^{-1} \circ F_2 = f_k$ also maps A to itself, giving a contradiction. ∎

Remark. The solution shows that the problem statement remains true even if the condition that A be a nonempty finite subset of $\mathbb{Z}$ is relaxed to the condition that A be a nonempty proper subset of $\mathbb{Z}$.

B1. (45, 26, 57, 0, 0, 0, 0, 0, 42, 28, 6, 2)

Find all positive integers that are within 250 of exactly 15 perfect squares.

Answer. An integer N is within 250 of exactly 15 perfect squares if and only if either $315 \leq N \leq 325$ or $332 \leq N \leq 350$.

Solution. The squares within 250 of a positive integer N form a set of consecutive squares. If N is such that there are 15 such squares, then they are m^2, $(m+1)^2, \dots,$ $(m+14)^2$ for some $m \geq 0$. If $m = 0$, then $14^2 \leq N + 250 < 15^2$, contradicting $N > 0$.

Now, given $N, m > 0$, the following two conditions are necessary and sufficient for m^2, $(m+1)^2, \dots, (m+14)^2$ to be the squares within 250 of N:

$$(m+14)^2 \leq N + 250 \leq (m+15)^2 - 1,$$

$$m^2 \geq N - 250 \geq (m-1)^2 + 1.$$

Subtraction shows that these imply

$$28m + 196 \leq 500 \leq 32m + 222,$$

which implies $m = 9$ or 10.

If $m = 9$, the two conditions $23^2 \leq N + 250 \leq 24^2 - 1$, $9^2 \geq N - 250 \geq 8^2 + 1$ are equivalent to $315 \leq N \leq 325$. If $m = 10$, the two conditions $24^2 \leq N + 250 \leq 25^2 - 1$, $10^2 \geq N - 250 \geq 9^2 + 1$ are equivalent to $332 \leq N \leq 350$. ∎

B2. (28, 8, 49, 0, 0, 0, 0, 0, 56, 10, 39, 16)

For which real numbers c is there a straight line that intersects the curve

$$y = x^4 + 9x^3 + cx^2 + 9x + 4$$

in four distinct points?

Answer. There exists such a line if and only if $c < 243/8$.

Remark. Since all vertical lines meet the curve in one point, we need only consider lines of the form $y = mx + b$.

Solution 1 (geometric). The constant and linear terms of

$$P(x) = x^4 + 9x^3 + cx^2 + 9x + 4$$

are irrelevant to the problem; $y = P(x)$ meets the line $y = mx + b$ in four points if and only if $y = P(x) + 9x + 4$ meets the line $y = (m+9)x + (b+4)$ in four points.

Also, $y = P(x)$ meets the line $y = mx + b$ in four points if and only if $y = P(x - \alpha)$ meets the line $y = m(x - \alpha) + b$ in four points, so we may replace the given quartic with $P(x - 9/4) = x^4 + (c - 243/8)x^2 + \cdots$ (where we ignore the linear and constant terms).

The problem is then to determine the values of c for which there is a straight line that intersects $y = x^4 + (c - 243/8)x^2$ in four distinct points. The result is now apparent from the shapes of the curves $y = x^4 + ax^2$. For example, when $a < 0$, this "W-shaped" curve has a relative maximum at $x = 0$, so the horizontal lines $y = -\epsilon$ for small positive ϵ intersect the curve in four points, while for $a \geq 0$, the curve is always concave upward, so no line can intersect it in more than two points; see Figure 28. ■

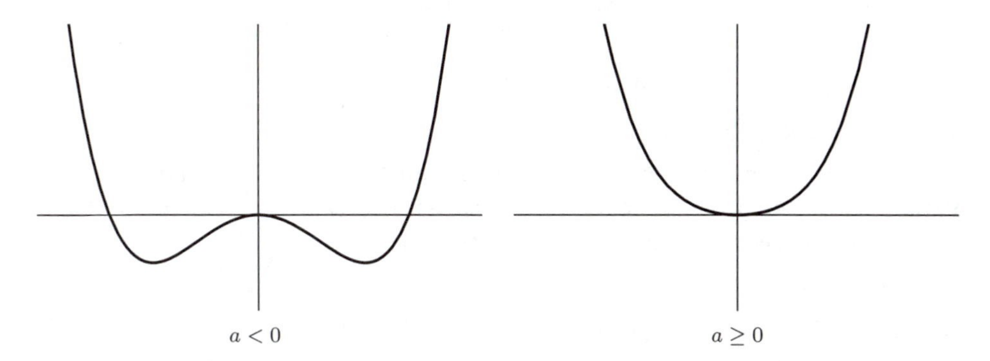

FIGURE 28.
Graphs of $y = x^4 + ax^2$ for $a < 0$ and $a \geq 0$.

Solution 2 (algebraic). We wish to know if we can choose m and b so that

$$q(x) = x^4 + 9x^3 + cx^2 + 9x + 4 - (mx + b)$$

has four distinct real solutions α_1, α_2, α_3, α_4. If we can find four distinct real numbers such that

$$\alpha_1 + \alpha_2 + \alpha_3 + \alpha_4 = -9 \tag{1}$$

$$\alpha_1\alpha_2 + \alpha_1\alpha_3 + \alpha_1\alpha_4 + \alpha_2\alpha_3 + \alpha_2\alpha_4 + \alpha_3\alpha_4 = c, \tag{2}$$

then we can choose m and b appropriately so that $q(x)$ has these four zeros (by the expansion of $\prod_{i=1}^4 (x - \alpha_i)$).

Then from

$$0 < \sum_{i<j} (\alpha_i - \alpha_j)^2 = 3(\alpha_1 + \alpha_2 + \alpha_3 + \alpha_4)^2 - 8c \tag{3}$$

we get $c < 243/8$. Conversely, we must show that if $c < 243/8$, then we can find distinct real α_1, α_2, α_3, α_4 satisfying (1) and (2). This is indeed possible; try $(\alpha_1, \alpha_2, \alpha_3, \alpha_4) = (9/4 - \mu, 9/4 + \mu, 9/4 - \nu, 9/4 + \nu)$ where μ and ν are distinct positive numbers. Then (1) is automatically satisfied, and we can choose μ and ν so that the right side of (3), which is $8(\mu^2 + \nu^2)$, is any desired positive number. ■

Remark. The previous two solutions highlight two ways of constructing a line for given c. Other possible tools to do this include Rolle's Theorem, the Mean Value Theorem [Spv, Ch. 11, Theorem 4], Descartes' Rule of Signs, and more. Also, another quick way to see that $c < 243/8$ is necessary is to consider the discriminant of the quadratic $P''(x)$.

Related question. A related problem is 1977A1 [PutnamII, p. 29]:

Consider all lines which meet the graph of

$$y = 2x^4 + 7x^3 + 3x - 5$$

in four distinct points, say (x_i, y_i), $i = 1, 2, 3, 4$. Show that

$$\frac{x_1 + x_2 + x_3 + x_4}{4}$$

is independent of the line and find its value.

B3. (27, 10, 8, 5, 0, 0, 0, 2, 45, 49, 15, 45)

Find the set of all real numbers k with the following property: For any positive, differentiable function f that satisfies $f'(x) > f(x)$ for all x, there is some number N such that $f(x) > e^{kx}$ for all $x > N$.

Answer. The desired set is $(-\infty, 1)$.

Solution. Let $h(x) = \ln f(x) - x$. Then the problem becomes that of determining for which k the following holds: if a real-valued function $h(x)$ satisfies $h'(x) > 0$ for all x, then there exists a number N such that $h(x) > (k-1)x$ for all $x > N$.

The function $-e^{-x}$ is always negative, but has positive derivative, so no number $k \geq 1$ is in the set. (This corresponds to the function $f(x) = e^{x-e^{-x}}$.) On the other hand any $k < 1$ *is* in the set: choose N such that $(k-1)N < h(0)$; then for $x > N$, $h(x) > h(0) > (k-1)N > (k-1)x$. ∎

B4. (15, 1, 4, 0, 0, 0, 0, 0, 5, 22, 71, 88)

For $n \geq 1$, let d_n be the greatest common divisor of the entries of $A^n - I$, where

$$A = \begin{pmatrix} 3 & 2 \\ 4 & 3 \end{pmatrix} \text{ and } I = \begin{pmatrix} 1 & 0 \\ 0 & 1 \end{pmatrix}.$$

Show that $\lim_{n\to\infty} d_n = \infty$.

Solution 1. Experimentation suggests and induction on n proves that there exist integers $a_n, b_n > 0$ such that

$$A^n = \begin{pmatrix} a_n & b_n \\ 2b_n & a_n \end{pmatrix}.$$

Since $\det A^n = 1$, we have $a_n^2 - 1 = 2b_n^2$. Thus $a_n - 1$ divides $2b_n^2$. By definition, $d_n = \gcd(a_n - 1, b_n)$, so $2d_n^2 = \gcd(2(a_n-1)^2, 2b_n^2) \geq a_n - 1$. From $A^{n+1} = A \cdot A^n$ we have $a_{n+1} > 3a_n$, so $\lim_{n\to\infty} a_n = \infty$. Hence $\lim_{n\to\infty} d_n = \infty$. ∎

Solution 2 (Robin Chapman). The set of matrices of the form $\begin{pmatrix} a & b \\ 2b & a \end{pmatrix}$ with $a, b \in \mathbb{Z}$ is closed under left multiplication by $\begin{pmatrix} 1 & 1 \\ 2 & 1 \end{pmatrix}$. It follows by induction on n

that $\begin{pmatrix} 1 & 1 \\ 2 & 1 \end{pmatrix}^n = \begin{pmatrix} a & b \\ 2b & a \end{pmatrix}$ for some $a, b \in \mathbb{Z}$ depending on n. Taking determinants shows $a^2 - 2b^2 = (-1)^n$. But $\begin{pmatrix} 1 & 1 \\ 2 & 1 \end{pmatrix}^2 = \begin{pmatrix} 3 & 2 \\ 4 & 3 \end{pmatrix}$, so

$$\begin{aligned} \begin{pmatrix} 3 & 2 \\ 4 & 3 \end{pmatrix}^n - \begin{pmatrix} 1 & 0 \\ 0 & 1 \end{pmatrix} &= \begin{pmatrix} a & b \\ 2b & a \end{pmatrix}^2 - \begin{pmatrix} 1 & 0 \\ 0 & 1 \end{pmatrix} \\ &= \begin{pmatrix} a^2 + 2b^2 - 1 & 2ab \\ 4ab & a^2 + 2b^2 - 1 \end{pmatrix}. \end{aligned}$$

If n is odd then $a^2 - 2b^2 = -1$, so $a^2 + 2b^2 - 1 = 2a^2$ and all entries are divisible by a. If n is even $a^2 - 2b^2 = 1$, so $a^2 + 2b^2 - 1 = 4b^2$ and all entries are divisible by b. Both a and b increase as $n \to \infty$ (by the same argument as in Solution 1), so we are done. ■

Solution 3. Define the sequence $r_0, r_1, r_2, \ldots$ by $r_0 = 0$, $r_1 = 1$, and $r_k = 6r_{k-1} - r_{k-2}$ for $k > 1$. Then $r_n > 5r_{n-1}$ for $n \geq 1$, so $\lim_{n\to\infty} r_n = \infty$. We first show by induction on k that

$$A^n - I = r_{k+1}(A^{n-k} - A^k) - r_k(A^{n-k-1} - A^{k+1}) \quad \text{for } k \geq 0. \tag{1}$$

This is clear for $k = 0$ and, for the inductive step, using $A^2 - 6A + I = 0$ (the characteristic equation), we have

$$\begin{aligned} &r_{k+1}(A^{n-k} - A^k) - r_k(A^{n-k-1} - A^{k+1}) \\ &\quad = r_{k+1}\left((6A^{n-k-1} - A^{n-k-2}) - (6A^{k+1} - A^{k+2})\right) - r_k(A^{n-k-1} - A^{k+1}) \\ &\quad = (6r_{k+1} - r_k)(A^{n-k-1} - A^{k+1}) - r_{k+1}(A^{n-k-2} - A^{k+2}) \\ &\quad = r_{k+2}(A^{n-k-1} - A^{k+1}) - r_{k+1}(A^{n-k-2} - A^{k+2}). \end{aligned}$$

Applying (1) with $k = \lfloor n/2 \rfloor$, we obtain

$$A^n - I = \begin{cases} r_{n/2}(A^{n/2+1} - A^{n/2-1}) & \text{if } n \text{ is even,} \\ \left(r_{(n+1)/2} + r_{(n-1)/2}\right)\left(A^{(n+1)/2} - A^{(n-1)/2}\right) & \text{if } n \text{ is odd.} \end{cases}$$

In either case, the entries of $A^n - I$ have a common factor that goes to ∞ since $\lim_{n\to\infty} r_n = \infty$. ■

Solution 4. The entries of A^n are each of the form $\alpha_1\lambda_1^n + \alpha_2\lambda_2^n$, where $\lambda_1 = 3 + 2\sqrt{2}$ and $\lambda_2 = 3 - 2\sqrt{2}$ are the eigenvalues of A. This follows from diagonalization (as suggested by our hint), or from the theory of linear recursive relations and the Cayley-Hamilton Theorem: the latter yields $A^2 - 6A + I = 0$, so $A^{n+2} - 6A^{n+1} + A^n = 0$ for all n, and each entry of A^n satisfies the recursion $x_{n+2} - 6x_{n+1} + x_n = 0$. Using the entries for $n = 1, 2$, we derive

$$A^n = \begin{pmatrix} \frac{\lambda_1^n + \lambda_2^n}{2} & \frac{\lambda_1^n - \lambda_2^n}{2\sqrt{2}} \\ \frac{\lambda_1^n - \lambda_2^n}{\sqrt{2}} & \frac{\lambda_1^n + \lambda_2^n}{2} \end{pmatrix}.$$

Since $\lambda_i = \mu_i^2$ where $\mu_1 = 1 + \sqrt{2}$ and $\mu_2 = 1 - \sqrt{2}$, we see

$$\begin{aligned} d_n &= \gcd\left(\frac{\lambda_1^n + \lambda_2^n}{2} - 1, \frac{\lambda_1^n - \lambda_2^n}{2\sqrt{2}}\right) \\ &= \gcd\left(\frac{(\mu_1^n - \mu_2^n)^2}{2}, \frac{(\mu_1^n - \mu_2^n)(\mu_1^n + \mu_2^n)}{2\sqrt{2}}\right) \\ &= \left(\frac{(\mu_1^n - \mu_2^n)}{2\sqrt{2}}\right) \gcd\left(\frac{(\mu_1^n - \mu_2^n)}{\sqrt{2}}, \frac{(\mu_1^n + \mu_2^n)}{2}\right) \end{aligned}$$

since $(\mu_1^n - \mu_2^n)/\sqrt{2}$ and $(\mu_1^n + \mu_2^n)/2$ are (rational) integers. Since $|\mu_1| > 1$ and $|\mu_2| < 1$, we conclude $\lim_{n\to\infty}(\mu_1^n - \mu_2^n) = \infty$. Hence $\lim_{n\to\infty} d_n = \infty$. ∎

Solution 5. The characteristic polynomial of A is $x^2 - 6x + 1$, so A has distinct eigenvalues λ, λ^{-1}, where $\lambda = 3+2\sqrt{2}$. Hence $A = CDC^{-1}$ where $D = \begin{pmatrix} \lambda & 0 \\ 0 & \lambda^{-1} \end{pmatrix}$ and C is an invertible matrix with entries in $\mathbb{Q}(\sqrt{2})$. Choose an integer $k \geq 1$ such that the entries of kC and kC^{-1} are in $\mathbb{Z}[\sqrt{2}]$. Then $k^2(A^n - I) = (kC)(D^n - I)(kC^{-1})$ and $D^n - I = (\lambda^n - 1)\begin{pmatrix} 1 & 0 \\ 0 & \lambda^{-n} \end{pmatrix}$ so $\lambda^n - 1$ divides $k^2 d_n$ in $\mathbb{Z}[\sqrt{2}]$. Taking norms, we find that the (rational) integer $(\lambda^n - 1)(\lambda^{-n} - 1)$ divides $k^4 d_n^2$. But $|\lambda| > 1$, so $|(\lambda^n - 1)(\lambda^{-n} - 1)| \to \infty$ as $n \to \infty$. Hence $\lim_{n\to\infty} d_n = \infty$. ∎

Remark. Each solution generalizes to establish the same result for integral matrices $A = \begin{pmatrix} a & b \\ c & d \end{pmatrix}$ of determinant 1 and $|\operatorname{trace}(A)| > 1$ (the latter to guarantee $r_n \to \infty$ where $r_n = \operatorname{trace}(A) r_{n-1} - r_{n-2}$).

B5. (11, 4, 4, 0, 0, 0, 0, 0, 32, 10, 15, 130)

For any real number α, define the function $f_\alpha(x) = \lfloor \alpha x \rfloor$. Let n be a positive integer. Show that there exists an α such that for $1 \leq k \leq n$,‡

$$f_\alpha^k(n^2) = n^2 - k = f_{\alpha^k}(n^2).$$

Solution 1. We will show that α satisfies the conditions of the problem if and only if

$$1 - \frac{1}{n^2} \leq \alpha < \left(\frac{n^2 - n + 1}{n^2}\right)^{1/n},$$

and then show that this interval is nonempty.

We have $f_\alpha^k(n^2) = n^2 - k$ for $k = 1, \ldots, n$ if and only if $\lfloor \alpha(n^2 - k + 1) \rfloor = n^2 - k$ for $k = 1, \ldots, n$, which holds if and only if

$$\frac{n^2 - k}{n^2 - k + 1} \leq \alpha < 1 \quad \text{for } k = 1, \ldots, n.$$

Since

$$\frac{n^2 - k}{n^2 - k + 1} = 1 - \frac{1}{n^2 - k + 1}$$

‡ Here $f_\alpha^k(n^2) = f_\alpha(\cdots(f_\alpha(n^2))\cdots)$, where f_α is applied k times to n^2.

decreases with k, these hold if and only if $1 - \frac{1}{n^2} \le \alpha < 1$. We assume this from now on.

Next we consider the conditions $f_{\alpha^k}(n^2) = n^2 - k$ for $k = 1, \dots, n$. Since $f_\alpha^k(n^2)$ is an integer less than $\alpha^k n^2$, we have $f_\alpha^k(n^2) \le f_{\alpha^k}(n^2)$. We have already arranged that $f_\alpha^k(n^2) = n^2 - k$, so $f_{\alpha^k}(n^2) = n^2 - k$ if and only if $\alpha^k n^2 < n^2 - k + 1$. Moreover, if the latter holds for $k = n$, then it holds for $k = 1, \dots, n$ by reverse induction, since multiplying $\alpha^k n^2 < n^2 - k + 1$ by

$$\alpha^{-1} \le \frac{n^2}{n^2-1} = 1 + \frac{1}{n^2-1} \le 1 + \frac{1}{n^2-k+1} \le \frac{n^2-(k-1)+1}{n^2-k+1}$$

yields the inductive step. Hence all the conditions hold if and only if

$$1 - \frac{1}{n^2} \le \alpha < \left(\frac{n^2-n+1}{n^2}\right)^{1/n}.$$

It remains to show that this interval is nonempty, or equivalently, that

$$\left(1 - \frac{1}{n^2}\right)^n < 1 - \frac{1}{n} + \frac{1}{n^2}. \tag{1}$$

First we will prove

$$(1-x)^n \le 1 - nx + \binom{n}{2}x^2 \qquad \text{for } 0 \le x \le 1. \tag{2}$$

The two sides are equal at $x = 0$, so it suffices to show that their derivatives satisfy

$$-n(1-x)^{n-1} \le -n + n(n-1)x \qquad \text{for } 0 \le x \le 1.$$

Again the two sides are equal at $x = 0$, so, differentiating again, it suffices to show

$$n(n-1)(1-x)^{n-2} \le n(n-1) \qquad \text{for } 0 \le x \le 1.$$

This is obvious. Taking $x = 1/n^2$ in (2) yields

$$\left(1 - \frac{1}{n^2}\right)^n \le 1 - \frac{1}{n} + \frac{n(n-1)/2}{n^4} < 1 - \frac{1}{n} + \frac{1}{n^2}. \blacksquare$$

Remark. Inequality (2) is a special case of the fact that for real $a > b > 0$ and integers $0 < k < n$, the partial binomial expansion

$$a^n - \binom{n}{1}a^{n-1}b + \cdots + (-1)^k \binom{n}{k} a^{n-k}b^k \tag{3}$$

is greater than or less than $(a-b)^n$, according to whether k is even or odd. It can be proved by the same argument used to prove (2). This is sometimes called the Inclusion-Exclusion Inequality, because if a, b are sizes of sets $A \supsetneq B \supsetneq \emptyset$, then (3) is the overestimate or underestimate for $\#(A-B)^n$ obtained by terminating the inclusion-exclusion argument prematurely.

Solution 2 (Dave Rusin). Let $\alpha = e^{-1/n^2}$. The same method used to prove (2) shows that

$$1 - r + r^2/2 > e^{-r} > 1 - r. \tag{4}$$

Substituting $r = k/n^2$ $(0 < k \leq n)$ and simplifying, we find

$$n^2 - k + \frac{1}{2}\left(\frac{k}{n}\right)^2 > \alpha^k n^2 > n^2 - k.$$

The right side is an integer, and the left side is at most 1/2 more, so $\lfloor \alpha^k n^2 \rfloor = n^2 - k$. Since $\alpha > 1 - 1/n^2$ (again by (4)), $f_\alpha^k(n^2) = n^2 - k$ for $1 \leq k \leq n$ by the same argument as in the previous solution. ∎

B6. (14, 11, 1, 0, 0, 0, 0, 0, 16, 10, 50, 104)

For any integer a, set

$$n_a = 101a - 100 \cdot 2^a.$$

Show that for $0 \leq a, b, c, d \leq 99$, $n_a + n_b \equiv n_c + n_d \pmod{10100}$ implies $\{a, b\} = \{c, d\}$.

The following lemma states that 2 is a *primitive root* modulo the prime 101.

Lemma. $2^a \equiv 1 \pmod{101}$ *if and only if a is divisible by* 100.

Proof. We need to show that the order m of the image of 2 in the group $(\mathbb{Z}/101\mathbb{Z})^*$ is 100. Since the group has order 100, m divides 100. If m were a proper divisor of 100, then m would divide either $100/2 = 50$ or $100/5 = 20$, so either 2^{50} or 2^{20} would be 1 modulo 101. But we compute

$$\begin{aligned} 2^{10} &= 1024 \equiv 14 \pmod{101} \\ 2^{20} &\equiv 14^2 \equiv -6 \pmod{101} \\ 2^{40} &\equiv (-6)^2 \equiv 36 \pmod{101} \\ 2^{50} &\equiv 36 \cdot 14 \equiv -1 \pmod{101}. \end{aligned}$$

□

Corollary 1. *If a and b are nonnegative integers such that $2^a \equiv 2^b \pmod{101}$, then $a \equiv b \pmod{100}$.*

Proof. Without loss of generality, assume $a \geq b$. If 101 divides $2^a - 2^b = 2^b(2^{a-b} - 1)$, then $2^{a-b} \equiv 1 \pmod{101}$, so $a - b \equiv 0 \pmod{100}$ by the lemma. □

Solution. By the Chinese Remainder Theorem, $n_a + n_b \equiv n_c + n_d \pmod{10100}$ is equivalent to

$$a + b \equiv c + d \pmod{100} \tag{1}$$

and

$$2^a + 2^b \equiv 2^c + 2^d \pmod{101}. \tag{2}$$

By Fermat's Little Theorem, (1) implies $2^{a+b} \equiv 2^{c+d} \pmod{101}$, or equivalently

$$2^a 2^b \equiv 2^c 2^d \pmod{101} \tag{3}$$

Solve for 2^b in (2) and substitute into (3) to obtain

$$2^a(2^c + 2^d - 2^a) \equiv 2^c 2^d \pmod{101},$$

or equivalently

$$0 \equiv (2^a - 2^c)(2^a - 2^d) \pmod{101}.$$

(That such a factorization exists could have been guessed from the desired conclusion that $a = c$ or $a = d$.) Hence $2^a \equiv 2^c \pmod{101}$ or $2^a \equiv 2^d \pmod{101}$. By the corollary above, a is congruent to c or d modulo 100. Then by (1), b is congruent to the other. But $0 \leq a, b, c, d \leq 99$, so these congruences are equalities. ■

Remark. A more conceptual way to get from (2) and (3) to the conclusion that 2^a, 2^b are congruent to 2^c, 2^d in some order, modulo 101, is to observe that $(x-2^a)(x-2^b)$ and $(x-2^c)(x-2^d)$ are equal as elements of the polynomial ring $\mathbb{F}_{101}[x]$ over the finite field $\mathbb{F}_{101}$. Then take the multiset of zeros of each.

The Fifty-Sixth William Lowell Putnam Mathematical Competition December 2, 1995

A1. (115, 64, 20, 0, 0, 0, 0, 0, 1, 0, 2, 2)

Let S be a set of real numbers which is closed under multiplication (that is, if a and b are in S, then so is ab). Let T and U be disjoint subsets of S whose union is S. Given that the product of any *three* (not necessarily distinct) elements of T is in T and that the product of any three elements of U is in U, show that at least one of the two subsets T, U is closed under multiplication.

Solution. Suppose on the contrary that there exist $t_1, t_2 \in T$ with $t_1t_2 \in U$ and $u_1, u_2 \in U$ with $u_1u_2 \in T$. Then $(t_1t_2)u_1u_2 \in U$ while $t_1t_2(u_1u_2) \in T$, contradiction. ■

Remark. It is possible for T to be closed and U not to be: for example, let T and U be the sets of integers congruent to 1 and 3, respectively, modulo 4.

A2. (26, 10, 14, 0, 0, 0, 0, 0, 14, 3, 56, 81)

For what pairs (a, b) of positive real numbers does the improper integral

$$\int_b^\infty \left(\sqrt{\sqrt{x+a}-\sqrt{x}} - \sqrt{\sqrt{x}-\sqrt{x-b}}\right) dx$$

converge?

Answer. The integral converges if and only if $a = b$.

Solution. Using $\sqrt{1+t} = 1 + t/2 + O(t^2)$ repeatedly, we obtain

$$\begin{aligned}
\sqrt{\sqrt{x+a}-\sqrt{x}} &= x^{1/4}\sqrt{\sqrt{1+\frac{a}{x}}-1} \\
&= x^{1/4}\sqrt{\frac{a}{2x}+O(x^{-2})} \\
&= \sqrt{\frac{a}{2}}\,x^{-1/4}(1+O(x^{-1})) \\
\sqrt{\sqrt{x}-\sqrt{x-b}} &= x^{1/4}\sqrt{1-\sqrt{1-\frac{b}{x}}} \\
&= x^{1/4}\sqrt{\frac{b}{2x}+O(x^{-2})} \\
&= \sqrt{\frac{b}{2}}\,x^{-1/4}(1+O(x^{-1})).
\end{aligned}$$

Thus the original integrand is $\left(\sqrt{a/2}-\sqrt{b/2}\right)x^{-1/4}+O(x^{-5/4})$. Since $\int_b^\infty x^{-5/4}\,dx$ converges, the original integral converges if and only if $\int_b^\infty \left(\sqrt{a/2}-\sqrt{b/2}\right)x^{-1/4}\,dx$ converges. But $\int_b^\infty x^{-1/4}\,dx$ diverges, so the original integral converges if and only if $a = b$. ■

Remark. See 1988A3 for a similar argument.

Remark. In the case $a = b$, one can show the integral converges by a "telescoping" argument. Write

$$\begin{aligned}
&\int_c^d \left(\sqrt{\sqrt{x+a}-\sqrt{x}} - \sqrt{\sqrt{x}-\sqrt{x-a}}\right) dx \\
&\qquad = \int_c^d \sqrt{\sqrt{x+a}-\sqrt{x}}\, dx - \int_{c-a}^{d-a} \sqrt{\sqrt{x+a}-\sqrt{x}}\, dx \\
&\qquad = \int_{d-a}^{d} \sqrt{\sqrt{x+a}-\sqrt{x}}\, dx - \int_{c}^{c-a} \sqrt{\sqrt{x+a}-\sqrt{x}}\, dx.
\end{aligned}$$

As $d \to \infty$ for fixed c, the second term is constant, while the first term tends to 0. (It is an integral over an interval of fixed length, and the integrand tends to 0 uniformly.)

A3. (95, 44, 39, 0, 0, 0, 0, 0, 12, 5, 3, 6)

The number $d_1d_2\dots d_9$ has nine (not necessarily distinct) decimal digits. The number $e_1e_2\dots e_9$ is such that each of the nine 9-digit numbers formed by replacing just one of the digits d_i in $d_1d_2\dots d_9$ by the corresponding digit e_i ($1 \le i \le 9$) is divisible by 7. The number $f_1f_2\dots f_9$ is related to $e_1e_2\dots e_9$ is the same way: that is, each of the nine numbers formed by replacing one of the e_i by the corresponding f_i is divisible by 7. Show that, for each i, $d_i - f_i$ is divisible by 7. [For example, if $d_1d_2\dots d_9 = 199501996$, then e_6 may be 2 or 9, since 199502996 and 199509996 are multiples of 7.]

Solution. Let D and E be the numbers $d_1\dots d_9$ and $e_1\dots e_9$, respectively. We are given that

$$\begin{aligned}
(e_i - d_i)10^{9-i} + D &\equiv 0 \pmod 7 \\
(f_i - e_i)10^{9-i} + E &\equiv 0 \pmod 7
\end{aligned}$$

for $i = 1, \dots, 9$. Sum the first relation over $i = 1, \dots, 9$ to get $E - D + 9D \equiv 0 \pmod 7$, or equivalently $E + D \equiv 0 \pmod 7$. Now add the first and second relations for any particular value of i to get $(f_i - d_i)10^{9-i} + E + D \equiv 0 \pmod 7$. Since $E + D$ is divisible by 7, and 10 is coprime to 7, we conclude $d_i - f_i \equiv 0 \pmod 7$. ■

A4. (39, 3, 13, 0, 0, 0, 0, 0, 19, 1, 83, 40)

Suppose we have a necklace of n beads. Each bead is labelled with an integer and the sum of all these labels is $n - 1$. Prove that we can cut the necklace to form a string whose consecutive labels $x_1, x_2, \dots, x_n$ satisfy

$$\sum_{i=1}^{k} x_i \le k - 1 \qquad \textbf{for} \quad k = 1, 2, \dots, n.$$

Solution 1. Number the beads 1, 2, ... starting from some arbitrary position, and let z_i be the label of bead i, with the convention that $z_{n+i} = z_i$. Let $S_j = z_1 + \dots + z_j - j(n-1)/n$, so $S_{n+j} = S_j$. Then the sum of the labels on beads $m+1, m+2, \dots, m+k$ is $S_{m+k} - S_m + k(n-1)/n$.

This suggests choosing m such that S_m is maximal, and cutting between m and $m+1$. Indeed, if we do so, then

$$S_{m+k} - S_m + \frac{k(n-1)}{n} \le k - \frac{k}{n},$$

but the left side is an integer, so we can replace the right side by $k-1$. ■

Remark. In fact, this argument shows that the location of the cut is unique, assuming that the orientation of the necklace is fixed. Namely, for any other choice of m, we would have $S_{m+k} - S_m + \frac{k(n-1)}{n} \ge \frac{k(n-1)}{n}$ for some k, but the left side is an integer, so we may replace the right side by k. (This implies in particular that the maximal S_m is unique up to replacing m by $m+in$, but this was already evident: the S_j have distinct fractional parts, so no two are equal.)

For a more visual reinterpretation, plot the points $(i, z_1 + \cdots + z_i)$ for $i \ge 1$. The set of these points is mapped into itself by translation by $(n, n-1)$, and the upper support line of slope $(n-1)/n$ of this set passes through $(m, z_1 + \cdots + z_m)$, where m is as chosen above.

Solution 2. Replace each label x with $1-x$, to obtain a necklace N_n with sum of labels 1. It suffices to prove that we can cut N_n so that the consecutive labels $y_1, \dots, y_n$ satisfy $\sum_{i=1}^k y_i \ge 1$ for $k = 1, \dots, n$. We prove this by induction on n, the case $n = 1$ being trivial. Suppose $n > 1$; then some bead b in N_n has a positive label, since otherwise the sum of labels would not be positive. Form the new necklace N_{n-1} by merging b with its successor, replacing the two labels by their sum. The induction hypothesis provides a cut point in N_{n-1} such that the partial sums are positive; the corresponding cut point for N_n has the same property. ■

Literature note. In the reformulation used in Solution 2, this problem appears in [GKP, Section 7.5], but it is much older: see [Tak] and the references listed there for some generalizations and applications. It is closely related to that of the "ballot problem": if m votes are cast for A and n votes are cast for B in some order, find the number of orderings of the $m+n$ votes in which A is never behind during the tally. When $m = n$, the solution to the ballot problem involves the Catalan numbers [St, Corollary 6.2.3]; see [St, Ex. 6.19 a.-nnn.] for more occurrences of the Catalan numbers.

A5. (1, 1, 2, 1, 0, 0, 0, 6, 18, 14, 43, 118)

Let $x_1, x_2, \dots, x_n$ be differentiable (real-valued) functions of a single variable t which satisfy

$$\begin{aligned}
\frac{dx_1}{dt} &= a_{11}x_1 + a_{12}x_2 + \cdots + a_{1n}x_n \\
\frac{dx_2}{dt} &= a_{21}x_1 + a_{22}x_2 + \cdots + a_{2n}x_n \\
&\vdots \\
\frac{dx_n}{dt} &= a_{n1}x_1 + a_{n2}x_2 + \cdots + a_{nn}x_n
\end{aligned}$$

for some constants $a_{ij} > 0$. Suppose that for all i, $x_i(t) \to 0$ as $t \to \infty$. Are the functions $x_1, x_2, \ldots, x_n$ necessarily linearly dependent?

Answer. Yes, the functions must be linearly dependent.

Solution. If $(v_1, \ldots, v_n)$ is a (complex) eigenvector of the matrix (a_{ij}) with (complex) eigenvalue λ, then the function $y = v_1 x_1 + \cdots + v_n x_n$ satisfies the differential equation

$$\frac{dy}{dt} = \sum_{i,j} v_i a_{i,j} x_j = \sum_j \lambda v_j x_j = \lambda y.$$

Therefore $y = ce^{\lambda t}$ for some $c \in \mathbb{C}$.

Now the trace of the matrix (a_{ij}) is $a_{11} + \cdots + a_{nn}$, which is positive. Since the trace is the sum of the eigenvalues of the matrix, there must be at least one eigenvalue $\lambda = \alpha + i\beta$ with positive real part. Setting $y = v_1 x_1 + \cdots + v_n x_n$, we have that

$$y = ce^{\lambda t} = ce^{\alpha t} e^{i\beta t}.$$

We are given that $|x_i| \to 0$ as $t \to \infty$ for $i = 1, \ldots, n$, which implies that $|y| \to 0$ as well. On the other hand, for $\alpha > 0$, $e^{\alpha t} e^{i\beta t}$ does not tend to 0 as $t \to \infty$. Therefore we must have $c = 0$, and hence $v_1 x_1 + \cdots + v_n x_n = 0$. Since the x_i are linearly dependent over $\mathbb{C}$, they are also linearly dependent over $\mathbb{R}$, as desired. ∎

Remark. We did not require the eigenvalue λ to be real. It turns out, however, that a matrix with nonnegative entries always has an eigenvector with nonnegative entries, and its corresponding eigenvalue is then also nonnegative. Moreover, if the matrix has all positive entries, it has exactly one positive real eigenvector: this assertion is the Perron-Frobenius Theorem, which is important in the theory of Markov processes and random walks. See the remarks following 1997A6 for further discussion.

Literature note. See Chapter 7 of [BD] for a treatment of systems of linear differential equations with constant coefficients, such as the one occurring here.

A6. (1, 0, 1, 0, 0, 0, 0, 0, 0, 0, 61, 141)

Suppose that each of n people writes down the numbers 1, 2, 3 in random order in one column of a $3 \times n$ matrix, with all orders equally likely and with the orders for different columns independent of each other. Let the row sums a, b, c of the resulting matrix be rearranged (if necessary) so that $a \le b \le c$. Show that for some $n \ge 1995$, it is at least four times as likely that both $b = a + 1$ and $c = a + 2$ as that $a = b = c$.

Solution 1. For integers x, y, z with $x + y + z = 0$ and n a positive integer, let $f(x, y, z, n)$ be the number of $3 \times n$ matrices whose columns are permutations of 1, 2, 3, and whose row sums p, q, r (in that order) satisfy $p - q = x$, $q - r = y$, $r - p = z$. Then we have the recursion

$$\begin{aligned} f(x, y, z, n + 1) &= f(x + 1, y - 1, z, n) + f(x + 1, y, z - 1, n) + f(x - 1, y + 1, z, n) \\ &\quad + f(x, y + 1, z - 1, n) + f(x - 1, y, z + 1, n) + f(x, y - 1, z + 1, n). \end{aligned}$$

When $x = y = z = 0$, the right-hand side counts the number of possible $3 \times n$ matrices satisfying both $b = a + 1$ and $c = a + 2$. Thus the problem requires us to show that $4f(0,0,0,n) \leq f(0,0,0,n+1)$ for some $n \geq 1995$. Suppose this were not the case; then we would have $f(0,0,0,n) = O(4^n)$ as $n \to \infty$. On the other hand, $f(0,0,0,6m)$ is at least the number of $3 \times 6m$ matrices such that each permutation of 1, 2, 3 appears in exactly m columns. The latter number equals the multinomial coefficient $\binom{6m}{m,m,m,m,m,m}$, so we would have $\binom{6m}{m,m,m,m,m,m} = O(4^{6m})$. This rate of growth contradicts the fact that

$$\begin{aligned}\lim_{m\to\infty} \frac{\binom{6(m+1)}{m+1,m+1,m+1,m+1,m+1,m+1}}{\binom{6m}{m,m,m,m,m,m}} &= \lim_{m\to\infty} \frac{(6m+6)(6m+5)(6m+4)(6m+3)(6m+2)(6m+1)}{(m+1)(m+1)(m+1)(m+1)(m+1)(m+1)} \\ &= 6^6.\end{aligned}$$

■

Remark. Instead of estimating the ratio of consecutive multinomial coefficients, we could obtain a direct contradiction from Stirling's approximation (see the remarks following 1996B2), which implies that

$$\binom{6m}{m,m,m,m,m,m} \sim (2\pi m)^{-5/2} 6^{6m+1/2}$$

as $m \to \infty$.

Solution 2. In fact, we can prove $4f(x,y,z,n) \leq f(x,y,z,n+1)$ for all $n \geq 3$ and all x, y, and z. Using the recursion in Solution 1, we can deduce this from the claim for $n = 3$. The verification for $n = 3$ can be made from the following table, which shows the results of computing $f(x,y,z,n)$ for small values of x, y, z, n from the recursion. (We list only those (x,y,z) with $x \geq 0 \geq z \geq y$ and $f(x,y,z,3) \neq 0$, since these suffice by symmetry.)

(x,y,z)	$f(x,y,z,0)$	$f(x,y,z,1)$	$f(x,y,z,2)$	$f(x,y,z,3)$	$f(x,y,z,4)$
$(0,0,0)$	1	0	6	12	60
$(1,-1,0)$	0	1	2	15	60
$(2,-2,0)$	0	0	1	6	34
$(2,-1,-1)$	0	0	2	6	48
$(3,-3,0)$	0	0	0	1	12
$(3,-2,-1)$	0	0	0	3	16

■

Reinterpretation. Consider a random walk on the triples of integers (x,y,z) with $x+y+z=0$, where a move increases one integer by 1 and decreases another one by 1. (More geometrically, one can think of the triples as the points of a triangular lattice, where a move leaves a lattice point and goes to an adjacent one.) Then $f(x,y,z,n)/6^n$ is the probability of ending up at (x,y,z) after n steps, and for fixed x, y, z, one can show that the ratio between these probabilities for successive n tends to 1. Greg Lawler points out that this is a special case of the Local Central Limit Theorem for two-dimensional walks. For more on random walks on lattice points of Euclidean space, see [Spt].

A related fact is that the random walk is recurrent: the expected number of returns to the origin is infinite, or equivalently, $\sum_{n=1}^{\infty} f(x,y,z,n)/6^n$ diverges. This is not true of the corresponding random walk on a lattice of three (or more) dimensions.

B1. (124, 26, 7, 0, 0, 0, 0, 0, 4, 10, 11, 22)

For a partition π of $\{1,2,3,4,5,6,7,8,9\}$, let $\pi(x)$ be the number of elements in the part containing x. Prove that for any two partitions π and π', there are two distinct numbers x and y in $\{1,2,3,4,5,6,7,8,9\}$ such that $\pi(x)=\pi(y)$ and $\pi'(x)=\pi'(y)$. [A *partition* of a set S is a collection of disjoint subsets (parts) whose union is S.]

Solution. For a given π, no more than three different values of $\pi(x)$ are possible. (Four would require one part each of size at least 1,2,3,4, which is already too many elements.) If no such x,y exist, each pair $(\pi(x),\pi'(x))$ occurs for at most one x, and since there are at most 3×3 possible pairs, each must occur exactly once. Moreover, there must be three different values of $\pi(x)$, each occurring 3 times. However, any given value of $\pi(x)$ occurs $k\pi(x)$ times, where k is the number of distinct parts of that size. Thus $\pi(x)$ can occur 3 times only if $\pi(x)$ equals 1 or 3, but we have three distinct values occurring 3 times, contradiction. ∎

Remark. Noam Elkies points out that the only n for which all pairs of partitions of $\{1,\dots,n\}$ satisfy the statement of the problem are $n=2,5,9$.

B2. (2, 54, 26, 0, 0, 0, 0, 0, 3, 13, 51, 55)

An ellipse, whose semi-axes have lengths a and b, rolls without slipping on the curve $y=c\sin\left(\frac{x}{a}\right)$. How are a,b,c related, given that the ellipse completes one revolution when it traverses one period of the curve?

Answer. The given condition forces $b^2=a^2+c^2$.

Solution 1. The assumption of rolling without slipping means that any point, the length of the curve already traversed equals the perimeter of the arc of the ellipse that has already touched the curve. For the ellipse to complete one revolution in one period of the sine curve, the perimeter of the ellipse must then equal the length of a period of the sine curve. Stating this in terms of integrals, we have

$$\int_0^{2\pi}\sqrt{(-a\sin\theta)^2+(b\cos\theta)^2}\,d\theta=\int_0^{2\pi a}\sqrt{1+(c/a\cos x/a)^2}\,dx.$$

Let $\theta=x/a$ in the second integral and write 1 as $\sin^2\theta+\cos^2\theta$; the result is

$$\int_0^{2\pi}\sqrt{a^2\sin^2\theta+b^2\cos^2\theta\,d\theta}=\int_0^{2\pi}\sqrt{a^2\sin^2\theta+(a^2+c^2)\cos^2\theta\,d\theta}.$$

Since the left side is increasing as a function of b, we have equality if and only if $b^2=a^2+c^2$. ∎

Solution 2. As in Solution 1, it suffices to find the relation between a,b,c that makes the length of one period of the sine curve equal to the perimeter of the ellipse. We will show that $b^2=a^2+c^2$. By scaling a,b,c, we may assume $a=1$.

In the strip $0\le\theta\le 2\pi$ in the (θ,z)-plane, graph one period S of the curve $z=c\sin\theta$. Curl the strip and glue its left and right edges to form the cylinder

$x^2+y^2=1$ in (x,y,z)-space so that a point (θ,z) on the strip maps to $(\cos\theta,\sin\theta,z)$ on the cylinder. Then S is mapped to the curve E described parametrically by $(x,y,z)=(\cos\theta,\sin\theta,c\sin\theta)$ for $0\le\theta\le 2\pi$, and S and E have the same length. The parameterization is of the form $(\cos\theta)\vec{v}+(\sin\theta)\vec{w}$ where $\vec{v}=(1,0,0)$ and $\vec{w}=(0,1,c)$ are orthogonal vectors, so E is an ellipse with semi-axes of lengths $|\vec{v}|=1$ and $|\vec{w}|=\sqrt{1+c^2}$. (Alternatively, this can be seen geometrically, since E is the intersection of the plane $z=cy$ with the cylinder.) Thus when $b^2=1+c^2$, the length of S equals the perimeter of the ellipse with semi-axes of lengths 1 and b.

On the other hand, the length of S is an increasing function of $|c|$. (If this is not clear geometrically, consider the arc length formula from calculus.) Hence for fixed b, the *only* values of c for which the length of S equals the perimeter of the ellipse with semi-axes of lengths 1 and b are those for which $b^2=1+c^2$. ∎

Remark. Noam Elkies points out that it is not obvious that the condition $b^2=a^2+c^2$ is sufficient for the rolling to be physically possible: it is conceivable that the ellipse could be too fat to touch the bottom of the sine curve. We show that the rolling is in fact physically possible if $b^2=a^2+c^2$, for an appropriately chosen starting point, by showing that the radius of curvature of the ellipse at any point is always less than that of the sine curve at the corresponding point.

The radius of curvature of the curve $y=c\sin(x/a)$ at (x,y) is

$$\frac{(1+(y')^2)^{3/2}}{|y''|}=\frac{(a^2+c^2\cos^2(x/a))^{3/2}}{ac|\sin(x/a)|},$$

while the radius of curvature of the ellipse $x=a\cos\theta$, $y=b\sin\theta$ at a given θ is

$$\frac{((x')^2+(y')^2)^{3/2}}{|x'y''-y'x''|}=\frac{(a^2\sin^2\theta+b^2\cos^2\theta)^{3/2}}{ab}.$$

(See [Ap2, pp. 538–539] for the basic properties of the radius of curvature.) If we start with the points $\theta=0$ of the ellipse and $x=0$ of the sine curve in contact, the parameters will be related by the equation $\theta=x/a$. Since $ac|\sin(x/a)|\le ac<ab$, the radius of curvature of the sine curve will always be larger than that of the ellipse at the corresponding point, so the rolling will be physically possible.

Remark. The result of the previous calculation can also be explained by the geometry of Solution 2. Since we obtain the ellipse by rolling up one period of the sine curve, it suffices to prove the following geometrically believable claim: Rolling up a curve does not increase the radius of curvature at any point. Here "rolling up a curve" means taking the image of a smooth curve in the (θ,z)-plane under the map $(\theta,z)\mapsto(\cos\theta,\sin\theta,z)$.

To prove the claim, parameterize the smooth curve by $(\theta(t),z(t))$ so that the speed for all t is 1. The radius of curvature is $1/|\vec{a}|$, where $\vec{a}$ is the acceleration [Ap2, p. 538]. Hence it remains to show that the magnitude of the acceleration is not decreased by rolling up.

Let $\vec{a}(t)$ and $\vec{A}(t)$ be the accelerations of the original and rolled-up curves, respectively. For the original curve, the velocity is (θ',z') and the acceleration is $\vec{a}=(\theta'',z'')$, so the magnitude of the acceleration is $|\vec{a}|=\sqrt{(\theta'')^2+(z'')^2}$.

For the image curve, the velocity is $(-\theta'\sin\theta, \theta'\cos\theta, z')$. The speed is $\sqrt{(\theta')^2 + (z')^2} = 1$. (We expect this, since "rolling up" is an *isometric embedding*: it does not involve stretching the paper.) The acceleration is

$$\vec{A} = (-\theta''\sin\theta - (\theta')^2\cos\theta, \theta''\cos\theta - (\theta')^2\sin\theta, z''),$$

so

$$|\vec{A}|^2 = (\theta'')^2 + (z'')^2 + (\theta')^4 = |\vec{a}|^2 + (\theta')^4 \geq |\vec{a}|^2,$$

as desired. Equality holds precisely when the original curve is moving only in the z-direction (i.e., when the curve has a vertical tangent vector); this never happens for the sine curve.

B3. (54, 15, 11, 0, 0, 0, 0, 0, 33, 15, 49, 27)

To each positive integer with n^2 decimal digits, we associate the determinant of the matrix obtained by writing the digits in order across the rows. For example, for $n = 2$, to the integer 8617 we associate $\det\begin{pmatrix} 8 & 6 \\ 1 & 7 \end{pmatrix} = 50$. Find, as a function of n, the sum of all the determinants associated with n^2-digit integers. (Leading digits are assumed to be nonzero; for example, for $n = 2$, there are 9000 determinants.)

Answer. The sum is 45 for $n = 1$, 20250 for $n = 2$, and 0 for $n \geq 3$.

Solution. The set of matrices in question can be constructed as follows: choose the first row from one set of possibilities, choose the second row from a second set, and so on. By the multilinearity of the determinant, the answer is the determinant of the matrix whose kth row is the sum of the possibilities for that row.

For $n = 1$, this sum matrix is the 1×1 matrix with entry 45, so the answer is 45 in this case. For $n = 2$, the sum matrix is $\begin{pmatrix} 450 & 405 \\ 450 & 450 \end{pmatrix}$, whose determinant is 20250. (More explicitly, the first row is the sum of the 90 vectors $\begin{pmatrix} a & b \end{pmatrix}$ with $a \in \{1, \dots, 9\}$ and $b \in \{0, \dots, 9\}$; the second row is the sum of the 100 vectors $\begin{pmatrix} a & b \end{pmatrix}$ with $a \in \{0, 1, \dots, 9\}$ and $b \in \{0, \dots, 9\}$.) For $n \geq 3$, all but the first row of the sum matrix are equal, so its determinant is 0. ∎

Remark. Alternatively, for $n \geq 3$, any matrix either has its last two columns equal, or cancels with the matrix obtained by switching these two columns. For $n = 2$, this argument eliminates all matrices from consideration except those with a 0 in the top right, and these can be treated directly.

B4. (9, 18, 62, 4, 0, 0, 0, 2, 21, 33, 10, 45)

Evaluate

$$\sqrt[8]{2207 - \frac{1}{2207 - \frac{1}{2207 - \cdots}}}.$$

Express your answer in the form $\frac{a+b\sqrt{c}}{d}$, where a, b, c, d are integers.

Answer. The expression equals $(3 + \sqrt{5})/2$.

Solution. The infinite continued fraction is defined as the limit L of the sequence defined by $x_0 = 2207$ and $x_{n+1} = 2207 - 1/x_n$; the limit exists because the sequence is decreasing (by a short induction). Also, L must satisfy $L = 2207 - 1/L$. Moreover, $x_i > 1$ for all i by induction, so $L \geq 1$. Let $r = L^{1/8}$, so $r^8 + \frac{1}{r^8} = 2207$. Then

$$\left(r^4 + \frac{1}{r^4}\right)^2 = 2207 + 2 = 47^2, \qquad \text{so } r^4 + \frac{1}{r^4} = 47;$$
$$\left(r^2 + \frac{1}{r^2}\right)^2 = 49, \qquad \text{so } r^2 + \frac{1}{r^2} = 7; \text{ and}$$
$$\left(r + \frac{1}{r}\right)^2 = 9, \qquad \text{so } r + \frac{1}{r} = 3.$$

Thus $r^2 - 3r + 1 = 0$, so $r = \frac{3\pm\sqrt{5}}{2}$. But $r = L^{1/8} \geq 1$, so $r = \frac{3+\sqrt{5}}{2}$. ■

Remark. This question deals with the asymptotic behavior of the dynamical system given by $x_0 = 2207$, $x_{n+1} = 2207 - 1/x_n$. See 1992B3 for another dynamical system.

Stronger result. Let L_n denote the nth Lucas number: that is, $L_0 = 2$, $L_1 = 1$, and $L_{n+2} = L_{n+1} + L_n$ for $n \geq 0$. Then it can be shown [Wo] that for any $n > 0$,

$$\sqrt[n]{L_{2n} - \cfrac{1}{L_{2n} - \cfrac{1}{L_{2n} - \cdots}}} = \frac{3+\sqrt{5}}{2}.$$

B5. (72, 11, 13, 0, 0, 0, 0, 0, 9, 6, 42, 51)

A game starts with four heaps of beans, containing 3, 4, 5, and 6 beans. The two players move alternately. A move consists of taking either

(a) one bean from a heap, provided at least two beans are left behind in that heap, or

(b) a complete heap of two or three beans.

The player who takes the last heap wins. To win the game, do you want to move first or second? Give a winning strategy.

Answer. The first player has a winning strategy, described below.

Solution. The first player wins by removing one bean from the pile of 3, leaving heaps of size 2, 4, 5, 6. Regarding heaps of size 2 and heaps of odd size as "odd", and heaps of even size other than 2 as "even", the total parity is now even. As long as neither player removes a single bean from a heap of size 3, the parity will change after each move. Thus the first player can always ensure that after his move, the total parity is even and there are no piles of size 3. (If the second player removes a heap of size 2, the first player can move in another odd heap; the resulting heap will be even and so cannot have size 3. If the second player moves in a heap of size greater than 3, the first player can move in the same heap, removing it entirely if it was reduced to size 3 by the second player.) ■

Remark. Because 3, 4, 5, 6 are not so large, it is also possible to describe a winning strategy with a state diagram, where a state consists of an unordered tuple of heap sizes, together with a label (1st or 2d) saying which player is to move next. The diagram would need to show, for each state labelled 1st, an arrow indicating a valid move to a state labelled 2nd, and for each state labelled 2nd, arrows to *all* states that can be reached in one move. If all paths eventually lead to the state in which there are no heaps left and the second player is to move, then the first player has a winning strategy. Figure 29 shows a state diagram for a smaller instance of the problem: it gives a winning strategy for the first player in the game with heaps of size 3, 3, and 5.

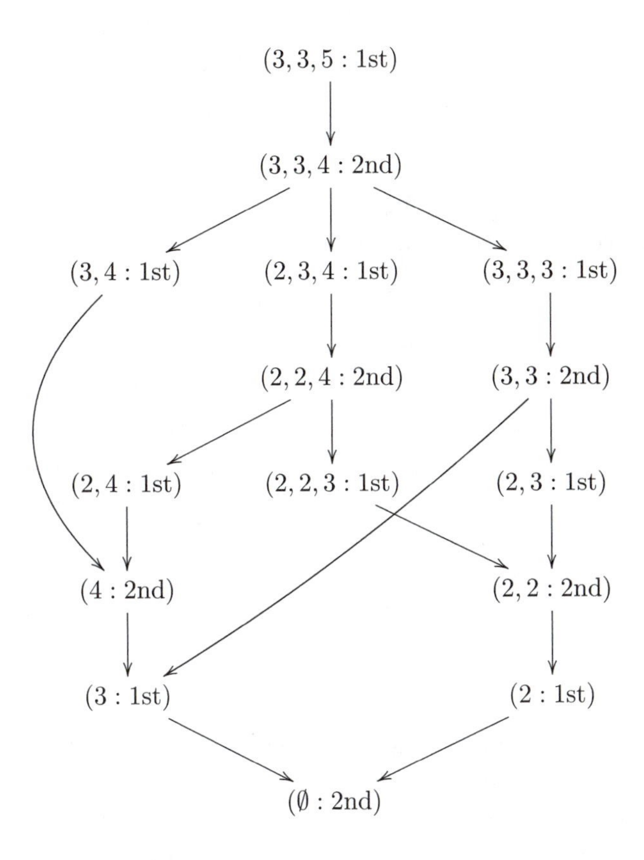

FIGURE 29.
An example of a state diagram.

Remark. In the language of game theory, this game is *normal* (the last player to move wins) and *impartial* (the options in any given position are the same for both players). The classic example of such a game is Nim: the game starts with some piles of sticks, and each player in turn can remove any number of sticks from a single pile. It has long been known that one can determine who wins in a given Nim position by adding the sizes of the piles in base 2 without carries: the position is a second player win if the sum is zero. The point is that every position with nonzero sum has a move to a zero position, but every zero position has moves only to nonzero positions. So from a nonzero position, the first player can arrange to move only to zero positions; when the game ends, it must do so after a move by the first player.

The main result of Sprague-Grundy theory [BCG, Volume 1, Ch. 3, p. 58] is that *every* normal impartial game works the same way! Explicitly, to each position one assigns a "nim-value" which is the smallest nonnegative integer not assigned to any position reachable from the given position in one move. In particular, the final position has value 0, and a position is a second player win if and only if it has value 0. The result then is that the value of a disjunction of two positions (in which a player may move in one position or the other) is obtained by adding the values of the positions in base 2 without carries.

In the original problem, piles of size 0, 2, 3, 4, 5, 6 have values 0, 1, 2, 0, 1, 0, so the initial position has value $2 \oplus 0 \oplus 1 \oplus 0 = 3$, and the winning first move creates a position of value $1 \oplus 0 \oplus 1 \oplus 0 = 0$, as expected.

B6. (3, 2, 0, 0, 0, 0, 0, 0, 2, 3, 61, 133)
For a positive real number α, define

$$S(\alpha) = \{ \lfloor n\alpha \rfloor : n = 1, 2, 3, \dots \}.$$

Prove that $\{1, 2, 3, \dots\}$ cannot be expressed as the disjoint union of three sets $S(\alpha), S(\beta)$ and $S(\gamma)$.

Solution 1. Suppose on the contrary that $S(\alpha), S(\beta), S(\gamma)$ form a partition of the positive integers. Then 1 belongs to one of these sets, say $S(\alpha)$, and so $n\alpha < 2$ for some $n \geq 1$. In particular $\alpha < 2$; moreover, $\alpha > 1$ or else $S(\alpha)$ contains every positive integer, contradicting our hypothesis.

Let $m \geq 2$ be the integer satisfying $1 + 1/m \leq \alpha < 1 + 1/(m-1)$. Then $\lfloor n\alpha \rfloor = n$ for $n = 1, \dots, m-1$, but $\lfloor m\alpha \rfloor = m+1$, so m is the smallest integer in the complement $S(\beta) \cup S(\gamma)$ of $S(\alpha)$. Moreover, any two consecutive elements of this complement differ by m or $m+1$.

Without loss of generality, suppose $m \in S(\beta)$, so $\lfloor \beta \rfloor = m$. Let n be an element of $S(\gamma)$; by the previous paragraph, the nearest elements of $S(\beta)$ on either side of n each differ from n by at least m, so they differ from each other by at least $2m$. This is a contradiction, because consecutive elements of $S(\beta)$ differ by at most $m + 1 < 2m$. ■

Related question. A similar argument arises in Problem 3 from the 1999 USA Mathematical Olympiad [USAMO]:

> Let $p > 2$ be a prime and let a, b, c, d be integers not divisible by p, such that
>
> $$\{ra/p\} + \{rb/p\} + \{rc/p\} + \{rd/p\} = 2$$
>
> for any integer r not divisible by p. Prove that at least two of the numbers $a+b$, $a+c$, $a+d$, $b+c$, $b+d$, $c+d$ are divisible by p. (Note: $\{x\} = x - \lfloor x \rfloor$ denotes the fractional part of x.)

Solution 2 (attributed to John H. Lindsey II). Let $a = 1/\alpha, b = 1/\beta, c = 1/\gamma$. For $n \geq 1$, let $f(n, x) = \#(S(x) \cap \{1, 2, \dots, n\})$. Since $S(\alpha) \neq \{1, 2, \dots\}$, $\alpha > 1$. Now $\lfloor k\alpha \rfloor \leq n$ is equivalent to $k\alpha < n+1$ and to $k < (n+1)a$, so $f(n, \alpha) = \lceil (n+1)a \rceil - 1$. Hence if $\{1, 2, 3, \dots\}$ is the disjoint union of $S(\alpha)$, $S(\beta)$ and $S(\gamma)$, then

$$n = f(n, \alpha) + f(n, \beta) + f(n, \gamma) = \lceil (n+1)a \rceil + \lceil (n+1)b \rceil + \lceil (n+1)c \rceil - 3. \quad (1)$$

Dividing by $n+1$ and taking the limit as $n \to \infty$ shows $1 = a+b+c$. Multiplying this by $n+1$ and subtracting the result from (1) yields

$$2 = g(n,a) + g(n,b) + g(n,c), \tag{2}$$

where $g(n,x) = \lceil (n+1)x \rceil - (n+1)x$.

We will contradict (2) by proving that for any $x > 0$, the average value of $g(i,x)$ for $1 \le i \le n$ tends to a limit less than or equal to $1/2$ as $n \to \infty$. If x is rational, say m/d in lowest terms, then any block of d consecutive integers contains exactly one solution z to $mz \equiv j \pmod{d}$ for each integer j, so the values of $g(i,x)$ for i in this block are $0/d$, $1/d$, $2/d$, ... , $(d-1)/d$ in some order; hence the average value of $g(i,x)$ for $1 \le i \le n$ tends to

$$\frac{1}{d}\left(\frac{0}{d} + \frac{1}{d} + \cdots + \frac{d-1}{d}\right) = \frac{d-1}{2d} < \frac{1}{2}.$$

If instead x is irrational, then the average value of $g(i,x)$ for $1 \le i \le n$ tends to $1/2$ by the equidistribution result mentioned in 1988B3. ∎

Stronger result. Solution 2 generalizes to show that if $k > 2$ and α_1, α_2, ... , α_k are real numbers, then $\{1, 2, \dots\}$ cannot be the disjoint union of $S(\alpha_1)$, $S(\alpha_2)$, ... , $S(\alpha_k)$. This fact also follows from the following paragraph.

Stronger result. The sets $S(\alpha)$, $S(\beta)$, $S(\gamma)$ cannot be disjoint. Proof (John Rickard): Let N be an integer larger than $\max\{\alpha, \beta, \gamma\}$ and consider the N^3+1 triples $v_n = (\{n/\alpha\}, \{n/\beta\}, \{n/\gamma\}) \in \mathbb{R}^3$ for $n = 0, \dots, N^3$, where $\{x\} = x - \lfloor x \rfloor$ denotes the fractional part of x. If we divide the unit cube $0 \le x, y, z \le 1$ in $\mathbb{R}^3$ into N^3 subcubes of side length $1/N$, then by the Pigeonhole Principle, two of the v_n, say v_i and v_j, lie in the same subcube. Let $k = |i-j|$. Then k/α is within $1/N$ of some integer m, and $\lfloor m\alpha \rfloor$ equals $k-1$ or k, depending on whether $k/\alpha > m$ or not. In other words, $S(\alpha)$ contains k or $k-1$. Similarly, $S(\beta)$ and $S(\gamma)$ each contain $k-1$ or k. We conclude that one of $k-1$ or k lies in at least two of $S(\alpha)$, $S(\beta)$, $S(\gamma)$, so these sets are not disjoint.

Warning. Not all authors use the notation $\{x\}$ for the fractional part of x. In fact, the same notation is used in Problem 1997B1 to denote the distance from x to the nearest integer. And sometimes it denotes the set whose only element is x!

Stronger result. If $S(\alpha)$ and $S(\beta)$ are disjoint, then there exist positive integers m and n such that $m/\alpha + n/\beta = 1$. This can be deduced from the following fact: if $x_1, \dots, x_k$ are real numbers, then the closure of the set of points $(\{nx_1\}, \dots, \{nx_k\})$ in the closed k-dimensional unit cube is the set of points $(y_1, \dots, y_k)$ in the cube with the property that for every sequence of integers $n_1, \dots, n_k$ such that $n_1x_1 + \cdots + n_kx_k \in \mathbb{Z}$, we have $n_1y_1 + \cdots + n_ky_k \in \mathbb{Z}$. (Hint: reduce to the special case where this set is the entire cube.)

The $k=1$ case of this fact is the denseness result proved in the remark in 1988B3. One can also generalize the equidistribution result given there: if $1, x_1, \dots, x_k$ are linearly independent over $\mathbb{Q}$, the k-tuples $(\{nx_1\}, \dots, \{nx_k\})$ are equidistributed in the k-dimensional unit cube. (That is, for any subcube of volume μ, the number of $n \le N$ such that the k-tuple $(\{nx_1\}, \dots, \{nx_k\})$ lies in the subcube is asymptotic to μN as $N \to \infty$.) This can be deduced from a multidimensional version of

Weyl's Equidistribution Theorem, which can be proved in the same way as the one-dimensional case in [Kör, Theorem 3.1′].

Reinterpretation. The multidimensional version of Weyl's Equidistribution Theorem can be deduced from the following fact: every continuous function on the n-dimensional unit cube can be uniformly approximated by finite linear combinations of characters. (A character is a continuous group homomorphism from $\mathbb{R}^n/\mathbb{Z}^k$ to $\mathbb{C}^*$; each such homomorphism is

$$(x_1, \ldots, x_k) \mapsto \exp(2\pi i(a_1 x_1 + \cdots + a_k x_k))$$

for some $a_1, \ldots, a_k \in \mathbb{Z}$.) The approximation assertion follows from basic properties of Fourier series, which are ubiquitous in physics and engineering as well as mathematics. See [Kör, Theorem 3.1′] for the proof in the $k = 1$ case, and [DMc, Section 1.10] for the general case.

Related question. Problem 1979A5 [PutnamII, p. 33] is related:

> Denote by $[x]$ the greatest integer less than or equal to x and by $S(x)$ the sequence $[x]$, $[2x]$, $[3x]$, $\ldots$. Prove that there are distinct real solutions α and β of the equation $x^3 - 10x^2 + 29x - 25 = 0$ such that infinitely many positive integers appear both in $S(\alpha)$ and $S(\beta)$.

Remark. If α and β are irrational and $1/\alpha + 1/\beta = 1$, then $S(\alpha)$ and $S(\beta)$ partition the positive integers (Beatty's Theorem). See 1993A6 for another instance of this fact, and the solution to that problem for further discussion.

The Fifty-Seventh William Lowell Putnam Mathematical Competition December 7, 1996

A1. (87, 26, 47, 0, 0, 0, 0, 0, 6, 9, 31, 0)

Find the least number A such that for any two squares of combined area 1, a rectangle of area A exists such that the two squares can be packed in the rectangle (without the interiors of the squares overlapping). You may assume that the sides of the squares will be parallel to the sides of the rectangle.

Answer. The least such A is $(1+\sqrt{2})/2$.

Solution 1. If x and y are the sides of two squares with combined area 1, then $x^2+y^2=1$. Suppose without loss of generality that $x \geq y$. Then the shorter side of a rectangle containing both squares without overlap must be at least x, and the longer side must be at least $x+y$. Hence the desired value of A is the maximum of $x(x+y)$ subject to the constraints $x^2+y^2=1$ and $x \geq y > 0$.

To find this maximum, we make the trigonometric substitution $x=\cos\theta, y=\sin\theta$ with $\theta \in (0, \pi/4]$. Then

$$\begin{aligned}\cos^2\theta + \sin\theta\cos\theta &= \frac{1}{2}(1+\cos 2\theta + \sin 2\theta)\\ &= \frac{1}{2} + \frac{\sqrt{2}}{2}\cos(2\theta - \pi/4)\\ &\leq \frac{1+\sqrt{2}}{2},\end{aligned}$$

with equality for $\theta = \pi/8$. Hence the least A is $(1+\sqrt{2})/2$. ■

Solution 2. As in Solution 1, we find the maximum value of $x(x+y)$ on the arc defined by the constraints $x^2+y^2=1$ and $x \geq y > 0$. By the theory of Lagrange multipliers [Ap2, Ch. 9.14], the maximum on the closure of the arc occurs either at an endpoint, or at a point on the arc where the gradient of $x(x+y)$ is a multiple of the gradient of x^2+y^2-1, i.e., where there exists λ such that $(2x+y, x) = \lambda(x,y)$.

If such a λ exists, we may substitute $x=\lambda y$ into $2x+y=\lambda x$ to obtain $(\lambda^2 - 2\lambda - 1)y = 0$. If $y=0$, then $x=\lambda y = 0$, contradicting $x^2+y^2=1$. Therefore $\lambda^2 - 2\lambda - 1 = 0$ and $\lambda = 1 \pm \sqrt{2}$. Only the plus sign gives $\lambda > 0$, and for this choice, there is a unique pair (x,y) with $x = \lambda y > y > 0$ and $x^2+y^2=1$. For this pair, we have $1 = x^2+y^2 = (\lambda^2+1)y^2 = (2\lambda+2)y^2$, so

$$\begin{aligned}x(x+y) &= (\lambda^2+\lambda)y^2\\ &= \frac{\lambda^2+\lambda}{2\lambda+2}\\ &= \frac{\lambda}{2}\\ &= \frac{1+\sqrt{2}}{2}.\end{aligned}$$

At the endpoints $(1,0)$ and $(\sqrt{2}/2, \sqrt{2}/2)$ of the closure of the arc, we have $x(x+y)=1$, so the maximum is $(1+\sqrt{2})/2$. ■

A2. (6, 11, 27, 0, 0, 0, 0, 0, 65, 23, 46, 28)

Let C_1 and C_2 be circles whose centers are 10 units apart, and whose radii are 1 and 3. Find, with proof, the locus of all points M for which there exists points X on C_1 and Y on C_2 such that M is the midpoint of the line segment XY.

Answer. Let O_1 and O_2 be the centers of C_1 and C_2, respectively. (We are assuming C_1 has radius 1 and C_2 has radius 3.) Then the desired locus is an annulus centered at the midpoint O of O_1O_2, with inner radius 1 and outer radius 2.

Solution. If X lies on C_1, Y lies on C_2, and M is the midpoint of XY, then

$$\begin{aligned}\overrightarrow{OM} &= \frac{1}{2}\left(\overrightarrow{OX} + \overrightarrow{OY}\right) \\ &= \frac{1}{2}\left(\overrightarrow{OO_1} + \overrightarrow{O_1X} + \overrightarrow{OO_2} + \overrightarrow{O_2Y}\right) \\ &= \frac{1}{2}\left(\overrightarrow{O_1X} + \overrightarrow{O_2Y}\right).\end{aligned}$$

By the triangle inequality, the length of $\overrightarrow{O_1X} + \overrightarrow{O_2Y}$ lies between $3-1$ and $3+1$. Conversely, every vector of length ℓ between $3-1$ and $3+1$ can be expressed as the sum of a vector of length 1 and a vector of length 3, by building a (possibly degenerate) triangle of side lengths $\ell, 1, 3$ having the given vector as one side. Thus the set of possible M is precisely the closed annulus centered at O with inner radius 1 and outer radius 2. ■

Related question. Given C_1, C_2, and a point M in the annulus, construct points X on C_1 and Y on C_2 using compass and straightedge such that M is their midpoint.

A3. (63, 18, 6, 0, 0, 0, 0, 0, 0, 0, 47, 72)

Suppose that each of 20 students has made a choice of anywhere from 0 to 6 courses from a total of 6 courses offered. Prove or disprove: there are 5 students and 2 courses such that all 5 have chosen both courses or all 5 have chosen neither course.

Answer. There need not exist 5 such students and 2 such courses.

Solution. The number of ways to choose 3 of the 6 courses is $\binom{6}{3} = 20$. Suppose each student chooses a different set of 3 courses. Then given any pair of courses, four students have chosen both and four students have chosen neither course. Thus there do not exist 5 students who have chosen both courses, nor do there exist 5 students who have chosen neither course. ■

Remark. It can be shown that in any example contravening the assertion of the problem, each student must choose 3 courses. To see this, note that for each pair of courses, there are at most $4+4=8$ students who choose both or neither. Thus there are at most $8\binom{6}{2} = 120$ pairs $(s, \{c, d\})$, where s is a student and $\{c, d\}$ is a set of two distinct courses, such that s either chooses both courses or chooses neither course. On the other hand, if a student chooses k courses, then there are $\binom{k}{2} + \binom{6-k}{2}$ pairs of courses such that the student chose both or chose neither. This sum is minimized for $k=3$, where it equals 6. Thus there are at least $6 \times 20 = 120$ pairs $(s, \{c, d\})$;

if any students choose more or fewer than 3 courses, we get strict inequality and a contradiction.

Remark. A consequence of the previous observation is that the example given above is unique if we assume no two students choose the same set of courses. On the other hand, there exist other counterexamples to the assertion of the problem in which more than one student selects the same set of courses. Here is an elegant construction attributed to Robin Chapman. Take an icosahedron, and label its vertices with the names of the 6 courses so that two vertices have the same label if and only if they are antipodal. For each of the 20 faces of the icosahedron, have one student select the three courses labelling the vertices of the face; then antipodal faces give rise to the same set of courses. Concretely, if the courses are A, B, C, D, E, F, the sets could be

$$ABC,\ ACD,\ ADE,\ AEF,\ AFB,\ BCE,\ CDF,\ DEB,\ EFC,\ FBD.$$

Literature note. The study of problems such as this one, asking for combinatorial structures not containing certain substructures, is known as *Ramsey theory*. See [Gr1] for an introduction.

A4. (3, 7, 7, 0, 0, 0, 0, 0, 7, 19, 67, 96)

Let S be a set of ordered triples (a,b,c) of distinct elements of a finite set A. Suppose that

(1) $(a,b,c) \in S$ if and only if $(b,c,a) \in S$;

(2) $(a,b,c) \in S$ if and only if $(c,b,a) \notin S$ [for a,b,c distinct];

(3) (a,b,c) and (c,d,a) are both in S if and only if (b,c,d) and (d,a,b) are both in S.

Prove that there exists a one-to-one function g from A to $\mathbb{R}$ such that $g(a) < g(b) < g(c)$ implies $(a,b,c) \in S$.

Solution. We may assume that A is nonempty. Pick some x in A. Let us define a binary relation on A as follows: we define $b < c$ if and only if $b = x \neq c$ or $(x,b,c) \in S$. For x, b, c distinct, we have

$$(x,b,c) \in S \iff (c,x,b) \in S \iff (b,x,c) \notin S \iff (x,c,b) \notin S$$

by (1),(2),(1), respectively. Thus $b < c$ if and only if $c \not< b$. We also have $b \not< x$ for any $b \neq x$, since $(x,b,x) \notin S$.

We now show that if $b < c$ and $c < d$, then $b < d$. These conditions imply $c, d \neq x$; if $b = x$, we are done, so assume $b \neq x$. Then (x,b,c) and (x,c,d) are in S; by (1), (c,d,x) is also in S, and then (3) implies $(d,x,b) \in S$. By (1) again, $(x,b,d) \in S$, and so $b < d$.

Thus $<$ is a total ordering of A, and so we can find the desired function g. For example, let $g(a)$ be the number of $y \in A$ such that $y < a$. ■

Remark. The construction of the total ordering did not use the finiteness of A. The finiteness was only used to construct the order-preserving injection of A into $\mathbb{R}$. In fact, such an injection also exists if A is countable: list the elements of A in an arbitrary fashion (without regard to the total ordering), and choose their images in $\mathbb{R}$ one at a time, making sure that the total ordering is preserved.

A5. (4, 4, 2, 0, 0, 0, 0, 0, 0, 13, 40, 143)

If p is a prime number greater than 3 and $k = \lfloor 2p/3 \rfloor$, prove that the sum

$$\binom{p}{1} + \binom{p}{2} + \cdots + \binom{p}{k}$$

of binomial coefficients is divisible by p^2.

Solution. For $1 \leq n \leq p-1$, p divides $\binom{p}{n}$ and

$$\frac{1}{p}\binom{p}{n} = \frac{1}{n} \cdot \frac{p-1}{1} \cdot \frac{p-2}{2} \cdots \frac{p-n+1}{n-1} \equiv \frac{(-1)^{n-1}}{n} \pmod{p},$$

where the congruence $x \equiv y \pmod{p}$ means that $x - y$ is a rational number whose numerator, in reduced form, is divisible by p. Therefore

$$\begin{aligned}\sum_{n=1}^{k} \frac{1}{p}\binom{p}{n} &\equiv \sum_{n=1}^{k} \frac{(-1)^{n-1}}{n} \\ &= \sum_{n=1}^{k} \frac{1}{n} - 2\sum_{n=1}^{\lfloor k/2 \rfloor} \frac{1}{2n} \\ &\equiv \sum_{n=1}^{k} \frac{1}{n} + \sum_{n=p-\lfloor k/2 \rfloor}^{p-1} \frac{1}{n} \pmod{p}.\end{aligned}$$

Any prime $p > 3$ is of one of the forms $6r+1$ or $6s+5$. In the former case, $k = 4r$ and $p - \lfloor k/2 \rfloor = 4r+1 = k+1$. In the latter case, $k = 4s+3$ and $p - \lfloor k/2 \rfloor = 4s+4 = k+1$. In either case, we conclude that

$$\sum_{n=1}^{k} \frac{1}{p}\binom{p}{n} \equiv \sum_{n=1}^{p-1} \frac{1}{n} \equiv \sum_{n=1}^{\lfloor p/2 \rfloor} \frac{1}{n} + \sum_{n=1}^{\lfloor p/2 \rfloor} \frac{1}{p-n} \equiv 0 \pmod{p},$$

which completes the proof. ∎

Related question. Compare with Problem 1 of the 1979 International Mathematical Olympiad [IMO79–85, p. 2]:

Let m and n be positive integers such that

$$\frac{m}{n} = 1 - \frac{1}{2} + \frac{1}{3} - \cdots - \frac{1}{1318} + \frac{1}{1319}.$$

Prove that m is divisible by 1979.

A6. (4, 4, 11, 0, 0, 0, 0, 0, 18, 13, 46, 110)

Let $c \geq 0$ be a constant. Give a complete description, with proof, of the set of all continuous functions $f : \mathbb{R} \to \mathbb{R}$ such that $f(x) = f(x^2 + c)$ for all $x \in \mathbb{R}$.

Answer. If $0 \leq c \leq 1/4$ then f is constant. If $c > 1/4$, then f is uniquely determined (as described below) by its restriction g to $[0, c]$, which may be any continuous function on $[0, c]$ with $g(0) = g(c)$.

Solution. The functional equation implies $f(x) = f(x^2 + c) = f((-x)^2 + c) = f(-x)$. Conversely, if f satisfies the functional equation for $x \geq 0$ and also satisfies

$f(x) = f(-x)$, then it satisfies the functional equation for all x. So we will consider only $x \geq 0$ hereafter.

Case 1: $0 \leq c \leq 1/4$. In this case, the zeros a, b of $x^2 - x + c$ are real and satisfy $0 \leq a \leq b$. We will show that $f(x) = f(a)$ for all $x \geq 0$. First suppose $0 \leq x < a$. Define $x_0 = x$ and $x_{n+1} = x_n^2 + c$ for $n \geq 0$. Then we have $x_n < x_{n+1} < a$ for all n, by induction: if $x_n < a$, then on one hand $x_{n+1} - x_n = x_n^2 - x_n + c > 0$, and on the other hand $x_{n+1} = x_n^2 + c < a^2 + c = a$. We conclude that x_n converges to a limit L not exceeding a, which must satisfy $L^2 + c = L$. Since the only real roots of this equation are a and b, we have $L = a$. Now since $f(x_n) = f(x)$ for all n and $x_n \to a$, we have $f(x) = f(a)$ by continuity.

If $a < x < b$, we argue similarly. Define x_n as above; then we have $a < x_{n+1} < x_n$ for all n, by induction: if $x_n < b$, then on one hand $x_{n+1} - x_n = x_n^2 - x_n + c < 0$, and on the other hand $x_{n+1} = x_n^2 + c > a^2 + c = a$. So again $f(x) = f(a)$ for $a < x < b$. By continuity, $f(b) = f(a)$.

Finally, suppose $x > b$. Now define $x_0 = x$ and $x_{n+1} = \sqrt{x_n - c}$ for $n \geq 0$. Then we have $b < x_{n+1} < x_n$ for all n, by induction: on one hand, $x_{n+1}^2 + c = x_n < x_n^2 + c$, and on the other hand $x_{n+1} > \sqrt{b - c} = b$. Thus x_n converges to a limit L not less than b, and $L = \sqrt{L - c}$, so $L = b$. So analogously, $f(x) = f(b) = f(a)$.

We conclude that $f(x) = f(a)$ is constant.

Case 2: $c > 1/4$. In this case, $x^2 + c > x$ for all x. Let g be any continuous function on $[0, c]$ with $g(0) = g(c)$. Then g extends to a continuous function f by setting $f(x) = g(x)$ for $x < c$, $f(x) = g(\sqrt{x - c})$ for $c \leq x < c^2 + c$, $f(x) = g(\sqrt{\sqrt{x - c} - c})$ for $c^2 + c \leq x < (c^2 + c)^2 + c$, and so on. Conversely, any f satisfying the functional equation is determined by its values on $[0, c]$ and satisfies $f(0) = f(c)$, and so arises in this fashion. ■

Stronger result. More generally, let p be a polynomial not equal to x or to $c - x$ for any $c \in \mathbb{R}$. One can imitate the above solution to classify continuous functions $f : \mathbb{R} \to \mathbb{R}$ such that $f(x) = f(p(x))$ for all $x \in \mathbb{R}$. The classification hinges on whether p has any fixed points.

Case 1: $p(x)$ has a fixed point. In this case, we show that f must be constant. We first reduce to the case where p has positive leading coefficient. If this is not the case and p has odd degree, we work instead with $p(p(x))$; if p has even degree, then $g(x) = f(-x)$ satisfies $g(x) = g(-p(-x))$, so we work instead with $-p(-x)$.

Let r be the largest fixed point of p; then $p(x) - x$ does not change sign for $x > r$, and it tends to $+\infty$ as $x \to +\infty$, so it is positive for all $x > r$. In particular, p maps $[r, \infty)$ onto itself. For any $x > r$, define the sequence $\{x_n\}$ by setting $x_0 = x$ and x_{n+1} is the smallest element of $[r, \infty)$ such that $p(x_{n+1}) = x_n$. The sequence $\{x_n\}$ is decreasing, so it converges to a limit $L \geq r$, and $L = \lim x_n = \lim p(x_{n+1}) = p(L)$. Thus $L = p(L)$, which can only occur if $L = r$. Since $f(x_n) = f(x_0)$ for all n and $x_n \to r$, we have $f(x_0) = f(r)$. Consequently, f is constant on $[r, \infty)$.

Suppose, in order to obtain a contradiction, that f is not constant on all of $\mathbb{R}$. Let

$$T = \inf\{\, t \in \mathbb{R} : f \text{ is constant on } [t, \infty)\}.$$

Then f is constant on $[T, \infty)$. It is also constant on $[p(T), \infty)$, so $p(T) \geq T$. It is also constant on $p^{-1}([T,\infty))$; if $p(T) > T$, this set also includes $T-\epsilon$ for small ϵ, contradiction. Thus $p(T) = T$.

Since p has finitely many fixed points, either $p(T-\epsilon) > T-\epsilon$ for small ϵ, or $p(T-\epsilon) < T-\epsilon$ for small ϵ. In the former case, the sequence with $x_0 = T-\epsilon$ and $x_{n+1} = p(x_n)$ converges to T, so $f(x_0) = f(T)$. In the latter case, the sequence with $x_0 = T-\epsilon$ and x_{n+1} the largest element of (x_n, T) with $q(x_{n+1}) = x_n$ converges to T, so $g(x_0) = T$. In either case, g is constant on a larger interval than $[T,\infty)$, a contradiction.

Case 2: p has no fixed point. In this case, p must have even degree, so as in Case 1, we may assume p has positive leading coefficient, which implies that $p(x) > x$ for all $x \in \mathbb{R}$. There exists $a_0 \in \mathbb{R}$ such that p is strictly increasing on $[a_0,\infty)$, since any a_0 larger than all zeros of p has this property. For $n \geq 0$, set $a_{n+1} = p(a_n)$, so $\{a_n\}$ is an increasing sequence. It cannot have a limit, since that limit would have to be a fixed point of p, so $a_n \to \infty$ as $n \to \infty$.

We claim f is determined by its values on $[a_0, a_1]$. Indeed, these values determine f on $[a_1, a_2]$, on $[a_2, a_3]$, and so on, and the union of these intervals is $[a_0,\infty)$. Furthermore, for any $x \in R$, if we set $x_0 = x$ and $x_{n+1} = p(x_n)$, then the sequence $\{x_n\}$ is increasing and unbounded, so some x_n lies in $[a_0,\infty)$, and $f(x_n) = f(x)$ is determined by the values of f on $[a_0, a_1]$.

Conversely, let $g(x)$ be any continuous function on $[a_0, a_1]$ with $g(a_0) = g(a_1)$; we claim there exists a continuous $f : \mathbb{R} \to \mathbb{R}$ with $f(x) = f(p(x))$ whose restriction to $[a_0, a_1]$ is g. For $x \in \mathbb{R}$, again set $x_0 = x$ and $x_{n+1} = p(x_n)$. If n is large enough, there exists $t \in [a_0, a_1)$ such that $x_n = p^k(t)$ for some integer k, and this t is independent of n. Hence we may set $f(x) = f(t)$.

It remains to show that f is continuous. We can define a continuous inverse p^{-1} of p mapping $[a_1,\infty)$ to $[a_0,\infty)$, and f is given on $[a_0,a_1]$ by $f(x) = g(x)$, on $[a_1, a_2]$ by $f(x) = g(p^{-1}(x))$, on $[a_2, a_3]$ by $f(x) = g(p^{-2}(x))$, and so on. Thus f is continuous on each of the intervals $[a_n, a_{n+1}]$, and hence on their union $[a_0,\infty)$. For x arbitrary, we can find a neighborhood of x and an integer k such that p^k maps the neighborhood into $[a_0,\infty)$, and writing $f(t) = f(p^k(t))$ for t in the neighborhood, we see that f is also continuous there.

We conclude that the continuous functions f satisfying $f(x) = f(p(x))$ correspond uniquely to the continuous functions g on $[a_0, a_1]$ such that $g(a_0) = g(a_1)$. Otherwise put, the quotient of $\mathbb{R}$ by the map p is isomorphic to the interval $[a_0, a_1]$ with the endpoints identified.

Remark. The ideas in this problem come from the theory of dynamical systems. For a problem involving a related dynamical system, see 1992B3.

B1. (113, 72, 17, 0, 0, 0, 0, 0, 3, 1, 0, 0)

Define a *selfish* set to be a set which has its own cardinality (number of elements) as an element. Find, with proof, the number of subsets of $\{1, 2, \ldots, n\}$ which are *minimal* selfish sets, that is, selfish sets none of whose proper subsets is selfish.

Answer. The number of subsets is F_n, the nth Fibonacci number. (See the definition at the end of 1988A5.)

Solution 1. Let f_n denote the number of minimal selfish subsets of $\{1,\dots,n\}$. We have $f_1 = 1$ and $f_2 = 2$. For $n > 2$, the number of minimal selfish subsets of $\{1,\dots,n\}$ not containing n is equal to f_{n-1}. On the other hand, for any minimal selfish set containing n, by removing n from the set and subtracting 1 from each remaining element, we obtain a minimal selfish subset of $\{1,\dots,n-2\}$. (Note that 1 could not have been an element of the set because $\{1\}$ is itself selfish.) Conversely, any minimal selfish subset of $\{1,\dots,n-2\}$ gives rise to a minimal selfish subset of $\{1,\dots,n\}$ containing n by the inverse procedure. Hence the number of minimal selfish subsets of $\{1,\dots,n\}$ containing n is f_{n-2}. Thus we obtain

$$f_n = f_{n-1} + f_{n-2},$$

which together with the initial values $f_1 = f_2 = 1$ implies that $f_n = F_n$. ■

Solution 2. A set is minimal selfish if and only if its smallest member is equal to its cardinality. (If a selfish set contains a member smaller than its cardinality, it contains a subset of that cardinality which is also selfish.) Thus we can compute the number S_n of minimal selfish subsets of $\{1,\dots,n\}$ by summing, over each cardinality k, the number of subsets of $\{k+1,\dots,n\}$ of size $k-1$ (since each of these, plus $\{k\}$, is minimal selfish, and vice versa). In other words, $S_n = \sum_k \binom{n-k}{k-1}$. We directly verify that $S_1 = S_2 = 1$, and that

$$\begin{aligned} S_{n-1} + S_n &= \sum_k \binom{n-k}{k-1} + \sum_k \binom{n-1-k}{k-1} \\ &= \sum_k \binom{n-k}{k-1} + \sum_k \binom{n-k}{k-2} \\ &= \sum_k \binom{n+1-k}{k-1} \\ &= S_{n+1}. \end{aligned}$$

Thus by induction, $S_n = F_n$ for all n. ■

Remark. The identity $F_n = \sum_k \binom{n-k}{k-1}$ can be interpreted as the fun fact that certain diagonal sums in Pascal's triangle yield the Fibonacci numbers. (See Figure 30.) It is common for a combinatorial enumeration problem to admit both a bijective and a recursive proof; see 1996B5 for another example.

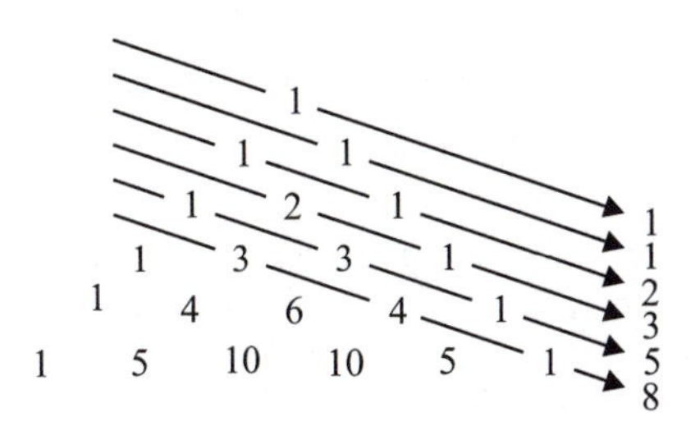

FIGURE 30.
Fibonacci in Pascal.

B2. (85, 13, 6, 0, 0, 0, 0, 0, 15, 7, 28, 52)

Show that for every positive integer n,

$$\left(\frac{2n-1}{e}\right)^{\frac{2n-1}{2}} < 1\cdot 3\cdot 5\cdots(2n-1) < \left(\frac{2n+1}{e}\right)^{\frac{2n+1}{2}}.$$

Solution 1. By estimating the area under the graph of $\ln x$ using upper and lower rectangles of width 2 (see Figure 31), we obtain

$$\int_1^{2n-1} \ln x\,dx < 2(\ln(3)+\cdots+\ln(2n-1)) < \int_3^{2n+1} \ln x\,dx.$$

Since $\int \ln x\,dx = x\ln x - x + C$, exponentiating and taking square roots yields the middle two inequalities in

$$\begin{aligned}\left(\frac{2n-1}{e}\right)^{\frac{2n-1}{2}} &< (2n-1)^{\frac{2n-1}{2}}e^{-n+1}\\ &\le 1\cdot 3\cdots(2n-1)\\ &\le (2n+1)^{\frac{2n+1}{2}}\frac{e^{-n+1}}{3^{3/2}} < \left(\frac{2n+1}{e}\right)^{\frac{2n+1}{2}},\end{aligned}$$

and the inequalities at the ends follow from $1 < e < 3$. ∎

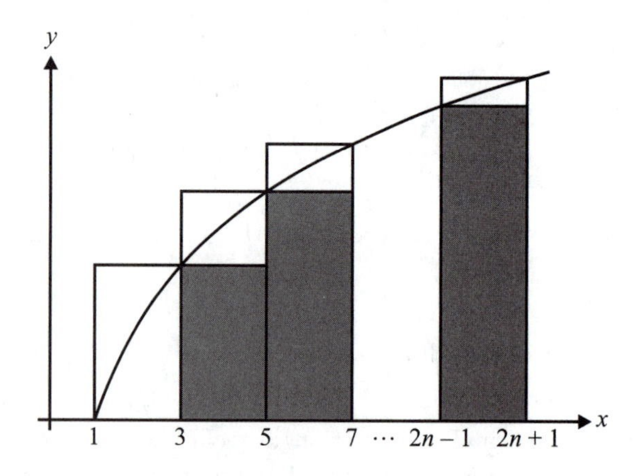

FIGURE 31.
Estimating $\int \ln x\,dx$.

Solution 2. We use induction on n. The $n = 1$ case follows from $1 < e < 3$. For the inductive step, we need to prove

$$\frac{\left(\frac{2n+1}{e}\right)^{\frac{2n+1}{2}}}{\left(\frac{2n-1}{e}\right)^{\frac{2n-1}{2}}} < 2n+1 < \frac{\left(\frac{2n+3}{e}\right)^{\frac{2n+3}{2}}}{\left(\frac{2n+1}{e}\right)^{\frac{2n+1}{2}}},$$

which is equivalent to

$$\left(\frac{2n+1}{2n-1}\right)^{\frac{2n-1}{2}} < e < \left(\frac{2n+3}{2n+1}\right)^{\frac{2n+3}{2}}.$$

The left side is equal to $(1+u)^{1/u}$ for $u = 2/(2n-1)$, and the right side is equal to $(1+u)^{1/u+1}$ for $u = 2/(2n+1)$, so it suffices to prove the inequalities

$$(1+u)^{1/u} < e < (1+u)^{1/u+1} \qquad \text{for } u > 0.$$

Apply ln and rearrange to rewrite this as

$$\frac{u}{1+u} < \ln(1+u) < u,$$

which follows by integrating

$$\frac{1}{(1+t)^2} < \frac{1}{1+t} < 1$$

from $t = 0$ to $t = u$. ∎

Related question. Use the method of Solution 1 to prove that

$$e(n/e)^n < n! < en(n/e)^n.$$

(This is a weak version of Stirling's approximation. The standard form of Stirling's approximation is stated in the following remark.)

Remark. Better estimates for $n!$ can be obtained using the *Euler-Maclaurin summation formula* [At, Section 5.4]: For any fixed $k > 0$,

$$\sum_{j=a}^{b} f(j) = \int_a^b f(t)\,dt + \frac{f(a)+f(b)}{2} + \sum_{i=1}^{k} \frac{B_{2i}}{(2i)!}\left(f^{(2i-1)}(b) - f^{(2i-1)}(a)\right) + R_k(a,b),$$

where the Bernoulli numbers B_{2i} are given by the power series

$$\frac{x}{e^x - 1} = -x/2 + \sum_{i=1}^{\infty} \frac{B_{2i}}{(2i)!} x^{2i},$$

(see [HW, Section 17.2]) and the error term $R_k(a,b)$ is given by

$$R_k(a,b) = \frac{-1}{(2k+2)!} \int_a^b B_{2k+2}(t - \lfloor t \rfloor) f^{(2k+2)}(t)\,dt.$$

(See 1993B3 for another appearance of Bernoulli numbers.) Specifically, this formula can be used to obtain Stirling's approximation to $n!$ [Ap2, Theorem 15.19],

$$n! \sim \sqrt{2\pi n}\left(\frac{n}{e}\right)^n,$$

where the tilde indicates that the ratio of the two sides tends to 1 as $n \to \infty$. See Solution 1 to 1995A6 for an application of Stirling's approximation.

The Euler-Maclaurin formula has additional applications in numerical analysis, as well as in combinatorics; see the remark following 1996B3 for an example.

B3. (20, 19, 8, 0, 0, 0, 0, 0, 14, 36, 56, 53)

Given that $\{x_1, x_2, \dots, x_n\} = \{1, 2, \dots, n\}$, find, with proof, the largest possible value, as a function of n (with $n \geq 2$), of

$$x_1x_2 + x_2x_3 + \cdots + x_{n-1}x_n + x_nx_1.$$

Answer. The maximum is $(2n^3 + 3n^2 - 11n + 18)/6$, and the unique arrangement (up to rotation and reflection) achieving the maximum is

$$\ldots, n-4, n-2, n, n-1, n-3, \ldots.$$

Solution 1. Since there are finitely many arrangements, at least one is optimal. In an optimal arrangement, if a, b, c, d occur around the circle in that order, with a adjacent to b, c adjacent to d and $a > d$, then $b > c$. Otherwise, reversing the segment from b to c would increase the sum by $(a-d)(c-b)$.

This implies first that in an optimal arrangement, n is adjacent to $n-1$, or else we could reverse the segment from $n-1$ to either neighbor of n. Likewise, n must be adjacent to $n-2$, or else we could reverse the segment from $n-2$ to the neighbor of n other than $n-1$.

Now $n-1$ is adjacent to n but not to $n-2$; we must have that $n-1$ is also adjacent to $n-3$, or else we could reverse the segment from $n-3$ to the neighbor of $n-1$ other than n. Likewise $n-2$ must be adjacent to $n-4$, and so on.

In this arrangement, the value of the function is

$$n(n-1) + n(n-2) + (n-1)(n-3) + \cdots + 4 \cdot 2 + 3 \cdot 1 + 2 \cdot 1.$$

Using the fact that $\sum_{k=1}^{n} k^2 = n(n+1)(2n+1)/6$, we can rewrite the sum as

$$\begin{aligned} n(n-1) + 2 + \sum_{k=1}^{n-2} k(k+2) &= n^2 - n + 2 + \sum_{k=1}^{n-2} (k+1)^2 - 1 \\ &= n^2 - n + 2 + \frac{(n-1)n(2n-1)}{6} - 1 - (n-2) \\ &= \frac{2n^3 + 3n^2 - 11n + 18}{6}. \end{aligned}$$

■

Solution 2. We prove that the given arrangement is optimal by induction on n. Suppose that the analogous arrangement of $n-1$ numbers is optimal. Given any arrangement of $n-1$ numbers, inserting n between a and b changes the sum by $na + nb - ab = n^2 - (n-a)(n-b) \leq n^2 - 2$. Thus the maximum sum for n numbers is at most $n^2 - 2$ plus the maximum sum for $n-1$ numbers, and this value is achieved by inserting n between $n-1$ and $n-2$ in the optimal arrangement of $n-1$ numbers. Thus the result is the optimal arrangement of n numbers; since the only possible arrangement for $n = 3$ has value $6 + 3 + 2 = 11$, the value for n numbers is

$$\begin{aligned} 11 + (4^2 - 2) + \cdots + (n^2 - 2) &= 11 + \frac{n(n+1)(2n+1)}{6} - 2(n-3) - (1^2 + 2^2 + 3^2) \\ &= \frac{2n^3 + 3n^2 - 11n + 18}{6}. \end{aligned}$$

■

Related question. Can you determine the arrangement that minimizes the function? Hint: the minimum value is $(n^3 + 3n^2 + 5n - 6)/6$ for n even and $(n^3 + 3n^2 + 5n - 3)/6$ for n odd.

Remark. The fact that $\sum_{k=1}^{n} k^2 = n(n+1)(2n+1)/6$, used above, can be generalized as follows: for any polynomial $P(x)$ of degree m, there exists a polynomial $Q(x)$ of degree $m+1$ such that $\sum_{k=1}^{n} P(k) = Q(n)$ for all positive integers n. This follows from the Euler-Maclaurin summation formula (see the remark in 1996B2), which can also be used to compute Q from P.

B4. (40, 4, 7, 0, 0, 0, 0, 0, 17, 4, 48, 86)

For any square matrix A, we can define $\sin A$ by the usual power series:

$$\sin A = \sum_{n=0}^{\infty} \frac{(-1)^n}{(2n+1)!} A^{2n+1}.$$

Prove or disprove: there exists a 2×2 matrix A with real entries such that

$$\sin A = \begin{pmatrix} 1 & 1996 \\ 0 & 1 \end{pmatrix}.$$

Answer. There does not exist such a matrix A.

Solution 1. Over the complex numbers, if A has distinct eigenvalues, it is diagonalizable. Since $\sin A$ is a convergent power series in A, eigenvectors of A are also eigenvectors of $\sin A$, so A having distinct eigenvalues would imply that $\sin A$ is diagonalizable. Since $\begin{pmatrix} 1 & 1996 \\ 0 & 1 \end{pmatrix}$ is not diagonalizable, it can be $\sin A$ only for a matrix A with equal eigenvalues. This matrix can be conjugated into the form $\begin{pmatrix} x & y \\ 0 & x \end{pmatrix}$ for some x and y. Using the power series for sin, we compute

$$\begin{aligned}
\sin \begin{pmatrix} x & y \\ 0 & x \end{pmatrix} &= \sum_{k=0}^{\infty} \frac{(-1)^k}{(2k+1)!} \begin{pmatrix} x & y \\ 0 & x \end{pmatrix}^{2k+1} \\
&= \sum_{k=0}^{\infty} \frac{(-1)^k x^{2k+1}}{(2k+1)!} \begin{pmatrix} 1 & y/x \\ 0 & 1 \end{pmatrix}^{2k+1} \\
&= \sum_{k=0}^{\infty} \frac{(-1)^k x^{2k+1}}{(2k+1)!} \begin{pmatrix} 1 & (2k+1)y/x \\ 0 & 1 \end{pmatrix} \\
&= \begin{pmatrix} \sum_{k=0}^{\infty} \frac{(-1)^k x^{2k+1}}{(2k+1)!} & y \sum_{k=0}^{\infty} \frac{(-1)^k x^{2k}}{(2k)!} \\ 0 & \sum_{k=0}^{\infty} \frac{(-1)^k x^{2k+1}}{(2k+1)!} \end{pmatrix} \\
&= \begin{pmatrix} \sin x & y \cos x \\ 0 & \sin x \end{pmatrix}.
\end{aligned}$$

Thus if $\sin x = 1$, then $\cos x = 0$ and $\sin \begin{pmatrix} x & y \\ 0 & x \end{pmatrix}$ is the identity matrix. In other words, $\sin A$ cannot equal a matrix whose eigenvalues are 1 but which is not the identity matrix. ■

Remark. The computation of $\sin \begin{pmatrix} x & y \\ 0 & x \end{pmatrix}$ can be simplified by taking $A = \begin{pmatrix} x & 0 \\ 0 & x \end{pmatrix}$ and $B = \begin{pmatrix} 0 & y \\ 0 & 0 \end{pmatrix}$ in the identity

$$\sin(A + B) = \sin A \cos B + \cos A \sin B,$$

which holds when A and B commute.

Solution 2. Put $\cos A = \sum_{n=0}^{\infty}(-1)^n A^{2n}/n!$. The identity $\sin^2 x + \cos^2 x = 1$ implies the identity

$$\left(\sum_{n=0}^{\infty} \frac{(-1)^n}{(2n+1)!} x^{2n+1}\right)^2 + \left(\sum_{n=0}^{\infty} \frac{(-1)^n}{(2n)!} x^{2n}\right)^2 = 1$$

of formal power series. The series converge absolutely if we substitute $x = A$, so $\sin^2 A + \cos^2 A$ equals the identity matrix I. But

$$I - \begin{pmatrix} 1 & 1996 \\ 0 & 1 \end{pmatrix}^2 = \begin{pmatrix} 0 & 3992 \\ 0 & 0 \end{pmatrix}$$

cannot be the square of a matrix; such a matrix would have to be nilpotent, and the kth power of a $k \times k$ nilpotent matrix is always zero, by the Cayley-Hamilton Theorem. ■

Stronger result. We will show that the image of $\sin : M_2(\mathbb{R}) \to M_2(\mathbb{R})$ is the set of matrices $B \in M_2(\mathbb{R})$ such that at least one of the following holds:

(1) B is conjugate to $\begin{pmatrix} \lambda & 0 \\ 0 & \lambda' \end{pmatrix}$ with $\lambda, \lambda' \in [-1, 1]$;

(2) B is conjugate to $\begin{pmatrix} \lambda & 1 \\ 0 & \lambda \end{pmatrix}$ with $\lambda \in (-1, 1)$;

(3a) $B = \lambda I$ for some $\lambda \in \mathbb{R}$; or

(3b) B has nonreal eigenvalues.

Recall that if $A, B \in M_n(\mathbb{R})$ are conjugate over $\mathbb{C}$, then they are also conjugate over $\mathbb{R}$. The entire function $\sin : \mathbb{C} \to \mathbb{C}$ is surjective, since it equals the composition of the surjection $e^{iz} : \mathbb{C} \to \mathbb{C}^*$ followed by the surjection $\frac{z - z^{-1}}{2i} : \mathbb{C}^* \to \mathbb{C}$.

Every $A \in M_2(\mathbb{R})$ is conjugate over $\mathbb{C}$ to a matrix of one of the following types:

(1′) $\begin{pmatrix} r & 0 \\ 0 & r' \end{pmatrix}$ with $r, r' \in \mathbb{R}$;

(2′) $\begin{pmatrix} x & 1 \\ 0 & x \end{pmatrix}$ with $x \in \mathbb{R}$; or

(3′) $\begin{pmatrix} z & 0 \\ 0 & \bar{z} \end{pmatrix}$ with $z \in \mathbb{C}$.

By definition, $\sin(CAC^{-1}) = C(\sin A)C^{-1}$ for any invertible C, so $B \in M_2(\mathbb{R})$ is in the image of $\sin : M_2(\mathbb{R}) \to M_2(\mathbb{R})$ if and only if it is conjugate over $\mathbb{C}$ to the sine of a matrix of one of types (1′), (2′), (3′).

If A is of type (1′), then $\sin A$ is of type (1), and conversely all type (1) matrices are conjugate to a matrix arising in this way. If A is of type (2′), then $\sin A = \begin{pmatrix} \sin x & \cos x \\ 0 & \sin x \end{pmatrix}$ is of type (1) or (2), according as $\cos x = 0$ or not, and conversely, any matrix of type (2) is conjugate to $\begin{pmatrix} \sin x & \cos x \\ 0 & \sin x \end{pmatrix}$ for some $x \in (-\pi/2, \pi/2)$. If A is of type (3′), then $\sin A = \begin{pmatrix} \sin z & 0 \\ 0 & \overline{\sin z} \end{pmatrix}$, which is of type (3a) if $\sin z \in \mathbb{R}$, and of type (3b) otherwise. Conversely, if $B = \lambda I$ if of type (3a), then $B = \sin(zI)$ where

$z \in \mathbb{C}$ satisfies $\sin z = \lambda$. And if B is of type (3b) with eigenvalues λ, $\bar{\lambda}$, choose $z \in \mathbb{C}$ with $\sin z = \lambda$, so that B is conjugate to $\sin A$ where $A \in M_2(\mathbb{R})$ is of type (3b) with eigenvalues z and $\bar{z}$.

B5. (34, 4, 7, 0, 0, 0, 0, 0, 3, 2, 58, 98)

Given a finite string S of symbols X and O, we write $\Delta(S)$ for the number of X's in S minus the number of O's. For example, $\Delta(XOOXOOX) = -1$. We call a string S *balanced* if every substring T of (consecutive symbols of) S has $-2 \le \Delta(T) \le 2$. Thus, $XOOXOOX$ is not balanced, since it contains the substring $OOXOO$. Find, with proof, the number of balanced strings of length n.

Answer. The number of balanced strings of length n is $3 \cdot 2^{n/2} - 2$ if n is even, and $2^{(n+1)/2} - 2$ if n is odd.

Solution 1. We give an explicit counting argument. Consider a $1 \times n$ checkerboard, in which we write an n-letter string, one letter per square. If the string is balanced, we can cover each pair of adjacent squares containing the same letter with a 1×2 domino, and these will not overlap (because no three in a row can be the same). Moreover, any domino is separated from the next by an even number of squares, since they must cover opposite letters, and the sequence must alternate in between.

Conversely, any arrangement of dominoes where adjacent dominoes are separated by an even number of squares corresponds to a unique balanced string, once we choose whether the string starts with X or O. In other words, the number of balanced strings is twice the number of acceptable domino arrangements.

We count these arrangements by numbering the squares $0, 1, \ldots, n-1$ and distinguishing whether the dominoes start on even or odd numbers. Once this is decided, one simply chooses whether or not to put a domino in each eligible position. Thus we have $2^{\lfloor n/2 \rfloor}$ arrangements in the first case and $2^{\lfloor (n-1)/2 \rfloor}$ in the second, but the case of no dominoes has been counted twice. Hence the number of balanced strings is $2\left(2^{\lfloor n/2 \rfloor} + 2^{\lfloor (n-1)/2 \rfloor} - 1\right)$, which equals the answer given above. ∎

Solution 2. Let b_n denote the number of balanced strings of length n. We establish the recursion $b_{n+2} = 2b_n + 2$ for $n \ge 1$. This recursion and the initial conditions $b_1 = 2, b_2 = 4$ yield the desired result by induction.

Let x_n, y_n, z_n denote the number of balanced strings of length n ending with XX, the number that end in XO and whose last doubled letter is X, and the number that end in XO and whose last doubled letter is O, respectively. (To keep things consistent, we count a purely alternating string as if its last doubled letter were both X and O.) The counts remain the same if X and O are interchanged. Thus $b_n = 2x_n + 2y_n + 2z_n - 2$ for $n \ge 1$ (we subtract 2 to avoid double-counting the purely alternating strings).

If a balanced string of length $n+2$ ends in XX, the remainder must end in OO, or end in XO and have last doubled letter O. Thus $x_{n+2} = x_n + z_n$. If a balanced string of length $n+2$ ends in XO and has last doubled letter X, the remainder (the first n symbols) must end in XO and have last doubled letter X, or end in OX and have last doubled letter O; conversely, appending XX to the end of such a balanced string results in a balanced string of length $n+2$. Thus $y_{n+2} = 2y_n$. Finally, if a balanced string of length $n+2$ ends in XO and has last doubled letter O, the remainder must end in OO, or end in XO and have last doubled letter O. Thus $z_{n+2} = x_n + z_n$ as well. Putting this together,

$$\begin{aligned} b_{n+2} &= 2x_{n+2} + 2y_{n+2} + 2z_{n+2} - 2 \\ &= 2(x_n + z_n) + 2(2y_n) + 2(x_n + z_n) - 2 \\ &= 2(b_n + 2) - 2, \end{aligned}$$

the desired recursion. ■

B6. (0, 1, 9, 0, 0, 0, 0, 0, 2, 0, 23, 171)

Let $(a_1, b_1), (a_2, b_2), \ldots, (a_n, b_n)$ be the vertices of a convex polygon which contains the origin in its interior. Prove that there exist positive real numbers x and y such that

$$(a_1, b_1)x^{a_1}y^{b_1} + (a_2, b_2)x^{a_2}y^{b_2} + \cdots + (a_n, b_n)x^{a_n}y^{b_n} = (0, 0).$$

Solution 1. Let $f(\vec{v}) = \sum_i e^{(a_i,b_i)\cdot\vec{v}}$. If $\nabla f(x_0, y_0) = \vec{0}$, then (e^{x_0}, e^{y_0}) is a solution of the original vector equation. Hence it suffices to show that f achieves a global minimum somewhere.

For $0 \le \theta \le 2\pi$, define

$$g(\theta) = \max_i\{(a_i, b_i) \cdot (\cos\theta, \sin\theta)\}.$$

Then $g(\theta)$ is always positive, because the origin lies in the interior of the convex hull of the points (a_i, b_i). Let $c > 0$ be the minimum value of $g(\theta)$ on the interval $[0, 2\pi]$; then

$$f(\vec{v}) \ge \max_i\{e^{(a_i,b_i)\cdot\vec{v}}\} \ge e^{c|\vec{v}|}$$

for all $\vec{v}$. In particular, f is greater than e^{cr} outside of a circle of radius r. Choose $r > 0$ such that $e^{cr} > f((0,0))$. Then the infimum of f on the entire plane equals its infimum within the disc of radius r centered at the origin. Since the closed disc is compact, the infimum of f there is the value of f at some point, which is the desired global minimum. ■

Solution 2. We retain the notation of the first solution. To prove that ∇f vanishes somewhere, it suffices to exhibit a simple closed contour over which ∇f has a nonzero winding number. In fact, any sufficiently large counterclockwise circle has this property. As in the first solution, there exists c such that $\max_i\{(a_i, b_i)\cdot\vec{v}\} \ge c|\vec{v}|$, so

$$\begin{aligned} \vec{v} \cdot \nabla f(\vec{v}) &= \sum_{i=1}^{n} e^{(a_i,b_i)\cdot\vec{v}}((a_i, b_i) \cdot \vec{v}) \\ &\ge c|\vec{v}|e^{c|\vec{v}|} + (n-1)\inf_x xe^x \\ &= c|\vec{v}|e^{c|\vec{v}|} - (n-1)/e, \end{aligned}$$

since xe^x is minimized over $x \in \mathbb{R}$ at $x = -1$.

Thus if $\vec{v}$ runs over a large enough circle, we will have $\vec{v} \cdot \nabla f(\vec{v}) > 0$ everywhere on the circle, implying that the vector field $\nabla f(\vec{v})$ has the same winding number as $\vec{v}$ on the circle, namely 1. (In particular, the number of solutions $\vec{v}$ to $\nabla f(\vec{v}) = 0$, counted with multiplicity, is odd.) ∎

Remark. The notion of winding number also occurs in complex analysis, for instance in the proof of Rouché's Theorem, which is stated in Solution 3 to 1989A3.

Remark. More generally, the following useful fact is true [Fu, p. 83]:

> Let $u_1, \ldots, u_r$ be points in $\mathbb{R}^n$, not contained in any affine hyperplane, and let K be their convex hull. Let $\varepsilon_1, \ldots, \varepsilon_r$ be any positive real numbers, and define $H : \mathbb{R}^n \to \mathbb{R}^n$ by
>
> $$H(x) = \frac{1}{f(x)} \sum_{k=1}^{r} \varepsilon_k e^{(u_k, x)} u_k,$$
>
> where $f(x) = \varepsilon_1 e^{(u_1,x)} + \cdots + \varepsilon_r e^{(u_r,x)}$, and $(\cdot, \cdot)$ is the usual inner product on $\mathbb{R}^n$. Then H defines a real analytic isomorphism of $\mathbb{R}^n$ onto the interior of K.

(We are grateful to Bernd Sturmfels for pointing this out.) In algebraic geometry, this can be used to show that "the moment map is surjective" for toric varieties.

The Fifty-Eighth William Lowell Putnam Mathematical Competition
December 6, 1997

A1. (138, 3, 8, 0, 0, 0, 0, 0, 1, 2, 21, 32)

A rectangle, $HOMF$, has sides $HO = 11$ and $OM = 5$. A triangle ABC has H as the intersection of the altitudes, O the center of the circumscribed circle, M the midpoint of BC, and F the foot of the altitude from A. What is the length of BC?

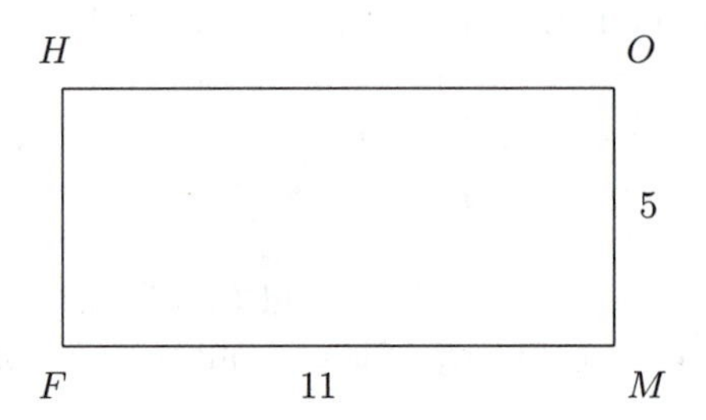

Answer. The length of BC is 28.

Solution 1 (Dave Rusin). (See Figure 32.) Introduce coordinates for which

$$O = (0,0), \quad H = (-11,0), \quad F = (-11,-5), \quad \text{and} \quad M = (0,-5).$$

Since B and C are both equidistant from M and O, the perpendicular bisector of BC is the y-axis. Since BC contains F, we have $B = (-x,-5)$ and $C = (x,-5)$ for some x. Also, $A = (-11, y)$ for some y.

The altitude through B passes through H, so its slope is $5/(x-11)$. It is perpendicular to the line AC, whose slope is $-(y+5)/(x+11)$, so these two slopes have product -1. That is, $5(y+5) = (x-11)(x+11)$.

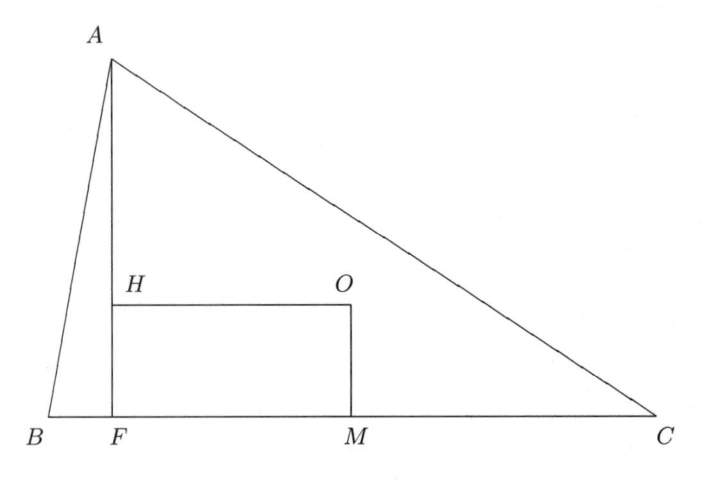

FIGURE 32.

On the other hand, A and B are also equidistant from O, so $y^2 + 11^2 = x^2 + 5^2$. Comparing this equality with the previous one, we find $y^2 - 5y - 50 = 0$, so $y = 10$ or $y = -5$. The latter is impossible, so $y = 10$ and $x = 14$, yielding $BC = 2x = 28$. ■

Solution 2. The centroid G of the triangle is collinear with H and O (Euler line), and G lies two-thirds of the way from A to M. Therefore H is two-thirds of the way from A to F, so $AF = 3OM = 15$. Triangles BFH and AFC are similar, since they are right triangles and

$$\angle HBF = \angle HBC = \pi/2 - \angle C = \angle CAF.$$

Hence $BF/FH = AF/FC$, and $BF \cdot FC = FH \cdot AF = 75$. Now $BC^2 = (BF+FC)^2 = (FC - BF)^2 + 4BF \cdot FC$, but

$$FC - BF = (FM + MC) - (BM - FM) = 2FM = 2HO = 22,$$

so

$$BC = \sqrt{22^2 + 4 \cdot 75} = \sqrt{784} = 28.$$ ■

Remark. For more on the Euler line and related topics, see [CG, Section 1.7].

Solution 3. Introduce a coordinate system with origin at O, and for a point X, let $\vec{X}$ denote the vector from O to X. Then $\vec{H} = \vec{A} + \vec{B} + \vec{C}$ (see remark below) and $\vec{M} = (\vec{B} + \vec{C})/2 = (\vec{H} - \vec{A})/2$; we now compute $AH = 2OM = 10$, $OC = OA = \sqrt{AH^2 + OH^2} = \sqrt{221}$, and

$$BC = 2MC = 2\sqrt{OC^2 - OM^2} = 2\sqrt{221 - 25} = 28.$$ ■

Remark. The vector equality $\vec{H} = \vec{A} + \vec{B} + \vec{C}$ holds for any triangle with circumcenter O at the origin, and can be proved either using the properties of the Euler line mentioned in the second solution, or directly as follows. Let $\vec{H}' = \vec{A}+\vec{B}+\vec{C}$; then

$$(\vec{H}' - \vec{A}) \cdot (\vec{B} - \vec{C}) = (\vec{B} + \vec{C}) \cdot (\vec{B} - \vec{C}) = \vec{B} \cdot \vec{B} - \vec{C} \cdot \vec{C} = OB^2 - OC^2 = 0.$$

Thus $H'A$ is perpendicular to BC, i.e., H' lies on the altitude through A. Similarly, H' lies on the other two altitudes, so $H' = H$.

A2. (42, 42, 44, 0, 0, 0, 0, 0, 18, 7, 19, 33)

Players $1, 2, 3, \ldots, n$ are seated around a table, and each has a single penny. Player 1 passes a penny to Player 2, who then passes two pennies to Player 3. Player 3 then passes one penny to Player 4, who passes two pennies to Player 5, and so on, players alternately passing one penny or two to the next player who still has some pennies. A player who runs out of pennies drops out of the game and leaves the table. Find an infinite set of numbers n for which some player ends up with all n pennies.

Solution. We show that the game terminates with one player holding all of the pennies if and only if $n = 2^m + 1$ or $n = 2^m + 2$ for some m; this assertion is clear for $n = 1$ and $n = 2$, so we assume $n \geq 3$. We begin by determining the state of the game after each player has moved once. The first player passes one penny and drops out, and the second player passes two pennies and drops out. Then the third player passes one penny and keeps two, the fourth player passes two pennies and drops out, the fifth player passes one penny and keeps two, and so on. The net result is that $\lfloor \frac{n-1}{2} \rfloor$ players remain, each of whom has two pennies except for the player to move next, who has 3 or 4 pennies.

Trying some small examples suggests the following induction argument. Suppose that for some $k \geq 2$, the game reaches a point where:

- Except for the player to move, each player has k pennies;
- The player to move has at least k pennies, and it is his turn to pass one penny.

We will show by induction that the game terminates if and only if the number of players remaining is a power of 2. If the number of players is odd, then two complete rounds leaves the situation unchanged (here we need $k \geq 2$), so the game terminates only if there is one player left to begin with. If the number of players is even, then after k complete rounds, the player who made the first move and every second player thereafter will have gained k pennies, and the other players will all have lost their k pennies. Thus we are in a situation of the same form with half as many players, and by induction the latter terminates with one player if and only if the number of players is a power of 2.

Returning to the original game, we see that if the player to move is to pass one penny, we are in the desired situation. If the player to move is to pass two pennies, after that move we end up in the desired situation. Thus the game terminates if and only if $\lfloor \frac{n-1}{2} \rfloor$ is a power of 2, that is, if and only if $n = 2^m + 1$ or $n = 2^m + 2$ for some m. ■

A3. (5, 28, 4, 0, 0, 0, 0, 0, 3, 36, 29, 100)

Evaluate

$$\int_0^\infty \left(x - \frac{x^3}{2} + \frac{x^5}{2 \cdot 4} - \frac{x^7}{2 \cdot 4 \cdot 6} + \cdots \right) \left(1 + \frac{x^2}{2^2} + \frac{x^4}{2^2 \cdot 4^2} + \frac{x^6}{2^2 \cdot 4^2 \cdot 6^2} + \cdots \right) dx.$$

Answer. The value of the integral is $\sqrt{e}$.

Solution 1. The series on the left is the Taylor series of $xe^{-x^2/2}$. Since the terms of the second sum are nonnegative, we may interchange the sum and integral by the Monotone Convergence Theorem [Ru, p. 319], so the expression becomes

$$\sum_{n=0}^{\infty} \int_0^\infty xe^{-x^2/2} \frac{x^{2n}}{2^{2n}(n!)^2}\, dx.$$

To evaluate these integrals, first note that

$$\int_0^\infty xe^{-x^2/2}\, dx = -e^{-x^2/2} \Big|_0^\infty = 1.$$

By integration by parts,

$$\int_0^\infty x^{2n} \left(xe^{-x^2/2} \right) dx = -x^{2n} e^{-x^2/2} \Big|_0^\infty + \int_0^\infty 2nx^{2n-1} e^{-x^2/2}\, dx$$

since both integrals converge absolutely, and (for $n \geq 1$) the first expression evaluates to 0 at both endpoints. Thus by induction,

$$\int_0^\infty x^{2n+1} e^{-x^2/2} dx = 2 \times 4 \times \cdots \times 2n.$$

Consequently, the desired integral is

$$\sum_{n=0}^\infty \frac{1}{2^n n!} = \sqrt{e}. \qquad \blacksquare$$

Remark. One can also first make the substitution $u = x^2/2$, in which case the integral becomes

$$\sum_{n=0}^\infty \frac{1}{2^n (n!)^2} \int_0^\infty u^n e^{-u}\, du.$$

The reader may recognize $\int_0^\infty u^n e^{-u}\, du$ as the integral defining the gamma function $\Gamma(n+1) = n!$.

Solution 2. Let $J_0(x)$ denote the function

$$J_0(x) = \sum_{n=0}^\infty (-1)^n \frac{x^{2n}}{2^{2n}(n!)^2};$$

the series converges absolutely for all $x \in \mathbb{C}$, by the Ratio Test. Applying the operator $x\frac{d}{dx}$ twice to $J_0(x)$ yields $-x^2 J_0(x)$, so $J_0(x)$ satisfies the differential equation

$$x^2 y'' + xy' + x^2 y = 0. \tag{1}$$

To justify certain calculations in the rest of the proof, we need some bounds on J_0 and its derivatives. For any $t \in \mathbb{C}$ and $\beta > 0$, we have $J_0(tx)/e^{\beta x^2} \to 0$ as $x \to +\infty$: this holds because for any $\epsilon > 0$, all but finitely many terms in the expansion of $J_0(tx)$ are bounded in absolute by ϵ times the corresponding term in $e^{\beta x^2}$; those finitely many terms also are negligible compared to $e^{\beta x^2}$ as $x \to \infty$. A similar argument shows that $J_0'(tx)/e^{\beta x^2} \to 0$ and $J_0''(tx)/e^{\beta x^2} \to 0$ as $x \to +\infty$, uniformly for t in any bounded subset of $\mathbb{C}$.

For $t \in \mathbb{C}$, define

$$F(t) = \int_0^\infty x e^{-x^2/2} J_0(tx)\, dx. \tag{2}$$

The previous paragraph shows that the integral converges and justifies the claim that integration by parts yields

$$\int_0^\infty e^{-x^2/2} J_0'(tx)\, dx = t^{-1}(F(t) - 1) \tag{3}$$

for $t \neq 0$. The rest of this paragraph justifies the claim that differentiating (3) with respect to t yields

$$\int_0^\infty x e^{-x^2/2} J_0''(tx)\, dx = -t^{-2}(F(t) - 1) + t^{-1}F'(t). \tag{4}$$

The nontrivial statement here is that if $g(x,t)$ is the integrand $e^{-x^2/2}J_0'(tx)$ on the left of (3), then

$$\frac{d}{dt}\int_0^\infty g(x,t)\,dx = \int_0^\infty \frac{\partial g}{\partial t}\,dx.$$

By definition of the derivative, the left side at $t = t_0$ equals

$$\lim_{t\to t_0} \frac{\int_0^\infty g(x,t)\,dx - \int_0^\infty g(x,t_0)\,dx}{t - t_0} = \lim_{t\to t_0} \int_0^\infty \frac{g(x,t) - g(x,t_0)}{t - t_0}\,dx, \tag{5}$$

so to obtain the right side, what we need is to interchange the limit and the integral on the right of (5). The interchange is justified by the Dominated Convergence Theorem [Ru, p. 321], provided that there exists $G(x)$ with $\int_0^\infty G(x)\,dx < \infty$ such that

$$\left|\frac{g(x,t) - g(x,t_0)}{t - t_0}\right| \le G(x)$$

for all t in a punctured neighborhood of t_0 and for all $x \ge 0$. For fixed x, the difference quotient $\frac{g(x,t)-g(x,t_0)}{t-t_0}$ is the average value of $\partial g/\partial t$ over the interval $[t_0, t]$, so we need only prove $\partial g/\partial t \le G(x)$. The limits in the previous paragraph show

$$\frac{\partial g}{\partial t} = x^2 e^{-x^2/2} J_0''(tx) \le C(t) e^{-x^2/4},$$

where $C(t)$ is uniformly bounded in a neighborhood of t_0. Hence we may take $G(x) = Ce^{-x^2}/4$ for some $C > 0$ depending on t_0 to complete the justification.

By (1), (2), (3), and (4),

$$\begin{aligned} -t^{-2}(F(t)-1) + t^{-1}F'(t) &= -t^{-1}\int_0^\infty e^{-x^2/2}J_0'(tx)\,dx - \int_0^\infty xe^{-x^2/2}J_0(tx)\,dx \\ &= -t^{-2}(F(t)-1) - F(t). \end{aligned}$$

Therefore $F'(t) = -tF(t)$. Separating variables and integrating, we get $F(t) = Ce^{-t^2/2}$. Here

$$C = F(0) = \int_0^\infty xe^{-x^2/2}J_0(0)\,dx = \int_0^\infty xe^{-x^2/2}\,dx = 1,$$

so $F(t) = e^{-t^2/2}$. The integral of the problem is

$$\int_0^\infty xe^{-x^2/2}\left(1 + \frac{x^2}{2^2} + \frac{x^4}{2^2\cdot 4^2} + \cdots\right)dx = F(i) = \sqrt{e}. \qquad \blacksquare$$

Remark. The above solution can be reinterpreted in terms of the Laplace transform, which takes $f(x)$ to $(Lf)(s) = \int_0^\infty e^{-sx}f(x)\,dx$. See Chapter 6 of [BD] for further applications of this technique.

Remark. The function $J_0(x)$ is an example of a Bessel function. See [O, Section 2.9] for the definitions and properties of these commonly occurring functions.

Remark. The trick of differentiating an integral with respect to an auxiliary parameter was a favorite of 1939 Putnam Fellow (and Nobel Laureate in physics) Richard Feynman, according to [Fe, p. 72]. A systematic treatment of the method can be found in [AZ].

A4. (100, 21, 1, 0, 0, 0, 0, 0, 3, 8, 24, 48)

Let G be a group with identity e and $\phi : G \to G$ a function such that

$$\phi(g_1)\phi(g_2)\phi(g_3) = \phi(h_1)\phi(h_2)\phi(h_3)$$

whenever $g_1g_2g_3 = e = h_1h_2h_3$. Prove that there exists an element $a \in G$ such that $\psi(x) = a\phi(x)$ is a homomorphism (that is, $\psi(xy) = \psi(x)\psi(y)$ for all $x, y \in G$).

Solution. Homomorphisms map e to e, so we take $a = \phi(e)^{-1}$ and define $\psi(x) = a\phi(x)$ in order to have $\psi(e) = e$. The hypothesis on ϕ implies

$$\phi(g)\phi(e)\phi(g^{-1}) = \phi(e)\phi(g)\phi(g^{-1}),$$

and cancelling $\phi(g^{-1})$ shows that $\phi(g)$ commutes with $\phi(e)$ for all g. The hypothesis also implies

$$\phi(x)\phi(y)\phi(y^{-1}x^{-1}) = \phi(e)\phi(xy)\phi(y^{-1}x^{-1}).$$

Since $\phi(e)$ commutes with anything in the image of ϕ, $\phi(e)^{-1}$ does too, so we deduce

$$\phi(e)^{-1}\phi(x)\phi(e)^{-1}\phi(y) = \phi(e)^{-1}\phi(xy)$$

or equivalently $\psi(xy) = \psi(x)\psi(y)$, as desired. ■

Remark. It is not necessarily true that $\phi(e)$ commutes with all elements of the group. For example, ϕ could be a constant map whose image is an element not in the center.

A5. (22, 10, 15, 0, 0, 0, 0, 0, 24, 11, 37, 86)

Let N_n denote the number of ordered n-tuples of positive integers $(a_1, a_2, \ldots, a_n)$ such that $1/a_1 + 1/a_2 + \cdots + 1/a_n = 1$. Determine whether N_{10} is even or odd.

Answer. The number N_{10} is odd.

Solution 1. Since we are only looking for the parity of the number of solutions, we may discard solutions in pairs. For example, any solution with $a_1 \neq a_2$ may be paired with the solution obtained from this one by interchanging a_1 and a_2. The solutions left unpaired are those with $a_1 = a_2$, so the parity of N_{10} equals the parity of the number of solutions with $a_1 = a_2$.

Similarly, we may restrict attention to solutions with $a_3 = a_4$, $a_5 = a_6$, $a_7 = a_8$, $a_9 = a_{10}$. Such solutions must satisfy the equation

$$2/a_1 + 2/a_3 + 2/a_5 + 2/a_7 + 2/a_9 = 1.$$

As above, we may restrict attention to solutions with $a_1 = a_3$ and $a_5 = a_7$, so we get $4/a_1 + 4/a_5 + 2/a_9 = 1$. We may restrict once more to solutions with $a_1 = a_5$, which satisfy $8/a_1 + 2/a_9 = 1$. This is equivalent to $(a_1 - 8)(a_9 - 2) = 16$, which has 5 solutions, corresponding to the factorizations of 16 as $2^i \times 2^{4-i}$ for $i = 0, \ldots, 4$. Thus N_{10} is odd. ■

Solution 2. Suppose we start with a solution in which the smallest a_i occurs p_1 times, the next smallest occurs p_2 times, and so on. Then the number of ordered 10-tuples which are rearrangements of this solution is given by the multinomial coefficient

$$\binom{10}{p_1, p_2, \ldots} = \frac{10!}{p_1!\,p_2!\,\cdots}.$$

The exponent of 2 in the prime factorization of this number is given by Kummer's Theorem.

Kummer's Theorem. *The exponent of the prime p in the factorization of the multinomial coefficient $\binom{a_1+\cdots+a_n}{a_1,\ldots,a_n}$ equals the number of carries that occur when $a_1, \ldots, a_n$ are added in base p.*

Proof. The exponent of p in the prime factorization of $n!$ can be written as

$$\sum_{i=1}^{\infty} \left\lfloor \frac{n}{p^i} \right\rfloor.$$

(This sum counts each multiple of p up to n once, each multiple of p^2 a second time, and so on.) Hence the exponent of p in $\binom{a_1+\cdots+a_n}{a_1,\ldots,a_n}$ is

$$\sum_{i=1}^{\infty} \left(\left\lfloor \frac{a_1+\cdots+a_n}{p^i} \right\rfloor - \left\lfloor \frac{a_1}{p^i} \right\rfloor - \cdots - \left\lfloor \frac{a_n}{p^i} \right\rfloor \right),$$

and the summand is precisely the number of carries into the p^i column when $a_1, \ldots, a_n$ are added in base p. □

In particular, $\binom{10}{p_1,p_2,\ldots}$ is even unless $p_1 = 10$ or $\{p_1, p_2\} = \{2, 8\}$. Up to rearrangement, there is one solution of the former shape, namely $(10, \ldots, 10)$, and four of the latter shape. Hence N_{10} is odd. ■

Reinterpretation. The symmetric group S_{10} on ten symbols acts on the set of ordered tuples. The number of elements in the orbit of a given tuple is $10!/|G|$, where $|G|$ is the order of the stabilizer G of the tuple. This orbit size is odd if and only if G contains a 2-Sylow subgroup. All 2-Sylow subgroups are conjugate, and one 2-Sylow subgroup contains the product of a disjoint 8-cycle and 2-cycle, so all 2-Sylow subgroups must contain such a product. But G is a product of symmetric groups, so G contains a 2-Sylow subgroup if and only if G contains a subgroup isomorphic to $S_8 \times S_2$. As seen above there are five solutions stabilized by $S_8 \times S_2$, up to rearrangement.

A6. (0, 1, 1, 0, 0, 0, 0, 0, 8, 8, 35, 152)

For a positive integer n and any real number c, define x_k recursively by $x_0 = 0$, $x_1 = 1$, and for $k \geq 0$,

$$x_{k+2} = \frac{cx_{k+1} - (n-k)x_k}{k+1}.$$

Fix n and then take c to be the largest value for which $x_{n+1} = 0$. Find x_k in terms of n and k, $1 \leq k \leq n$.

Answer. For $1 \leq k \leq n$, we have $x_k = \binom{n-1}{k-1}$.

Solution 1. Introduce the generating function

$$p(t) = \sum_{k \geq 0} x_{k+1} t^k.$$

The condition $x_{n+1} = 0$ forces $x_m = 0$ for all $m \geq n+1$, so $p(t)$ is a polynomial of degree at most $n-1$.

Now $p'(t) = \sum_{k \geq 0} (k+1) x_{k+2} t^k$, so the recursion implies that $p(t)$ satisfies the differential equation

$$p'(t) = cp(t) - (n-1)tp(t) + t^2 p'(t). \tag{1}$$

Since $p(0) = x_1 = 1$, $p(t)$ is not identically zero. Rearranging (1), we get

$$(\ln p(t))' = \frac{p'(t)}{p(t)} = \frac{c-(n-1)t}{1-t^2} = \frac{A}{1+t} + \frac{B}{1-t} \tag{2}$$

for some A and B depending on c and n. This implies that the only linear factors of the polynomial $p(t)$ over $\mathbb{C}$ are $1-t$ and $1+t$. (See the discussion of logarithmic differentiation in 1991A3.) Moreover, $p(0) = 1$, so

$$p(t) = (1+t)^r (1-t)^s$$

for some integers $r, s \geq 0$. Then (2) implies $A = r$, $B = -s$, and $c - (n-1)t = r(1-t) - s(1+t)$ as polynomials in t, so $r = n-1-s$ and $c = r - s = n-1-2s$. Since $s \geq 0$, $c \leq n-1$.

On the other hand, if we take $c = n-1$, then reversing our steps shows that the sequence x_k defined by $\sum_{k \geq 0} x_{k+1} t^k = p(t) = (1+t)^{n-1}$ satisfies the original recursion. Therefore the theoretical maximum $c = n-1$ is attained, and in that case $x_k = \binom{n-1}{k-1}$. ■

Solution 2. For fixed n, induction on k shows that x_k is a polynomial in c of degree exactly $k-1$, so there are at most n values of c making $x_{n+1} = 0$. We will prove by induction on n that the possible values of c are

$$-(n-1),\ -(n-3),\ \dots,\ n-5,\ n-3,\ n-1$$

and that if $c = n-1$ the sequence is $x_k = \binom{n-1}{k-1}$ for $k \geq 1$.

If $n = 1$, then the recursion gives $x_2 = c$. Thus the only possible value of c is 0. In this case, the recursion says $x_{k+2} = \left(\frac{k-1}{k+1}\right) x_k$ for $k \geq 0$. Starting from $x_0 = 0$ and $x_1 = 1$, this implies $x_k = 0$ for all $k \geq 2$. In other words, $x_k = \binom{0}{k-1}$ for all $k \geq 1$, as claimed.

Now suppose that $n \geq 1$, and that the claim has been verified for n. Suppose that c is one of

$$-(n-1),\ -(n-3),\ \dots,\ n-5,\ n-3,\ n-1$$

and that $\{y_k\}$ satisfies the hypotheses of the problem for n and c. Thus $y_{n+1} = 0$ and

$$(k+1)y_{k+2} + (n-k)y_k = cy_{k+1} \qquad \text{for } k \geq 0. \tag{3}$$

The recursion holds for $k=-1$ as well, if we define $y_{-1}=0$. Let $x_k = y_k + y_{k-1}$ for $k \geq 0$. Then for $k \geq 0$,

$$\begin{aligned}(k+1)x_{k+2}+(n+1-k)x_k &= (k+1)(y_{k+2}+y_{k+1})+(n+1-k)(y_k+y_{k-1}) \\ &= ((k+1)y_{k+2}+(n-k)y_k) \\ &\qquad + (ky_{k+1}+(n-k+1)y_{k-1})+y_k+y_{k+1} \\ &= cy_{k+1}+cy_k+y_k+y_{k+1} \quad \text{(by the } y \text{ recursion (3))} \\ &= (c+1)(y_{k+1}+y_k) \\ &= (c+1)x_{k+1},\end{aligned}$$

so $\{x_k\}$ satisfies the recursion for $n+1$ and $c+1$. Taking $k=n$ in the y recursion (3) shows that $y_{n+1}=0$ implies $y_{n+2}=0$, so $x_{n+2}=y_{n+2}+y_{n+1}=0$. Thus $\{x_k\}$ is a solution sequence for $n+1$ and $c+1$. Hence adding 1 to each value of c for n gives a possible value of c for $n+1$, so the inductive hypothesis implies that

$$-(n-2),\ -(n-4),\ \dots,\ n-4,\ n-2,\ n$$

are possible values of c for $n+1$. If $\{x_k\}$ is the sequence for $c=n$, then $\{(-1)^{k-1}x_k\}$ is a solution sequence for $c=-n$, so

$$-n,\ -(n-2),\ -(n-4),\ \dots,\ n-4,\ n-2,\ n$$

are $n+1$ possible values of c for $n+1$. By the first sentence of this solution, there cannot be any others.

Finally, the inductive hypothesis implies that $y_k=\binom{n-1}{k-1}$, $k \geq 0$, is a solution for n and $c=n-1$ (we interpret $\binom{n-1}{-1}$ as 0), so

$$x_k = y_k + y_{k-1} = \binom{n-1}{k-1} + \binom{n-1}{k-2} = \binom{n}{k-1} \quad \text{for } k \geq 1,$$

gives the solution for $n+1$ and $c=n$. This completes the inductive step. ∎

Remark. Both of the previous solutions can be used to show that the sequences for which $x_{n+1}=0$ are the sequences of coefficients of $t(1+t)^j(1-t)^{n-1-j}$ for some $j \in \{0,1,\dots,n-1\}$.

Solution 3 (Byron Walden). Rewrite the recursion as

$$(k+1)x_{k+2}-(k-1)x_k = cx_{k+1}-(n-1)x_k.$$

If $m \geq 0$, then summing from $k=0$ to m yields

$$(m+1)x_{m+2}+mx_{m+1} = cx_{m+1}+(c-(n-1))(x_1+x_2+\cdots+x_m),$$

since $x_0=0$. Thus, for $m \geq 0$,

$$x_{m+2} = \frac{c-m}{m+1}x_{m+1} + \frac{c-(n-1)}{m+1}(x_1+x_2+\cdots+x_m). \tag{4}$$

If $c > n-1$, then applying (4) for $m=0,\ 1,\ \dots,\ n-1$ in turn shows that x_2, x_3, $\dots$, x_{n+1} are all positive, since at each step, all terms on the right-hand side of (4) are positive. Hence the assumption $x_{n+1}=0$ forces $c \leq n-1$.

Suppose we take $c = n - 1$. Then (4) becomes $x_{m+2} = \frac{n-m-1}{m+1} x_{m+1}$ for $m \geq 0$. In particular, if $x_{m+1} = \binom{n-1}{m}$, then

$$x_{m+2} = \frac{n-m-1}{m+1} \frac{(n-1)!}{m!\,(n-m-1)!} = \frac{(n-1)!}{(m+1)!\,(n-m-2)!} = \binom{n-1}{m+1}.$$

Hence by induction, starting from $x_1 = \binom{n-1}{0}$, we can prove that $x_k = \binom{n-1}{k-1}$ for all $k \geq 1$. In particular, $x_{n+1} = \binom{n-1}{n} = 0$. Thus $n-1$ is a possible value for c, and by the previous paragraph it is the largest possible value. ■

Solution 4 (Greg Kuperberg). The condition $x_{n+1} = 0$ states that $(x_1, \ldots, x_n)$ is an eigenvector of the $n \times n$ matrix

$$A_{ij} = \begin{cases} i & \text{if } j = i+1 \\ n-j & \text{if } j = i-1 \\ 0 & \text{otherwise,} \end{cases}$$

with eigenvalue c. In particular, A has nonnegative entries. By the Perron-Frobenius Theorem (see remark below), A has an eigenvector with positive entries, unique up to a positive scalar multiple, and the corresponding eigenvalue has multiplicity one and has absolute value greater than or equal to that of any other eigenvalue. Using $\binom{n-1}{k+1} = \frac{n-k-1}{k+1}\binom{n-1}{k}$ and $\binom{n-1}{k-1} = \frac{k}{n-k}\binom{n-1}{k}$ (which follow from the definition of binomial coefficients in terms of factorials), one shows that the vector with $x_k = \binom{n-1}{k-1}$ is an eigenvector with eigenvalue $n-1$, and it has positive entries, so $c = n-1$ is the desired maximum. ■

Remark (the Perron-Frobenius Theorem). In its simplest form, this theorem states that any matrix with positive entries has a unique eigenvector with positive entries, and that the corresponding eigenvalue has multiplicity one and has absolute value strictly greater than that of any other eigenvalue. In the study of random walks, the theorem implies that any random walk on finitely many states, in which there is a positive transition probability from any state to any other state, has a unique steady state. One proof of the existence of the eigenvector, due to G. Birkhoff [Bi], is that a matrix with positive entries acts on the set of lines passing through the first orthant as a contraction mapping under the Hilbert metric:

$$d((v_1, \ldots, v_n), (w_1, \ldots, w_n)) = \ln\left(\max_i \{v_i/w_i\} / \min_i \{v_i/w_i\}\right).$$

In many applications to random walks, a more refined version of the Perron-Frobenius Theorem is needed. Namely, let A be a matrix with *nonnegative* entries, and assume that for some $k > 0$, A^k has strictly positive entries. The refined theorem states that A then has a unique eigenvector with positive entries, whose eigenvalue has multiplicity one and has absolute value strictly greater than that of any other eigenvalue. This assertion can be deduced from the simpler form of the theorem as follows. If v is a positive eigenvector of A, then it is also a positive eigenvector of A^k, to which we may apply the simpler form; thus the positive eigenvector of A, if it exists, is unique. To show existence, let w be a positive eigenvector of A^k with eigenvalue $\lambda > 0$; then $w + \lambda^{-1}Aw + \cdots + \lambda^{-k+1}A^{k-1}w$ is a positive eigenvector of A with eigenvalue λ. Finally, if λ is the eigenvalue of the positive eigenvector of A, and ρ is any other eigenvalue, then $|\lambda^k| > |\rho^k|$ by the simpler form of the theorem, so $|\lambda| > |\rho|$.

There is also a formulation of the theorem that applies directly to matrices like the one occurring in Solution 4, in which A^n may not have positive entries for any n. Let A be a matrix with nonnegative entries. Let G be the directed graph with vertices $\{1, \ldots, n\}$ and with an edge from i to j if and only if $A_{ij} > 0$. Call A *irreducible* if G is strongly connected (see 1990B4 for the definition). In this case, A has a unique eigenvector with positive entries, whose eigenvalue has multiplicity one and has absolute value greater than *or equal to* that of any other eigenvalue. This can be deduced by applying the version in the previous paragraph to $A + \epsilon I$ for each $\epsilon > 0$.

In fact, the matrix in Solution 4 shows that "greater than or equal to" cannot be replaced by "strictly greater than" in this final version: the matrix A there has $1-n$ as an eigenvalue in addition to the eigenvalue $n-1$ of the positive eigenvector.

Related question. The following is an example of a problem that can be solved by an appropriate application of the Perron-Frobenius Theorem. (We are grateful to Mira Bernstein for passing it on to us.) It can also be solved by other means.

> The Seven Dwarfs are sitting around the breakfast table; Snow White has just poured them some milk. Before they drink, they perform a little ritual. First, Dwarf #1 distributes all the milk in his mug equally among his brothers' mugs (leaving none for himself). Then Dwarf #2 does the same, then Dwarf #3, #4, etc., finishing with Dwarf # 7. At the end of the process, the amount of milk in each dwarf's mug is the same as at the beginning! If the total amount of milk is 42 ounces, how much milk did each of them originally have?

See 1995A5 for another application of the Perron-Frobenius Theorem, and see 1993B4 for an application of a continuous analogue of the theorem. For more on matrices with nonnegative entries, including the Perron-Frobenius Theorem and related results, see [HJ, Chapter 8].

B1. (171, 6, 8, 0, 0, 0, 0, 0, 1, 8, 10, 1)

Let $\{x\}$ denote the distance between the real number x and the nearest integer. For each positive integer n, evaluate

$$S_n = \sum_{m=1}^{6n-1} \min\left(\left\{\frac{m}{6n}\right\}, \left\{\frac{m}{3n}\right\}\right).$$

(Here $\min(a, b)$ denotes the minimum of a and b.)

Answer. We have $S_n = n$.

Solution 1. It is trivial to check that $\frac{m}{6n} = \{\frac{m}{6n}\} \leq \{\frac{m}{3n}\}$ for $1 \leq m \leq 2n$, that $1 - \frac{m}{3n} = \{\frac{m}{3n}\} \leq \{\frac{m}{6n}\}$ for $2n \leq m \leq 3n$, that $\frac{m}{3n} - 1 = \{\frac{m}{3n}\} \leq \{\frac{m}{6n}\}$ for $3n \leq m \leq 4n$, and that $2 - \frac{m}{6n} = \{\frac{m}{6n}\} \leq \{\frac{m}{3n}\}$ for $4n \leq m \leq 6n$. (See Figure 33 for a graph.) Therefore

$$S_n = \sum_{m=1}^{2n-1} \frac{m}{6n} + \sum_{m=2n}^{3n-1} \left(1 - \frac{m}{3n}\right) + \sum_{m=3n}^{4n-1} \left(\frac{m}{3n} - 1\right) + \sum_{m=4n}^{6n-1} \left(2 - \frac{m}{6n}\right).$$

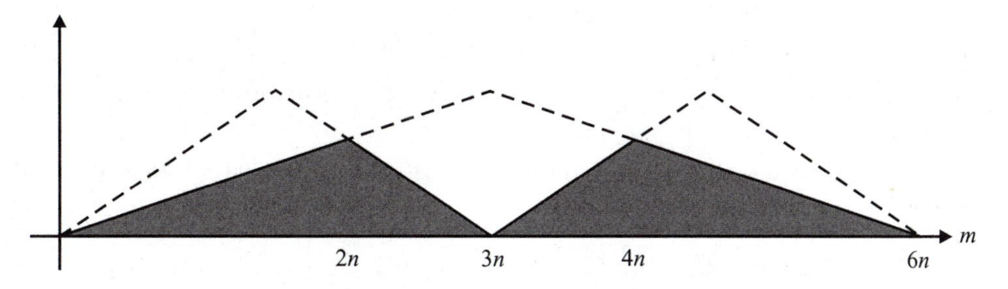

FIGURE 33.
Graph of $y = \min\left(\left\{\frac{m}{6n}\right\}, \left\{\frac{m}{3n}\right\}\right)$.

Each of the four sums is an arithmetic progression, which can be evaluated as the number of terms times the average of the first and last terms. This yields

$$S_n = (2n-1)\left(\frac{2n}{12n}\right) + n\left(\frac{n+1}{6n}\right) + n\left(\frac{n-1}{6n}\right) + 2n\left(\frac{2n+1}{12n}\right) = n. \qquad \blacksquare$$

Remark. Solution 1 can be simplified by exploiting symmetry: since $\{1-x\} = \{x\}$ and $\{2-x\} = \{x\}$, the mth term in the series S_n equals the $(6n-m)$th term.

Solution 2. The series S_n is the approximation given by the Trapezoid Rule to the area A under the graph of the function

$$f(x) = \min\left(\left\{\frac{x}{6n}\right\}, \left\{\frac{x}{3n}\right\}\right)$$

for $0 \le x \le 6n$, using the sampling points $0, 1, \dots, 6n$. But $f(x)$ is piecewise linear and the break points $x = 2n, 3n, 4n$ are also sample points, so $S_n = A$. The area A consists of the triangle with vertices $(0,0)$, $(3n,0)$, $(2n,1/3)$, and the congruent triangle with vertices $(6n,0)$, $(3n,0)$, $(4n,1/3)$. Thus $A = n/2 + n/2 = n$, and $S_n = n$. $\blacksquare$

B2. (28, 1, 1, 0, 0, 0, 0, 0, 1, 3, 38, 133)

Let f be a twice-differentiable real-valued function satisfying

$$f(x) + f''(x) = -xg(x)f'(x),$$

where $g(x) \ge 0$ for all real x. Prove that $|f(x)|$ is bounded.

Solution. Multiplying both sides of the given differential equation by $2f'(x)$, we have

$$2f(x)f'(x) + 2f'(x)f''(x) = -2xg(x)f'(x)^2.$$

The left side of the equation is the derivative of $f(x)^2 + f'(x)^2$, whereas the right side is nonnegative for $x < 0$ and nonpositive for $x > 0$. Thus $f(x)^2 + f'(x)^2$ increases to its maximum value at $x = 0$ and decreases thereafter. In particular, it is bounded, so $f(x)$ and $f'(x)$ are bounded. $\blacksquare$

Reinterpretation. This problem has a physical interpretation: $f(x)$ is the amplitude of an oscillator with time-dependent damping $xg(x)$. Since the damping is negative for $x < 0$ and positive for $x > 0$, the oscillator gains energy before time 0 and loses energy thereafter.

B3. (1, 4, 7, 0, 0, 0, 0, 0, 22, 20, 89, 62)

For each positive integer n, write the sum $\sum_{m=1}^{n} \frac{1}{m}$ in the form $\frac{p_n}{q_n}$, where p_n and q_n are relatively prime positive integers. Determine all n such that 5 does not divide q_n.

Answer. The only such n are the 19 integers in the ranges 1–4, 20–24, 100–104, and 120–124; i.e., the set of such n is

$$\{1, 2, 3, 4, 20, 21, 22, 23, 24, 100, 101, 102, 103, 104, 120, 121, 122, 123, 124\}.$$

Solution. To simplify the discussion, we introduce some terminology. For s a positive integer, we say the rational number a/b, in lowest terms, is *s-integral* if b is coprime to s, and *divisible by s* if a is divisible by s.

Let $H_n = \sum_{m=1}^{n} m^{-1}$. (By convention, the empty sum H_0 is 0.) Let S_n be the sum of m^{-1} over those $m \in \{1, \ldots, n\}$ not divisible by 5. Then

$$H_n = S_n + \frac{1}{5} H_{\lfloor n/5 \rfloor}. \tag{1}$$

By design, S_n is always 5-integral, so H_n is 5-integral if and only if $H_{\lfloor n/5 \rfloor}$ is divisible by 5.

In particular, if H_n is 5-integral, then $H_{\lfloor n/5 \rfloor}$ must have been 5-integral. Thus the nonnegative integers n such that H_n is 5-integral are precisely the integers in the set $\mathcal{S}$ after carrying out the following algorithm.

(a) Start with the set $\mathcal{S} = \{0\}$ and the integer $n = 0$.

(b) If H_n is divisible by 5, add $5n, 5n+1, 5n+2, 5n+3, 5n+4$ to $\mathcal{S}$.

(c) If n is the largest element of $\mathcal{S}$, stop; otherwise, increase n to the smallest element of $\mathcal{S}$ greater than the current n, and go to (b).

To carry out the algorithm, we first take $n = 0$ and add $1, 2, 3, 4$ to $\mathcal{S}$. Now $H_1 = 1, H_2 = 3/2, H_3 = 11/6$ are not divisible by 5, but $H_4 = 25/12$ is. Thus we add $20, 21, 22, 23, 24$ to $\mathcal{S}$ and keep going.

We next must determine which of $H_{20}, \ldots, H_{24}$ are divisible by 5. From (1), we have $H_{20+i} = S_{20+i} + \frac{1}{5} H_4$ for $i = 0, \ldots, 4$. Before proceeding further, we verify that

$$\frac{1}{5k+1} + \frac{1}{5k+2} + \cdots + \frac{1}{5k+j} \tag{2}$$

is not divisible by 5 if $j = 1, 2, 3$ but is divisible by 25 if $j = 4$. For $j = 1, 2, 3$, the sum (2) differs from H_j by a rational number divisible by 5 (since this already holds termwise), and H_j is not divisible by 5. For $j = 4$, we combine terms to rewrite (2) as

$$\frac{(10k+5)(50k^2+50k+10)}{(5k+1)(5k+2)(5k+3)(5k+4)};$$

in this expression, the numerator is divisible by 25 while the denominator is coprime to 5.

Summing (2) for $k = 0, 1, 2, 3$ and $j = 4$, we find that S_{20} is divisible by 25. Adding to this (2) for $k = 4$ and appropriate j, we find that S_{21}, S_{22}, and S_{23} are not divisible by 5, but S_{24} is. Thus we add $100, 101, 102, 103, 104$, $120, 121, 122, 123, 124$ to $\mathcal{S}$.

To complete the proof, we must check that H_n is not divisible by 5 for any of the ten integers n just added to $\mathcal{S}$. Suppose $n = 100+j$ with $j = 0, \ldots, 4$. Applying (1) twice gives $H_n = S_{100+j} + \frac{1}{5}S_{20} + \frac{1}{25}H_4$. We know $\frac{1}{5}S_{20}$ is divisible by 5, and $H_4 = 25/12$, so it suffices to show that $S_{100+j} + 1/12$ is not divisible by 5. For $j = 0, \ldots, 4$,

$$S_{100+j} + \frac{1}{12} = S_{100} + \left(H_j + \frac{1}{12}\right) + \sum_{i=1}^{j}\left(\frac{1}{100+i} - \frac{1}{i}\right).$$

Here S_{100} is a sum of expressions of the form (2), hence divisible by 5, and $\frac{1}{100+i} - \frac{1}{i} = \frac{-100}{i(100+i)}$ is divisible by 5, but one can check (for $j = 0, \ldots, 4$ individually) that $H_j + 1/12$ is not divisible by 5. Thus S_{100+j} is not divisible by 5, and neither is H_{100+j}. By a similar argument, if $n = 120 + j$ with $j = 0, \ldots, 4$, then H_n is not divisible by 5.

Thus the algorithm terminates, and the list of integers n such that H_n is 5-integral is as given in the answer above. ■

Reinterpretation. Arguments of this sort are often more easily expressed in terms of the p-adic valuation on the rational numbers, for p a prime. Given $r \in \mathbb{Q}$ nonzero, let $v_p(r)$ be the largest integer m (positive, negative or zero) such that $p^{-m}r$ is p-integral. Then the problem is to determine when $v_5(H_n) \geq 0$, and the first step is to note that this is equivalent to $v_5(H_{\lfloor n/5 \rfloor}) \geq 1$. The rest of the solution amounts to computing some low order terms in the base 5 expansions of $1/m$ for some small m.

Remark. The fact that (2) is divisible by 25 for $j = 4$ is a special case of the fact that for a prime $p > 3$ and an integer x,

$$\frac{1}{px+1} + \frac{1}{px+2} + \cdots + \frac{1}{px+(p-1)} \equiv 0 \pmod{p^2}.$$

The case $x = 0$ is known as Wolstenholme's Theorem, which arises in a different context in 1991B4. To prove the general case, first combine opposite terms in the sum to rewrite it as

$$(2px + p) \sum_{i=1}^{(p-1)/2} \frac{1}{(px+i)(px+(p-i))}.$$

Hence it suffices to show that

$$\sum_{i=1}^{(p-1)/2} \frac{1}{(px+i)(px+(p-i))} \equiv 0 \pmod{p}.$$

This sum is congruent to $-S$ where $S = \sum_{i=1}^{(p-1)/2} i^{-2}$. If a^2 is a perfect square not congruent to 0 or 1 modulo p (e.g. $a^2 = 4$), then the terms of a^2S are congruent modulo p to a permutation of the terms of S, and we may solve $a^2S \equiv S \pmod{p}$ to obtain $S \equiv 0 \pmod{p}$, as desired.

By a similar argument, one can prove that for a prime $p > 3$, positive integer n, and integer x,

$$\sum_{i=1}^{p-1} \sum_{j=0}^{p^{n-1}-1} \frac{1}{p^n x + pj + i} \equiv 0 \pmod{p^{2n}}.$$

See the remarks following 1991B4 for an application of this generalization.

B4. (23, 6, 7, 0, 0, 0, 0, 0, 5, 4, 35, 125)

Let $a_{m,n}$ denote the coefficient of x^n in the expansion of $(1+x+x^2)^m$. Prove that for all integers $k \geq 0$,

$$0 \leq \sum_{i=0}^{\lfloor \frac{2k}{3} \rfloor} (-1)^i a_{k-i,i} \leq 1.$$

Solution 1. Let $s_k = \sum_i (-1)^i a_{k-i,i}$ be the given sum. (Note that $a_{k-i,i}$ is nonzero precisely for $i = 0, \dots, \lfloor \frac{2k}{3} \rfloor$.) Since

$$a_{m+1,n} = a_{m,n} + a_{m,n-1} + a_{m,n-2},$$

we have

$$\begin{aligned} s_k - s_{k+1} + s_{k+2} &= \sum_i (-1)^i (a_{k-i,i} + a_{k-i,i+1} + a_{k-i,i+2}) \\ &= \sum_i (-1)^i a_{k-i+1,i+2} = s_{k+3}. \end{aligned}$$

After computing $s_0 = 1, s_1 = 1, s_2 = 0$, we may prove by induction that $s_{4j} = s_{4j+1} = 1$ and $s_{4j+2} = s_{4j+3} = 0$ for all $j \geq 0$. ■

Reinterpretation. The characteristic polynomial of the linear recursion $s_{k+3} - s_{k+2} + s_{k+1} - s_k$ is $x^3 - x^2 + x - 1$, which divides $x^4 - 1$. Thus the sequence also satisfies $s_{k+4} - s_k = 0$, which explains the periodicity. See the remark following 1988A5 for more on linear recursive sequences.

Solution 2. We use a generating function. Define s_k as in Solution 1. Then

$$\begin{aligned} \sum_{k=0}^{\infty} s_k x^k &= \sum_{i,k} (-1)^i a_{k-i,i} x^k \\ &= \sum_{i,j} x^j a_{j,i} (-x)^i \quad \text{(change of variable: } k = i+j) \\ &= \sum_j x^j (1 - x + x^2)^j \quad \left(\text{since } \sum_i a_{j,i} z^i = (1+z+z^2)^j\right) \\ &= \frac{1}{1 - x + x^2 - x^3} \quad \text{(geometric series)} \\ &= \frac{1+x}{1-x^4} \\ &= (1+x) + (x^4 + x^5) + (x^8 + x^9) + \cdots, \end{aligned}$$

so $0 \leq s_k \leq 1$ for all $k \geq 0$. ■

Remark. If one does not notice that

$$\frac{1}{1 - x + x^2 - x^3} = \frac{1+x}{1-x^4},$$

one can instead use partial fractions to recover its coefficients.

B5. (9, 4, 3, 0, 0, 0, 0, 0, 17, 24, 44, 104)

Prove that for $n \geq 2$,

$$2^{2^{\cdot^{\cdot^{2}}}}\Big\}^{n} \equiv 2^{2^{\cdot^{\cdot^{2}}}}\Big\}^{n-1} \pmod{n}.$$

Solution. Define a sequence by $x_0 = 1$ and $x_m = 2^{x_{m-1}}$ for $m > 0$. We are asked to prove $x_n \equiv x_{n-1} \pmod{n}$ for all $n \geq 2$. We will use strong induction on n to prove the stronger result $x_{n-1} \equiv x_n \equiv x_{n+1} \equiv \cdots \pmod{n}$ for $n \geq 1$. The $n = 1$ case is obvious.

Now suppose $n \geq 2$. Write $n = 2^a b$, where b is odd. It suffices to show that $x_{n-1} \equiv x_n \equiv \cdots$ modulo 2^a and modulo b. For the former, we only need $x_{n-1} \geq a$, which holds, since $x_{n-1} \geq n$ by induction on n. For the latter, note that $x_m \equiv x_{m+1} \equiv \cdots \pmod{b}$ as long as $x_{m-1} \equiv x_m \equiv \cdots \pmod{\phi(b)}$, where $\phi(n)$ is the Euler ϕ-function. By the inductive hypothesis, the latter holds for $m = \phi(b)$; but $m = \phi(b) \leq n - 1$, so $x_{n-1} \equiv x_n \equiv \cdots \pmod{b}$, as desired. ∎

Remark. This solution also yields a solution to Problem 3 on the 1991 USA Mathematical Olympiad [USAMO]:

> Show that, for any fixed integer $n \geq 1$, the sequence
>
> $$2,\ 2^2,\ 2^{2^2},\ 2^{2^{2^2}},\ \ldots \pmod{n}$$
>
> is eventually constant.

See 1985A4 for a similar problem and for the properties of $\phi(n)$.

Remark. For $n \geq 1$, let $f(n)$ be the smallest integer $k \geq 0$ such that $x_k \equiv x_{k+1} \equiv x_{k+2} \equiv \cdots \pmod{n}$. The above solution shows that $f(n) \leq n - 1$, but the actual value of $f(n)$ is much smaller.

We now prove $f(n) \leq \lceil \log_2 n \rceil$ by strong induction on n. The base case $n = 1$ is trivial. If $n = 2^a$ for some $a \geq 1$, then 2^a divides $x_a, x_{a+1}, \ldots$ (since $x_{a-1} \geq a$ was proved in the solution above), so $f(n) \leq a$. If $n = 2^a b$ for some $a \geq 1$ and odd $b > 1$, then $f(n) \leq \max\{f(2^a), f(b)\}$, so we apply the inductive hypothesis at 2^a and b. Finally, if $n > 1$ is odd, write $\phi(n) = 2^c d$ with d odd, and observe that

$$\begin{aligned} f(n) &\leq 1 + f(\phi(n)) && \text{(as in the solution above)} \\ &\leq 1 + \max\{f(2^c), f(d)\} \\ &\leq 1 + \max\{c, \lceil \log_2 d \rceil\} && \text{(by the inductive hypothesis)} \\ &\leq \lceil \log_2 n \rceil, \end{aligned}$$

since $c \leq \log_2 \phi(n) < \log_2 n$ and $d \leq \phi(n)/2 < n/2$. This completes the inductive step.

When n is a power of 3, the bound just proved is best possible up to a constant factor: we now show that $f(3^m) = m + 1$ for $m \geq 1$ by induction on m. First, $x_{k+1} \equiv (-1)^{x_k} = 1 \pmod{3}$, if and only if $k \geq 1$, so $f(3) = 2$. Now suppose $m > 1$. Since the image of 2 generates the cyclic group $(\mathbb{Z}/3^m\mathbb{Z})^*$ (i.e., 2 is a primitive root of 3^m [NZM, Theorem 2.40]), we have $x_k \equiv x_{k+1} \equiv \cdots \pmod{3^m}$ if and only if $x_{k-1} \equiv x_k \equiv \cdots \pmod{2 \cdot 3^{m-1}}$, but all the x_k except x_0 are even, so this is equivalent to $x_{k-1} \equiv x_k \equiv \cdots \pmod{3^{m-1}}$ if $k \geq 2$. Hence $f(3^m) = f(3^{m-1}) + 1$, so $f(3^m) = m + 1$ by induction.

It is not true in general that $f(n)$ is bounded below by a constant multiple of $\ln n$, because there are some values of n, such as large powers of 2, for which $f(n)$ is much, much less than $\ln n$. We can at least say that $f(n) \to \infty$ as $n \to \infty$, though: we have $x_{f(n)+1} \geq x_{f(n)+1} - x_{f(n)} \geq n$, since n divides $x_{f(n)+1} - x_{f(n)}$.

B6. (5, 4, 5, 0, 0, 0, 0, 0, 3, 0, 51, 137)

The dissection of the 3–4–5 triangle shown below has diameter $5/2$.

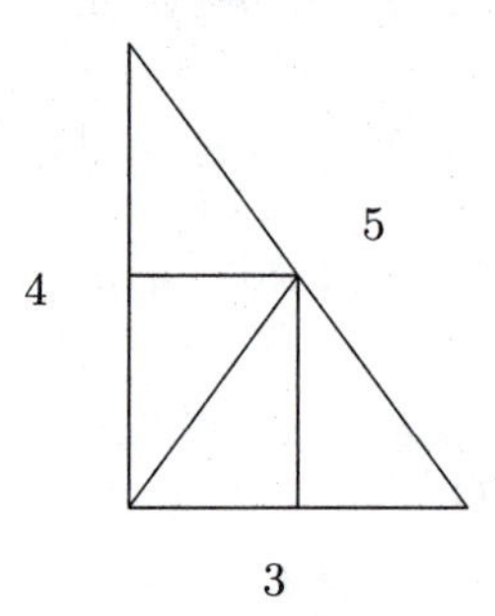

Find the least diameter of a dissection of this triangle into four parts. (The diameter of a dissection is the least upper bound of the distances between pairs of points belonging to the same part.)

Answer. The minimum diameter is $25/13$.

Solution. (See Figure 34.) Place the triangle on the cartesian plane so that its vertices are $A = (4,0)$, $B = (0,0)$, $C = (0,3)$. The five points A, B, C, $D = (27/13, 0)$, $E = (20/13, 24/13)$ lie at distance at least $25/13$ apart from each other (note that $AD = DE = EC = 25/13$). By the Pigeonhole Principle, any dissection of the triangle into four parts has a part containing at least two of these points, and hence has diameter at least $25/13$.

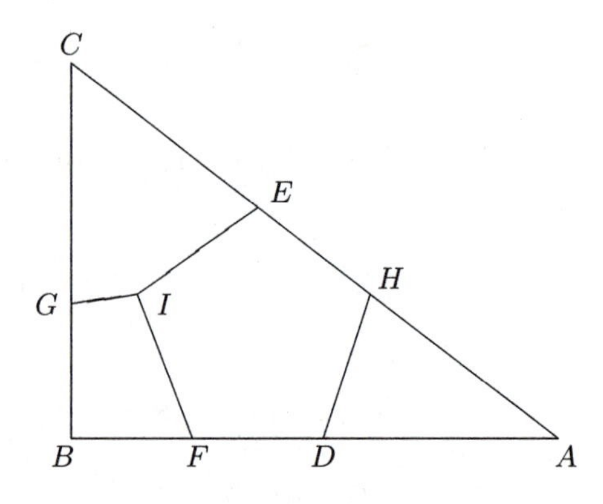

FIGURE 34.

On the other hand, we now construct a dissection into four parts in which each part has diameter $25/13$. Let $H = (32/13, 15/13)$, so that $AH = 25/13$. Next, construct a point I inside quadrilateral $BDEC$ so that BI, CI, DI, EI, HI have length less than $25/13$. For example, we may take $I = (7/13, 15/13)$ such that $DECI$ and $ADIH$ are parallelograms. Take $F = (1, 0)$ on AB so that $EF = 25/13$, and take $G = (0, 14/13)$ on BC so that $CG = 25/13$.

Our parts are the triangle ADH, the pentagon $DHEIF$, the quadrilateral $CEIG$, and the quadrilateral $BFIG$. Verifying that each has diameter $25/13$ entails checking that the distance between any two vertices of a polygon is at most $25/13$. For starters,

$$AD = AH = CE = CG = CI = DE = DI = EF = HI = 25/13.$$

Next, we see without any calculations that

$$BF, BG < FG \qquad FI, FD < DI \qquad DH < FH \qquad GI < CI.$$

Finally, we compute the remaining distances:

$$BI^2 = \frac{274}{169}, \qquad EG^2 = \frac{500}{169}, \qquad EH^2 = \frac{225}{169},$$

$$EI^2 = \frac{260}{169}, \qquad FG^2 = \frac{365}{169}, \qquad FH^2 = \frac{586}{169}. \qquad \blacksquare$$

Remark. Once the "skeleton" is in place, some variations in the placements of the points are possible.

The Fifty-Ninth William Lowell Putnam Mathematical Competition December 5, 1998

A1. (156, 23, 4, 0, 0, 0, 0, 0, 0, 0, 16, 0)

A right circular cone has base of radius 1 and height 3. A cube is inscribed in the cone so that one face of the cube is contained in the base of the cone. What is the side-length of the cube?

Answer. The side length of the cube is $(9\sqrt{2} - 6)/7$.

Solution. Consider a plane cross-section through a vertex of the cube and the axis of the cone as shown in Figure 35.

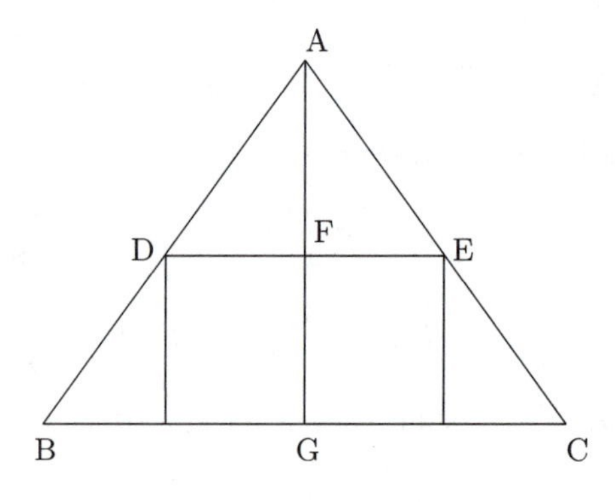

FIGURE 35.

Let s be the side length of the cube. Segment DE is a diagonal of the top of the cube, so its length is $s\sqrt{2}$. From the similar triangles $ADE \sim ABC$, we have $\frac{DE}{BC} = \frac{AF}{AG}$, or equivalently $\frac{s\sqrt{2}}{2} = \frac{3-s}{3}$. Solving for s, we get $s = (9\sqrt{2} - 6)/7$. ■

A2. (103, 35, 26, 0, 0, 0, 0, 0, 9, 8, 12, 6)

Let s be any arc of the unit circle lying entirely in the first quadrant. Let A be the area of the region lying below s and above the x-axis and let B be the area of the region lying to the right of the y-axis and to the left of s. Prove that $A + B$ depends only on the arc length, and not on the position, of s.

Solution 1. Let O be the center of the circle, K and H the endpoints of the arc (with K above H), C and F the projections of K onto the x-axis and y-axis, respectively, D and E the projections of K onto the x-axis and y-axis, respectively, and G the intersection of KC and HE, as in Figure 36.

Denote by $[XYZ]$ the area of the region with vertices X, Y, Z. In this notation,

$$\begin{aligned} A + B &= [EFGK] + [CDHG] + 2[GHK] \\ &= 2([OGK] + [OGH] + [GHK]) \\ &= 2[OHK] = \theta, \end{aligned}$$

where θ is the length of the arc s. ■

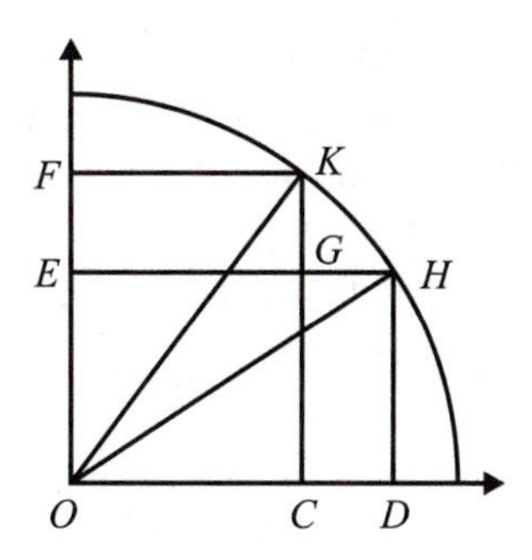

FIGURE 36.

Solution 2. We obtain A and B by integrating $-y\,dx$ and $x\,dy$, respectively, counterclockwise along the arc. Thus $A+B$ is the integral of $x\,dy - y\,dx$ over the arc. If we parameterize the arc by arc length, that is, by setting $x = \cos\theta$ and $y = \sin\theta$, the integrand becomes $\cos^2\theta + \sin^2\theta = 1$. Thus the integral is precisely the arc length, and so does not depend on the position of the arc. ■

Remark. For a straightforward but less elegant solution, one can also compute the integrals separately. Alternatively, for a fixed arc length, one can express the sum of areas as a function $F(\theta)$ of the argument θ at one endpoint of s, and show that $F'(\theta) = 0$.

A3. (82, 34, 2, 0, 0, 0, 0, 0, 5, 0, 39, 37)

Let f be a real function on the real line with continuous third derivative. Prove that there exists a point a such that

$$f(a) \cdot f'(a) \cdot f''(a) \cdot f'''(a) \geq 0.$$

Solution. If at least one of $f(a)$, $f'(a)$, $f''(a)$, or $f'''(a)$ vanishes at some point a, then we are done. Otherwise by the Intermediate Value Theorem, each of $f(x)$, $f'(x)$, $f''(x)$, and $f'''(x)$ is either strictly positive or strictly negative on the real line, and we only need to show that their product is positive for a single value of x. By replacing $f(x)$ by $-f(x)$ if necessary, we may assume $f''(x) > 0$; by replacing $f(x)$ by $f(-x)$ if necessary, we may assume $f'''(x) > 0$. Notice that these substitutions do not change the sign of $f(x)f'(x)f''(x)f'''(x)$.

Now $f'''(x) > 0$ implies that $f''(x)$ is increasing. Thus for $a > 0$,

$$f'(a) = f'(0) + \int_0^a f''(t)\,dt \geq f'(0) + af''(0).$$

In particular, $f'(a) > 0$ for large a. Similarly, since $f''(x) > 0$, $f(a)$ is positive for large a. Therefore $f(x)f'(x)f''(x)f'''(x) > 0$ for sufficiently large x. ■

Reinterpretation. More succinctly, f cannot both be positive and strictly concave-down everywhere, nor negative and strictly concave-up everywhere. So $f(x)f''(x)$ must be positive for some range of x, as must be $f'(x)f'''(x)$ by the same reasoning applied to f' instead of f.

A4. (39, 27, 52, 0, 0, 0, 0, 0, 49, 7, 14, 11)

Let $A_1 = 0$ and $A_2 = 1$. For $n > 2$, the number A_n is defined by concatenating the decimal expansions of A_{n-1} and A_{n-2} from left to right. For example $A_3 = A_2A_1 = 10$, $A_4 = A_3A_2 = 101$, $A_5 = A_4A_3 = 10110$, and so forth. Determine all n such that 11 divides A_n.

Answer. The number 11 divides A_n if and only if $n \equiv 1 \pmod 6$.

Solution. The number of digits in the decimal expansion of A_n is the nth Fibonacci number F_n. It follows that the sequence $\{A_n\}$ modulo 11 satisfies a recursion:

$$\begin{aligned} A_n &= 10^{F_{n-2}} A_{n-1} + A_{n-2} \\ &\equiv (-1)^{F_{n-2}} A_{n-1} + A_{n-2} \pmod{11}. \end{aligned}$$

By induction, F_n is even if and only if 3 divides n; hence $(-1)^{F_{n-2}}$ is periodic with period 3.

Computing A_n modulo 11 for small n using the recursion, we find

$$A_1, \dots, A_8 \equiv 0, 1, -1, 2, 1, 1, 0, 1 \pmod{11}.$$

By induction, we deduce that $A_{n+6} \equiv A_n \pmod{11}$ for all n, and so A_n is divisible by 11 if and only if $n \equiv 1 \pmod 6$. ■

Remark. See 1988A5 for more on linear recursions and Fibonacci numbers.

A5. (85, 24, 15, 0, 0, 0, 0, 0, 5, 2, 8, 60)

Let $\mathcal{F}$ be a finite collection of open discs in $\mathbb{R}^2$ whose union contains a set $E \subseteq \mathbb{R}^2$. Show that there is a pairwise disjoint subcollection $D_1, \dots, D_n$ in $\mathcal{F}$ such that

$$E \subseteq \bigcup_{j=1}^{n} 3D_j.$$

Here, if D is the disc of radius r and center P, then $3D$ is the disc of radius $3r$ and center P.

Solution. Define the sequence D_i by the following greedy algorithm: let D_1 be the disc of largest radius (breaking ties arbitrarily), let D_2 be the disc of largest radius not meeting D_1, let D_3 be the disc of largest radius not meeting D_1 or D_2, and so on, up to some final disc D_n (since F is finite). To see that $E \subseteq \cup_{j=1}^{n} 3D_j$, consider a point P in E; if P lies in one of the D_i, we are done. Otherwise, P lies in a disc $D \in \mathcal{F}$, say of radius r, which meets one of the D_i having radius $s \geq r$. (This is the only reason a disc can be skipped in our algorithm.) Choose Q in $D \cap D_i$. If O and O_i are the centers of D and D_i, respectively, then by the triangle inequality

$$PO_i \leq PO + OQ + QO_i < r + r + s \leq 3s.$$

Thus $P \in 3D_i$. ■

A6. (12, 1, 3, 0, 0, 0, 0, 0, 3, 9, 92, 79)

Let A, B, C denote distinct points with integer coordinates in $\mathbb{R}^2$. Prove that if

$$(|AB| + |BC|)^2 < 8 \cdot [ABC] + 1$$

then A, B, C are three vertices of a square. Here $|XY|$ is the length of segment XY and $[ABC]$ is the area of triangle ABC.

Solution 1. Given that A, B, C have integer coordinates, it follows that $|AB|^2, |BC|^2, |CA|^2$ are integers, and that $2[ABC]$ is an integer (see the first remark in 1990A3). These quantities are related via the AM-GM Inequality (see the end of 1985A2) in the form $|AB|^2 + |BC|^2 \geq 2|AB||BC|$, and the area formula

$$2[ABC] = |AB||BC| \sin \angle ABC \leq |AB||BC|.$$

The quantity $|AB|^2 + |BC|^2 + 4[ABC]$ is thus an integer which, by the above inequalities and the given condition, "sandwiches" $8[ABC]$ as follows:

$$\begin{aligned} |AB|^2 + |BC|^2 + 4[ABC] &\leq (|AB| + |BC|)^2 \\ &< 8[ABC] + 1 \\ &\leq |AB|^2 + |BC|^2 + 4[ABC] + 1, \end{aligned}$$

We may thus conclude that $|AB|^2 + |BC|^2 + 4[ABC] = 8[ABC]$, since these quantities are integers and their difference is at least 0 but less than 1. In particular, we have equality both in AM-GM and the area formula, so $AB = BC$ and $\angle ABC$ is a right angle, as desired. ∎

Solution 2. Set up a new coordinate system with the same unit length as the original system, but with $B = (0, 0)$ and $A = (s, 0)$ for some $s > 0$. We may assume that C is in the upper half plane in these coordinates, and put $C = (x, y + s)$. Let C' be the image of A under $90°$ counterclockwise rotation around B, so $C' = (0, s)$. The given inequality is equivalent to all of the following:

$$\begin{aligned} \left(s + \sqrt{x^2 + (y+s)^2}\right)^2 &< 4s(y+s) + 1 \\ s^2 + x^2 + (y+s)^2 + 2s\sqrt{x^2 + (y+s)^2} &< 4s(y+s) + 1 \\ 2s\left(\sqrt{x^2 + (y+s)^2} - (y+s)\right) &< 1 - x^2 - y^2. \end{aligned}$$

The left-hand side in the last inequality is nonnegative, so $x^2 + y^2 < 1$. But $x^2 + y^2 = |CC'|^2$, and C, C' have integer coordinates in the old coordinate system, so $C = C'$. Thus A, B, C are three consecutive vertices of a square. ∎

B1. (112, 30, 30, 0, 0, 0, 0, 0, 3, 3, 12, 9)

Find the minimum value of

$$\frac{(x + 1/x)^6 - (x^6 + 1/x^6) - 2}{(x + 1/x)^3 + (x^3 + 1/x^3)}$$

for $x > 0$.

Answer. The minimum value is 6.

Solution. If we put $a = (x + 1/x)^3$ and $b = x^3 + 1/x^3$, the given expression can be rewritten as

$$\frac{a^2 - b^2}{a + b} = a - b = 3(x + 1/x).$$

The minimum of $x + 1/x$ for $x > 0$ is 2 by the AM-GM inequality (see the end of 1985A2), by calculus, or by the identity $x + 1/x = (\sqrt{x} - \sqrt{1/x})^2 + 2$. Thus the minimum value of the original expression is 6, attained when $x = 1$. ■

B2. (82, 10, 9, 0, 0, 0, 0, 0, 3, 11, 38, 46)

Given a point (a, b) with $0 < b < a$, determine the minimum perimeter of a triangle with one vertex at (a, b), one on the x-axis, and one on the line $y = x$. You may assume that a triangle of minimum perimeter exists.

Answer. The minimum perimeter of such a triangle is $\sqrt{2a^2 + 2b^2}$.

Solution. Consider a triangle as described by the problem. Label its vertices A, B, C so that $A = (a, b)$, B lies on the x-axis, and C lies on the line $y = x$; see Figure 37. Further let $D = (a, -b)$ be the reflection of A in the x-axis, and let $E = (b, a)$ be the reflection of A in the line $y = x$. Then $AB = DB$ and $AC = CE$, so the perimeter of ABC is

$$DB + BC + CE \geq DE = \sqrt{(a - b)^2 + (a + b)^2} = \sqrt{2a^2 + 2b^2}.$$

This lower bound can be achieved: set B (resp., C) to be the intersection between the segment DE and the x-axis (resp., the line $x = y$). Thus the minimum perimeter is in fact $\sqrt{2a^2 + 2b^2}$. ■

Remark. The reflection trick comes up in various geometric problems where the sum of certain distances is to be minimized. For many applications of this principle, see [CG, Sections 4.4–4.6] and [Pó, Section IX].

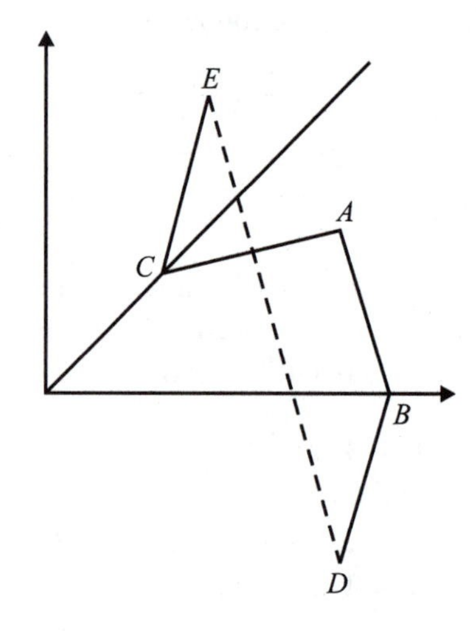

FIGURE 37.
The reflection trick: $AB + BC + CA = DB + BC + CD \geq DE$.

Remark. Our proof did not use the assumption that a triangle of minimum perimeter exists. This assumption might be slightly useful in a "brute force" approach using calculus to find the minimum of a two-variable function. Such an approach is painful to execute, however.

B3. (44, 4, 3, 0, 0, 0, 0, 0, 7, 2, 62, 77)

Let H be the unit hemisphere $\{(x,y,z) : x^2+y^2+z^2=1, z \geq 0\}$, C the unit circle $\{(x,y,0) : x^2+y^2=1\}$, and P the regular pentagon inscribed in C. Determine the surface area of that portion of H lying over the planar region inside P, and write your answer in the form $A\sin\alpha + B\cos\beta$, where A, B, α, β are real numbers.

Answer. The surface area is $5\pi\cos\frac{\pi}{5} - 3\pi$. In particular, we may take

$$A = -3\pi, \quad \alpha = \pi/2, \quad B = 5\pi, \quad \text{and} \quad \beta = \pi/5.$$

Solution. The surface area of the spherical cap

$$\{(x,y,z) : x^2+y^2+z^2=1,\ z \geq z_0\}$$

can be computed in spherical coordinates as

$$\int_{\phi=0}^{\cos^{-1} z_0} \int_{\theta=0}^{2\pi} \sin\phi \, d\theta \, d\phi = 2\pi(1-z_0).$$

The desired surface area is the area of a hemisphere minus the surface areas of five identical halves of spherical caps; these caps, up to isometry, correspond to z_0 being the distance from the center of the pentagon to any of its sides, i.e., $z_0 = \cos\frac{\pi}{5}$. Since the area of a hemisphere is 2π (take $z_0 = 0$ above), the desired area is

$$2\pi - \frac{5}{2}\left(2\pi\left(1-\cos\frac{\pi}{5}\right)\right) = 5\pi\cos\frac{\pi}{5} - 3\pi. \qquad \blacksquare$$

Remark. The surface area of a spherical cap can also be computed by noting that the cap is the surface of revolution obtained by rotating part of the graph of $y = f(x)$ around the x-axis, where $f(x) = \sqrt{1-x^2}$. This gives

$$\begin{aligned}
\int_{x=z_0}^{1} 2\pi y\sqrt{dx^2+dy^2} &= \int_{z_0}^{1} 2\pi f(x)\sqrt{1+f'(x)^2}\,dx \\
&= \int_{z_0}^{1} 2\pi\sqrt{1-x^2}\sqrt{1+\frac{x^2}{1-x^2}}\,dx \\
&= \int_{z_0}^{1} 2\pi\sqrt{1-x^2}\sqrt{\frac{1}{1-x^2}}\,dx \\
&= 2\pi(1-z_0).
\end{aligned}$$

Notice an amazing consequence of the formula: the area of a slice of a unit sphere depends only on the width of the slice, and not on its position on the sphere! This was known already to Archimedes, and forms the basis of the *Lambert equal-area cylindrical projection* and the *Gall-Peters equal-area projection* used in some maps.

This fact can be used to give a beautiful solution to the following problem [She]:

> Consider a disk of diameter d drawn on a plane and a number of very long white paper strips of different widths. Show that the disk can be covered by the strips if and only if the sum of the widths is at least d.

B4. (42, 9, 22, 0, 0, 0, 0, 0, 21, 28, 24, 53)

Find necessary and sufficient conditions on positive integers m and n so that

$$\sum_{i=0}^{mn-1} (-1)^{\lfloor i/m \rfloor + \lfloor i/n \rfloor} = 0.$$

Answer. We have $S(m,n) = 0$ if and only if $\frac{m}{\gcd(m,n)}$ and $\frac{n}{\gcd(m,n)}$ are not both odd.

Solution. For convenience, define $f_{m,n}(i) = \lfloor \frac{i}{m} \rfloor + \lfloor \frac{i}{n} \rfloor$, so that the given sum is $S(m,n) = \sum_{i=0}^{mn-1} (-1)^{f_{m,n}(i)}$. If m and n are both odd, then $S(m,n)$ is the sum of an odd number of ± 1's, and thus cannot be zero. If m and n are of opposite parity,

$$\begin{aligned} f_{m,n}(i) + f_{m,n}(mn-1-i) &= \left\lfloor \frac{i}{m} \right\rfloor + \left\lfloor \frac{i}{n} \right\rfloor + \left\lfloor \frac{mn-1-i}{m} \right\rfloor + \left\lfloor \frac{mn-1-i}{n} \right\rfloor \\ &= (n-1) + (m-1), \end{aligned}$$

since $\lfloor \frac{c}{a} \rfloor + \lfloor b - \frac{c+1}{a} \rfloor = b - 1$ for all integers a, b, c with $a > 0$. Thus the terms of $S(m,n)$ cancel in pairs and so the sum is zero.

Now suppose that $m = dk$ and $n = dl$ for some d. For $i = 0, \dots, d-1$, we have $\lfloor \frac{dj+i}{dk} \rfloor = \lfloor \frac{j}{k} \rfloor$ and $\lfloor \frac{dj+i}{dl} \rfloor = \lfloor \frac{j}{l} \rfloor$, so $f_{m,n}(dj+i) = f_{k,l}(j)$. Hence

$$\begin{aligned} S(dk, dl) &= d \sum_{j=0}^{dkl-1} (-1)^{f_{k,l}(j)} \\ &= d \sum_{j=0}^{kl-1} (-1)^{f_{k,l}(j)} + (-1)^{f_{k,l}(j+kl)} + \cdots + (-1)^{f_{k,l}(j+(d-1)kl)} \\ &= d \sum_{j=0}^{kl-1} (-1)^{f_{k,l}(j)} \sum_{i=0}^{d-1} (-1)^{i(k+l)} \\ &\qquad \text{(since } f_{k,l}(j+ikl) = i(k+l) + f_{k,l}(j)) \\ &= dS(k,l) \sum_{i=0}^{d-1} (-1)^{i(k+l)}. \end{aligned}$$

Thus we have $S(dk, dl) = d^2 S(k,l)$ if $k + l$ is even; this equality also holds if $k + l$ is odd, since then $S(k,l) = 0$. Thus $S(dk, dl)$ vanishes if and only if $S(k,l)$ vanishes. Piecing together the various cases gives the desired result. ■

Remark. There are many equivalent ways to state the answer. Here are some other possibilities, mostly from student papers:

- The highest powers of 2 dividing m and n are different.
- There exist integers $a, b, k \geq 0$ such that $m = 2^k a$, $n = 2^k b$, and a and b have opposite parity.

- $\operatorname{lcm}(m,n)$ is an even integer times m, or an even integer times n, but not both.
- Exactly one of $\frac{m}{\gcd(m,n)}$ and $\frac{n}{\gcd(m,n)}$ is even.
- $\frac{m+n}{\gcd(m,n)}$ is odd.
- $\frac{mn}{\gcd(m,n)^2}$ is even.
- $\frac{\operatorname{lcm}(m,n)}{\gcd(m,n)}$ is even.
- When expressed in lowest terms, m/n has even numerator or even denominator.
- The 2-adic valuation of m/n is nonzero. (See the remark following 1997B3 for a definition of the p-adic valuation for any prime p.)

Remark. If $S(m,n)$ is nonzero, it equals $\gcd(m,n)^2$. To show this, we use the equation $S(dk,dl) = S(k,l)$ from the above solution to reduce to the case where m,n are coprime and odd. In this case,

$$f_{m,n}(i) \equiv m\lfloor i/m \rfloor + n\lfloor i/n \rfloor \equiv i - m\lfloor i/m \rfloor + i - n\lfloor i/n \rfloor \pmod 2.$$

Now $i - m\lfloor i/m \rfloor$ is the remainder upon dividing i by m. By the Chinese Remainder Theorem, the pairs $(i - m\lfloor i/m \rfloor, i - n\lfloor i/n \rfloor)$ run through all pairs (j,k) of integers with $0 \le j \le m-1$, $0 \le k \le n-1$ exactly once. Hence

$$S(m,n) = \sum_{j=0}^{m-1}(-1)^j \sum_{k=0}^{n-1}(-1)^k = 1.$$

B5. (55, 2, 29, 0, 0, 0, 0, 0, 8, 0, 29, 76)

Let N be the positive integer with 1998 decimal digits, all of them 1; that is,

$$N = 1111\cdots 11.$$

Find the thousandth digit after the decimal point of $\sqrt{N}$.

Answer. The 1000th digit is 1.

Solution 1. Write $N = (10^{1998}-1)/9$. Then

$$\sqrt{N} = \frac{10^{999}}{3}\sqrt{1-10^{-1998}} = \frac{10^{999}}{3}(1 - \frac{1}{2}10^{-1998} + r),$$

where r can be bounded using Taylor's Formula with remainder: see Solution 2 to 1992A4. Recall that Taylor's Formula says that $f(x) = f(0) + xf'(0) + \frac{x^2}{2}f''(c)$ for some c between 0 and x. For $f(x) = \sqrt{1+x}$, $f''(x) = -\frac{1}{4}(1+x)^{-3/2}$; therefore for $x = -10^{-1998}$, $|r| < 10^{-3996}/8$.

The digits after the decimal point of $10^{999}/3$ are given by $.3333\ldots$, while the digits after the decimal point of $\frac{1}{6}10^{-999}$ are given by $.00000\ldots1666666\ldots$. Thus the first 1000 digits of $\sqrt{N}$ after the decimal point are given by $.33333\ldots3331$; in particular, the thousandth digit is 1. ∎

Solution 2 (Hoeteck Wee). The 1000th digit after the decimal point of $\sqrt{N}$ is the unit digit of

$$10^{1000}\sqrt{N} = 1000^{1000}\frac{\sqrt{10^{1998}-1}}{3} = \frac{\sqrt{10^{3998}-10^{2000}}}{3}.$$

Now

$$(10^{1999}-7)^2 < 10^{3998} - 10^{2000} < (10^{1999}-4)^2,$$

so

$$\frac{10^{1999}-7}{3} < 10^{1000}\sqrt{N} < \frac{10^{1999}-4}{3}.$$

Hence

$$33\ldots31 < 10^{1000}\sqrt{N} < 33\ldots32,$$

and the answer is 1. ■

Related question. What are the next thousand digits? (Hint: extend Solution 1.)

B6. (25, 8, 8, 0, 0, 0, 0, 0, 4, 5, 31, 118)

Prove that, for any integers a, b, c, there exists a positive integer n such that $\sqrt{n^3 + an^2 + bn + c}$ is not an integer.

Solution 1. Recall that all perfect squares are congruent to 0 or 1 modulo 4. Suppose $P(n) = n^3 + an^2 + bn + c$ is a square for $n = 1, 2, 3, 4$. Since $P(2)$ and $P(4)$ are perfect squares of the same parity, their difference $56 + 12a + 2b$ must be a multiple of 4; that is, b must be even. On the other hand, since $P(1)$ and $P(3)$ are also perfect squares of the same parity, their difference $26 + 8a + 2b$ must be a multiple of 4; that is, b must be odd. This is a contradiction. ■

Solution 2. If $4b - a^2 = 0$ and $c = 0$, then $n^3 + an^2 + bn + c = n\,(n + a/2)^2$, which is not a perfect square if n is not a perfect square and $n \neq -a/2$.

So suppose that $4b - a^2$ and c are not both zero. Take $n = 4m^2$. Then

$$n^3 + an^2 + bn + c = (8m^3 + am)^2 + (4b - a^2)m^2 + c.$$

If $4b - a^2 > 0$, or $4b - a^2 = 0$ and $c > 0$, then for m sufficiently large, $(4b - a^2)m^2 + c > 0$, and

$$|(4b - a^2)m^2 + c| < 2(8m^3 + am) - 1,$$

so $n^3 + an^2 + bn + c$ lies between $(8m^3 + am)^2$ and $(8m^3 + am + 1)^2$ and is hence not a perfect square. Similarly, if $4b - a^2 < 0$, or $4b - a^2 = 0$ and $c < 0$, then $n^3 + an^2 + bn + c$ lies between $(8m^3 + am - 1)^2$ and $(8m^3 + am)^2$ for sufficiently large m. ■

Solution 3. We will use the theory of height functions on elliptic curves (see Chapter 8 of [Sil] for this theory). Suppose first that $P(n) = n^3 + an^2 + bn + c$ is not squarefree as a polynomial in $\mathbb{Q}[n]$. Then by Gauss's Lemma (see the remark below), $P(n)$ factors as $(n - d)^2(n - e)$ for some $d, e \in \mathbb{Z}$. Choose a positive integer $n \neq d$ such that $n - e$ is not a square. Then $P(n)$ is not a square.

Now suppose instead that $P(n)$ is squarefree. Then $y^2 = P(x)$ is an elliptic curve E over $\mathbb{Q}$. Let $N(B)$ be the number of rational numbers $x = u/v$ in lowest terms with $|u|, |v| \leq B$ such that $P(x)$ is the square of a rational number. The theory of height functions implies that $N(B) = O((\ln B)^{r/2})$ as $B \to \infty$, where r is the rank of $E(\mathbb{Q})$. In particular, if B is a sufficiently large integer, then $N(B) < B$, so not all integers 1, 2, ... , B can be x-values making $P(x)$ a square. ■

Remark (Gauss's Lemma). Define the *content* $c(f)$ of a polynomial $f \in \mathbb{Z}[x]$ as the greatest common divisor of its coefficients. "Gauss's Lemma" refers to any of the three following statements:

1. For any $f, g \in \mathbb{Z}[x]$, we have $c(fg) = c(f)c(g)$.
2. If a monic polynomial f with integer coefficients factors into monic polynoimals with *rational coefficients*, say $f = gh$ with $g, h \in \mathbb{Q}[x]$, then g and h have integer coefficients [NZM, Theorem 9.7].
3. If a polynomial $f \in \mathbb{Z}[x]$ factors nontrivially into polynomials with rational coefficients, then it also factors nontrivially into (possibly different) polynomials with integer coefficients [Lar1, Problem 4.2.16].

The first statement immediately implies the other two. All three statements can be generalized to the case where $\mathbb{Z}$ is replaced by an arbitrary unique factorization domain (UFD). For the generalization of the first statement, see [Lan1, p. 181].

Stronger result. For any polynomial P with integer coefficients that is not the square of a polynomial with integer coefficients, there are arbitrarily large n such that $P(n)$ is not a perfect square. We present three proofs, using quite different techniques.

Proof 1 (Paul Cohen). Suppose that $P(n)$ is the square of a nonnegative integer for all $n \geq n_0$. Express $f(n) = \sqrt{P(n)}$ as an infinite *Puiseux series* (see the first remark after this solution)

$$f(n) = c_0 n^{k/2} + c_1 n^{k/2-1} + \cdots$$

with $c_0 > 0$, convergent for n sufficiently large. Define the difference operator Δ taking the function g to the new function $(\Delta g)(n) = g(n+1) - g(n)$. For $i \geq 1$, the function $(\Delta^i f)(n)$ is represented by the series $c_0^{(i)} n^{k/2-i} + \cdots$ for some coefficients $c_j^{(i)}$, and this series converges for n sufficiently large. In particular, if $i > k/2$, then $|(\Delta^i f)(n)| < 1$ for sufficiently large n, but $(\Delta^i f)(n)$ is also an integer, so it is zero. By the lemma in the Lagrange interpolation remark below, $f(n)$ agrees with a polynomial $F(n)$ for sufficiently large integers n. Then $P(n) - F(n)^2$ is a polynomial with infinitely many zeros, so $P(n) = F(n)^2$ as polynomials. Since $F(n)$ is an integer for sufficiently large integers n, Lagrange interpolation (described in the second remark below) implies that F has rational coefficients. Finally, Gauss's Lemma (given in the remark after Solution 3) implies that P, being the square of a polynomial with rational coefficients, is also the square of a polynomial with integer coefficients. □

Remark (Puiseux series). A Puiseux series in the variable x over a field k is a series of the form $\sum_{j=M}^{\infty} a_j x^{j/r}$ for some $M \in \mathbb{Z}$ (possibly negative), some integer $r \geq 1$, and some elements $a_j \in k$. In other words, it is a Laurent series in the parameter $x^{1/r}$ for some integer $r \geq 1$. (In Proof 1 above, we expressed $\sqrt{f(n)}$ as a Puiseux series in $x = n^{-1}$.) The set of all Puiseux series over k forms a field.

If k is an algebraically closed field of characteristic zero, such as $\mathbb{C}$, then the field of Puiseux series is an algebraic closure of the field $k((x))$ of Laurent series; in particular, under these hypotheses, if $f \in k(x)[t]$ is a one-variable polynomial with coefficients in the field of rational functions $k(x)$, then the zeros of f can be expressed as Puiseux series. Such zeros are called *algebraic functions*; for instance $\sqrt[3]{7+x^2}$ is an algebraic function. The fact that algebraic functions can be expanded in Puiseux series was observed by Newton; hence the term *Newton-Puiseux series* is also sometimes used. The statements in this paragraph are false when k is an algebraically closed field of characteristic $p > 0$, but see [Ke] for an analogue.

Remark (Lagrange interpolation). Suppose one is given $n+1$ pairs of numbers (a_i, b_i) such that $a_1, \ldots, a_{n+1}$ are distinct. The theorem of Lagrange interpolation states that there exists a unique polynomial $f(x)$ of degree less than or equal to n such that $f(a_i) = b_i$ for $1 \le i \le n+1$. Uniqueness follows from the observation that the difference of two such polynomials f is a polynomial of degree at most n, having at least $n+1$ zeros (the a_i): such a difference must be identically zero. Existence follows from the explicit formula

$$f(x) = \sum_{i=1}^{n+1} \frac{(x-a_1)\cdots\widehat{(x-a_i)}\cdots(x-a_{n+1})}{(a_i-a_1)\cdots\widehat{(a_i-a_i)}\cdots(a_i-a_{n+1})} b_i,$$

where $\widehat{\cdot}$ indicates that the corresponding term is omitted from the product. (It is easy to check that this $f(x)$ is a polynomial of degree less than or equal to n such that $f(a_i) = b_i$ for all i.) As an immediate consequence, if $a_i, b_i \in \mathbb{Q}$, then $f(x)$ has rational coefficients.

One application of Lagrange interpolation is the lemma below, used in Proof 1 above. For further applications, see [Lar1, Section 4.3].

Lemma. *If $n_0 \in \mathbb{Z}$, and $g(n)$ is a function defined on integers $n \ge n_0$ such that $(\Delta^k g)(n) = 0$ for all $n \ge n_0$, then there is a polynomial $G(x)$ of degree at most $k-1$ such that $g(n) = G(n)$ for all $n \ge n_0$.*

Proof of Lemma. First, if $h(x)$ is a polynomial of degree $m \ge 1$, then $(\Delta h)(x)$ is a polynomial of degree $m-1$. It follows that $(\Delta^k h)(n) = 0$ for all $k > m$.

Use Lagrange interpolation to find the polynomial $G(x)$ of degree at most $k-1$ such that $G(n) = g(n)$ for $n = n_0, n_0+1, \ldots, n_0+k-1$. Let $H(n) = G(n) - g(n)$ for integers $n \ge n_0$. Then $(\Delta^k G)(n) = 0$ by the previous paragraph, and $(\Delta^k g)(n) = 0$ for integers $n \ge n_0$ is given, so $(\Delta^k H)(n) = 0$ for integers $n \ge n_0$.

It remains to prove that if H is a function such that $(\Delta^k H)(n) = 0$ for integers $n \ge n_0$ and $H(n_0) = H(n_0+1) = \cdots = H(n_0+k-1) = 0$, then $H(n) = 0$ for all $n \ge n_0$. We use induction on k. The case $k = 0$ is trivial, since $\Delta^0 H$ is just H. For $k \ge 1$, applying the inductive hypothesis to ΔH, which satisfies the assumptions with $k-1$ instead of k, shows that $(\Delta H)(n) = 0$ for integers $n \ge n_0$. But $H(n_0) = 0$ too, so induction on n proves $H(n) = 0$ for all integers $n \ge n_0$. □

Proof 2. We may reduce to the case that P has no repeated factors by dividing by squares of factors, so that the discriminant D of P is nonzero. We construct a prime p not dividing D and an integer n such that p divides $P(n)$. Pick an integer m such that $P(m) \ne 0$, and for each prime p_i dividing D, let e_i be the exponent to which p_i divides $P(m)$. As n varies over integers congruent to m modulo $\prod p_i^{e_i+1}$, $P(n)$ assumes arbitrarily large values (since P is nonconstant), and none of these values is divisible by $p_i^{e_i+1}$, since $P(m)$ is not. Thus there exist arbitrarily large integers n for which $P(n)$ is divisible by a prime p not dividing D.

If p^2 does not divide $P(n)$, then $P(n)$ is not a square. Otherwise,

$$P(n+p) \equiv P(n) + pP'(n) \equiv pP'(n) \pmod{p^2},$$

and p does not divide $P'(n)$ since p does not divide the discriminant of P. Thus $P(n+p)$ is divisible by p but not p^2, so $P(n+p)$ is not a square. □

Remark. The end of the argument is related to Hensel's Lemma; a similar idea arises in 1986B3.

Proof 3. We show that for any sufficiently large prime p, there exist arbitrarily large integers n such that $P(n)$ is not a square modulo p. Again we reduce to the case where P is squarefree and nonconstant, say of degree d. Then the discriminant Δ of P is a nonzero integer; let p be any odd prime not dividing Δ. Suppose that $P(n)$ is a square modulo p for all sufficiently large n. Then for each x in the field $\mathbb{F}_p$ of p elements, the equation $y^2 = P(x)$ has two solutions y, unless x is a zero of P modulo p in which case there is only one solution y. The number of zeros of P modulo p is at most d, so $y^2 = P(x)$ has at least $2p - d$ solutions $(x, y) \in \mathbb{F}_p \times \mathbb{F}_p$. On the other hand, the Weil Conjectures (see the end of 1991B5) for the curve $y^2 = P(x)$ imply that the number of solutions is at most $p + c\sqrt{p}$, where the constant c depends only on d. This yields a contradiction for sufficiently large p. □

Remark. A fourth proof of the stronger result can be given using the Thue-Siegel Theorem [HiS, Theorem D.8.3], which asserts that if P is a polynomial of degree at least 3 with no repeated factors, then $P(n)$ is a perfect square for only finitely many integers n. For an effective bound on the size of such n, see Section 4.3 of [Bak2]. (The Thue-Siegel Theorem can fail if P has degree 2. For example, $2n^2+1$ is a square for infinitely many integers n. See Solution 2 to 2000A2.)

Related question. A similar problem is 1971A6 [PutnamII, p. 15]:

> Let c be a real number such that n^c is an integer for every positive integer n. Show that c is a nonnegative integer.

The Sixtieth William Lowell Putnam Mathematical Competition
December 4, 1999

A1. (124, 17, 34, 0, 0, 0, 0, 0, 10, 4, 11, 5)
Find polynomials $f(x)$, $g(x)$, and $h(x)$, if they exist, such that, for all x,

$$|f(x)| - |g(x)| + h(x) = \begin{cases} -1 & \textbf{if } x < -1 \\ 3x+2 & \textbf{if } -1 \le x \le 0 \\ -2x+2 & \textbf{if } x > 0. \end{cases}$$

Answer. Take

$$f(x) = \frac{3x+3}{2}, \quad g(x) = \frac{5x}{2}, \quad \text{and} \quad h(x) = -x + \frac{1}{2}.$$

Solution. Let

$$F(x) = \begin{cases} -1 & \text{if } x < -1 \\ 3x+2 & \text{if } -1 \le x \le 0 \\ -2x+2 & \text{if } x > 0. \end{cases}$$

The function $G(x) = \max\{-1, 3x+2\}$ agrees with $F(x)$ for $x \le 0$, but

$$F(x) - G(x) = -5x \quad \text{for } x > 0.$$

Thus

$$\begin{aligned} F(x) &= \max\{-1, 3x+2\} - \max\{0, 5x\} \\ &= (3x+1+|-3x-3|)/2 - (5x+|5x|)/2 \\ &\qquad \text{(by the identity } \max\{r,s\} = (r+s+|r-s|)/2) \\ &= |(3x+3)/2| - |5x/2| - x + \frac{1}{2}, \end{aligned}$$

so we may take

$$f(x) = (3x+3)/2, \qquad g(x) = 5x/2, \qquad h(x) = -x + 1/2. \qquad ■$$

Remark. Alternatively, based on Figure 38, one may guess that f, g, h are linear, and that f changes sign at -1 and g changes sign at 0. We may assume that f and g have positive leading coefficients. Then one has the linear equations

$$\begin{aligned} -1 &= -f(x) + g(x) + h(x) \\ 3x+2 &= f(x) + g(x) + h(x) \\ -2x+2 &= f(x) - g(x) + h(x) \end{aligned}$$

from which one solves for f, g, h as given above.

In fact, both guesses can be justified, and we can prove that the solution is unique up to the signs of f and g. Of the four polynomials $\pm f \pm g + h$, three must be equal to -1, $3x+2$, $-2x+2$ in some order. But of any three of these functions, two sum to $2h$, and the difference between some two is $2f$ or $2g$. Thus f, g, h are each linear. As for the sign changes, assume that the graphs of f and g have positive slope. The

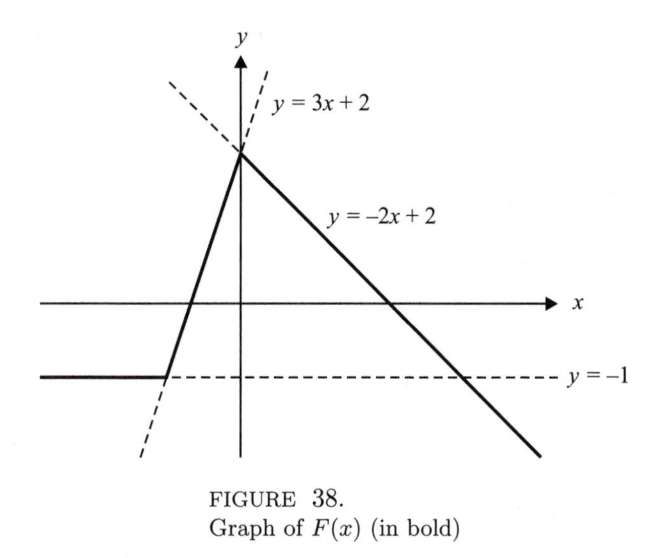

FIGURE 38.
Graph of $F(x)$ (in bold)

slope of the graph of $|f(x)| - |g(x)| + h(x)$ jumps up at the zero of f and jumps down at the zero of g, so these points must be $x = -1$ and $x = 0$, respectively.

A2. (61, 16, 28, 0, 0, 0, 0, 0, 2, 4, 52, 42)

Let $p(x)$ be a polynomial that is nonnegative for all real x. Prove that for some k, there are polynomials $f_1(x), \ldots, f_k(x)$ such that

$$p(x) = \sum_{j=1}^{k} (f_j(x))^2.$$

Solution 1. If $p(x)$ is identically zero, we are done. Otherwise factor $p(x)$ into linear and quadratic factors over the real numbers. Each linear factor must occur with even multiplicity, or else $p(x)$ would have a sign change at the zero of the linear factor. Thus $p(x)$ can be factored into squares of linear factors and irreducible quadratic factors. The latter can be written as sums of squares (namely, $x^2 + ax + b = (x + a/2)^2 + (b - a^2/4)$), so $p(x)$ is the product of polynomials which can be written as sums of squares, so is itself a sum of squares. ■

Solution 2. We proceed by induction on the degree of p, with base case where p has degree 0. As noted in Solution 1, real zeros of p occur with even multiplicity, so we may divide off the linear factors having such zeros to reduce to the case where p has no real zeros (and is nonconstant, or else we are already done). Then $p(x) > 0$ for all real x, and since $p(x) \to +\infty$ as $x \to \pm\infty$, p attains a minimum value $c > 0$. Now $p(x) - c$ has real zeros, so as above, we deduce that $p(x) - c$ is a sum of squares. Adding one more square, namely $(\sqrt{c})^2$, to $p(x) - c$ expresses $p(x)$ as a sum of squares. ■

Stronger result. In fact, only two polynomials are needed. This can be deduced from Solution 1, provided that one knows the identity (1) in Solution 3 to 2000A2, which says that a sum of two squares times a sum of two squares can be expressed as a sum of two squares.

Here is another proof that $k = 2$ suffices. Factor $p(x) = q(x)r(x)$, where q has all real zeros and r has all nonreal zeros and r is monic. Each zero of q has even multiplicity, since otherwise p would have a sign change at that zero. Also q has positive leading coefficient, since p must. Thus $q(x)$ has a square root $s(x)$ with real coefficients. Now write $r(x) = \prod_{j=1}^{k}(x - a_j)(x - \overline{a_j})$ (possible because r has zeros in complex conjugate pairs). Write $\prod_{j=1}^{k}(x - a_j) = t(x) + iu(x)$ with t and u having real coefficients. Then for x real,

$$\begin{aligned} p(x) &= q(x)r(x) \\ &= s(x)^2\,(t(x) + iu(x))\,(\overline{t(x) + iu(x)}) \\ &= (s(x)t(x))^2 + (s(x)u(x))^2. \end{aligned}$$

Literature note. This problem appeared as [Lar1, Problem 4.2.21], and is credited there to [MathS].

Remark. A polynomial in more than one variable with real coefficients taking nonnegative values need not be a sum of squares. (The reader may check that $1 - x^2y^2(1 - x^2 - y^2)$ is a counterexample.) Hilbert's Seventeenth Problem asked whether such a polynomial can always be written as the sum of squares of *rational functions*; this was answered affirmatively by E. Artin and O. Schreier. See [J, Section 11.4] for a proof and for generalizations.

A3. (103, 10, 2, 0, 0, 0, 0, 0, 1, 18, 27, 44)

Consider the power series expansion

$$\frac{1}{1 - 2x - x^2} = \sum_{n=0}^{\infty} a_n x^n.$$

Prove that, for each integer $n \geq 0$, there is an integer m such that

$$a_n^2 + a_{n+1}^2 = a_m.$$

Solution 1. Note that

$$\frac{1}{1 - 2x - x^2} = \frac{1}{2\sqrt{2}}\left(\frac{1 + \sqrt{2}}{1 - (1 + \sqrt{2})x} - \frac{1 - \sqrt{2}}{1 - (1 - \sqrt{2})x}\right)$$

and

$$\frac{1}{1 - (1 \pm \sqrt{2})x} = \sum_{n=0}^{\infty}(1 \pm \sqrt{2})^n x^n,$$

so

$$a_n = \frac{1}{2\sqrt{2}}\left((1 + \sqrt{2})^{n+1} - (1 - \sqrt{2})^{n+1}\right).$$

A simple computation now shows that $a_n^2 + a_{n+1}^2 = a_{2n+2}$. ∎

Solution 2. As in Solution 1, we see that $a_n = C\alpha^n + D\beta^n$ for some constants C, D, α, β with $\alpha\beta = -1$ (since α, β are the zeros of $x^2 - 2x - 1$). Thus

$$a_n^2 = E\alpha^{2n} + F(-1)^n + G\beta^{2n}$$

for some constants E, F, G, and

$$a_n^2 + a_{n+1}^2 = H\alpha^{2n} + I\beta^{2n}$$

for some constants H, I. But a_{2n+2} has the same form, so $a_n^2 + a_{n+1}^2$ and a_{2n+2} satisfy the same second-order linear recursion. (See the remarks in 1988A5 for more on linear recursions.) Hence we can prove $a_n^2 + a_{n+1}^2 = a_{2n+2}$ for all $n \geq 0$ by checking it for $n = 0$ and $n = 1$, which is easy. ■

Remark. From the recursion $a_{n+1} = 2a_n + a_{n-1}$, one can also find the recursion satisfied by a_{2n+2}, then directly prove that $a_n^2 + a_{n+1}^2$ satisfies the same recursion.

Solution 3 (Richard Stanley). Let A be the matrix $\begin{pmatrix} 0 & 1 \\ 1 & 2 \end{pmatrix}$. By induction, the recursion $a_{n+1} = 2a_n + a_{n-1}$ implies

$$A^{n+2} = \begin{pmatrix} a_n & a_{n+1} \\ a_{n+1} & a_{n+2} \end{pmatrix}.$$

The desired result follows from equating the top left entries in the matrix equality $A^{n+2}A^{n+2} = A^{2n+4}$. ■

Related question. This sequence, translated by one, also appeared in the following proposal by Bulgaria for the 1988 International Mathematical Olympiad [IMO88, p. 62].

> An integer sequence is defined by $a_n = 2a_{n-1} + a_{n-2}$ $(n > 1)$, $a_0 = 0$, $a_1 = 1$. Prove that 2^k divides a_n if and only if 2^k divides n.

A4. (33, 27, 3, 0, 0, 0, 0, 0, 3, 2, 14, 123)
Sum the series

$$\sum_{m=1}^{\infty}\sum_{n=1}^{\infty} \frac{m^2 n}{3^m(n3^m + m3^n)}.$$

Answer. The series converges to $9/32$.

Solution. Denote the series by S, and let $a_n = 3^n/n$. Note that

$$S = \sum_{m=1}^{\infty}\sum_{n=1}^{\infty} \frac{1}{a_m(a_m + a_n)} = \sum_{m=1}^{\infty}\sum_{n=1}^{\infty} \frac{1}{a_n(a_m + a_n)},$$

where the second equality follows by interchanging m and n. Thus

$$\begin{aligned} 2S &= \sum_m \sum_n \left(\frac{1}{a_m(a_m + a_n)} + \frac{1}{a_n(a_m + a_n)} \right) \\ &= \sum_m \sum_n \frac{1}{a_m a_n} = \left(\sum_{n=1}^{\infty} \frac{n}{3^n} \right)^2. \end{aligned}$$

Finally, if

$$A = \sum_{n=1}^{\infty} \frac{n}{3^n}, \tag{1}$$

then

$$3A = \sum_{n=1}^{\infty} \frac{n}{3^{n-1}} = \sum_{n=0}^{\infty} \frac{n+1}{3^n}, \tag{2}$$

so subtracting (1) from (2) gives

$$2A = 1 + \sum_{n=1}^{\infty} \frac{1}{3^n} = \frac{3}{2}.$$

Hence

$$A = \frac{3}{4}, \tag{3}$$

so $S = 9/32$. ■

Remark. Equation (3) also follows by differentiating both sides of

$$\sum_{n=0}^{\infty} \frac{x^n}{3^n} = \frac{3}{3-x},$$

and then evaluating at $x = 1$. Either method can be generalized to evaluate arbitrary sums of the form $\sum_{n=0}^{\infty} P(n)\alpha^n$ where $P(n)$ is a polynomial, and $|\alpha| < 1$.

Remark. The rearrangement of terms in various parts of the solution is justified, since all terms are nonnegative.

A5. (5, 1, 4, 0, 0, 0, 0, 0, 3, 10, 63, 119)

Prove that there is a constant C such that, if $p(x)$ is a polynomial of degree 1999, then

$$|p(0)| \leq C \int_{-1}^{1} |p(x)|\, dx.$$

Solution 1. Let P denote the set of polynomials of degree at most 1999. Identify P with $\mathbb{R}^{2000}$ by identifying $\sum_{i=0}^{1999} a_i x^i$ with $(a_0, a_1, \ldots, a_{1999})$. Let S be the set of polynomials $\sum_{i=0}^{1999} a_i x^i$ such that $\max\{|a_i|\} = 1$. Then S is a closed and bounded subset of $P \approx \mathbb{R}^{2000}$, so S is compact. The function $P \times \mathbb{R} \to \mathbb{R}$ mapping (p, x) to $|p(x)|$ is the absolute value of a polynomial in all 2001 coordinates, so it is continuous. Therefore the function $g : P \to \mathbb{R}$ defined by $g(p) = \int_{-1}^{1} |p(x)|\, dx$ is continuous, as is its restriction to S. Similarly the function $f : S \to \mathbb{R}$ defined by $f(p) = |p(0)|$ is continuous. Since $g(p) \neq 0$ for $p \in S$, the quotient $f/g : S \to \mathbb{R}$ is continuous. By the Extreme Value Theorem, there exists a constant C such that $f(p)/g(p) \leq C$ for all $p \in S$. An arbitrary $p \in P$ can be written as cq for some $c \in \mathbb{R}$ and $q \in S$; then $f(p) = |c| f(q) \leq |c| C g(q) = C g(p)$, as desired. ■

Remark. The same method proves the standard result that any two norms on a finite-dimensional vector space are equivalent [Hof, p. 249]. In fact, another way to solve the problem would be to apply this result to the two norms

$$\sup_{x \in [-1,1]} |p(x)| \qquad \text{and} \qquad \int_{-1}^{1} |p(x)|\, dx$$

on P.

Solution 2 (Reid Barton). We exhibit an explicit constant C, by showing that $|p(x)|$ must be large compared to $|p(0)|$ on some subinterval of $[-1,1]$ whose length is bounded below. Assume $p(0)=1$ without loss of generality.

Then $p(x)=\prod_{i=1}^{1999}(1-x/r_i)$, where $r_1,\dots,r_{1999}$ are the complex zeros of p, listed with multiplicity. Fix $\epsilon<1/3998$, and draw closed discs of radius ϵ centered at the r_i in the complex plane. These discs intersect $(-1/2,1/2)$ in at most 1999 intervals of total length at most 3998ϵ, so their complement consists of at most 2000 intervals of total length at least $1-3998\epsilon$. By the Pigeonhole Principle, one of these intervals, say (c,d), has length at least $\delta=(1-3998\epsilon)/2000>0$. For $x\in(c,d)$, if $|r_i|\le 1$ then $|1-x/r_i|\ge|x-r_i|>\epsilon$, whereas if $|r_1|\ge 1$ then $|1-x/r_i|\ge 1-|x/r_i|\ge 1/2>\epsilon$. Hence

$$\int_{-1}^{1}|p(x)|\,dx\ge\int_{c}^{d}\prod_{i=1}^{1999}|1-x/r_i|\,dx\ge\delta\epsilon^{1999},$$

so

$$|p(0)|\le C\int_{-1}^{1}|p(x)|\,dx$$

with $C=1/(\delta\epsilon^{1999})$. Taking $\epsilon=1/4000$ yields the explicit value $C=2^{1999}2000^{2001}$. ∎

Related question. Can you improve the constant in Solution 2? (We do not know what the smallest possible C is.)

A6. (31, 11, 9, 0, 0, 0, 0, 0, 1, 5, 28, 120)

The sequence $(a_n)_{n\ge1}$ is defined by $a_1=1, a_2=2, a_3=24$, and, for $n\ge4$,

$$a_n=\frac{6a_{n-1}^2a_{n-3}-8a_{n-1}a_{n-2}^2}{a_{n-2}a_{n-3}}.$$

Show that, for all n, a_n is an integer multiple of n.

Solution. Rearranging the given equation yields the much more tractable equation

$$\frac{a_n}{a_{n-1}}=6\frac{a_{n-1}}{a_{n-2}}-8\frac{a_{n-2}}{a_{n-3}}.$$

Let $b_n=a_n/a_{n-1}$. With the initial conditions $b_2=2, b_3=12$, one obtains $b_n=2^{n-1}(2^{n-1}-1)$, by induction or by the theory of linear recursive sequences: see the remark in 1988A5. Thus

$$a_n=a_1b_2b_3\dots b_n=2^{n(n-1)/2}\prod_{i=1}^{n-1}(2^i-1).\tag{1}$$

If $n=1$, then n divides a_n. Otherwise factor n as 2^km, with m odd. Then $k\le n-1\le n(n-1)/2$, and there exists $i\le n-1$ such that m divides 2^i-1, namely $i=\phi(m)$. (Here ϕ denotes the Euler ϕ-function: see 1985A4.) Hence n divides a_n for all $n\ge1$. ∎

Remark. Alternatively, the result for $n\ge3$ can be proved from the following two facts:

(a) The right side of the formula (1) for a_n equals $2^{n-1}\#\mathrm{GL}_{n-1}(\mathbb{F}_2)$. (See page xi for the definition of $\mathrm{GL}_{n-1}(\mathbb{F}_2)$.)

(b) If $n\ge3$, $\mathrm{GL}_{n-1}(\mathbb{F}_2)$ contains an element of exact order n.

These suffice because of Lagrange's Theorem, which states that the order of an element of a finite group G divides the order of G.

Let us prove (a). Matrices in $\mathrm{GL}_{n-1}(\mathbb{F}_2)$ can be constructed one row at a time: the first row may be any nonzero vector, and then each successive row may be any vector not in the span of the previous rows, which by construction are independent. Hence the number of possibilities for the jth row, given the previous ones, is $2^{n-1} - 2^{j-1}$. Thus

$$\#\mathrm{GL}_{n-1}(\mathbb{F}_2) = \prod_{j=1}^{n-1}(2^{n-1} - 2^{j-1}) = 2^{(n-1)(n-2)/2}\prod_{j=1}^{n-1}(2^{n-j} - 1).$$

Setting $i = n - j$ and comparing with (1) proves (a).

It remains to prove (b). Let $V = \{(x_1, \dots, x_n) \in (\mathbb{F}_2)^n : \sum x_i = 0\}$. Since $\dim_{\mathbb{F}_2} V = n - 1$, the group of automorphisms of the vector space V is isomorphic to $\mathrm{GL}_{n-1}(\mathbb{F}_2)$. Let $T : V \to V$ be the automorphism $(x_1, x_2, \dots, x_n) \mapsto (x_n, x_1, \dots, x_{n-1})$. Then T^n is the identity. If $m < n$ and $n \geq 3$, then there exists $v = (v_1, \dots, v_n) \in V$ with $v_1 = 1$ and $v_{m+1} = 0$: make just one other v_i equal to 1, to make the sum zero. Then $T^m v \neq v$, so T^m is not the identity. Hence T has exact order n.

(The matrix of T with respect to the basis

$$\epsilon_1 = (1, 1, 0, 0, \dots, 0),\ \epsilon_2 = (0, 1, 1, 0, \dots, 0),\ \dots,\ \epsilon_{n-1} = (0, 0, \dots, 0, 1, 1)$$

of V is

$$\begin{pmatrix} 0 & 0 & 0 & \cdots & 0 & 1 \\ 1 & 0 & 0 & \cdots & 0 & 1 \\ 0 & 1 & 0 & \cdots & 0 & 1 \\ \vdots & \vdots & \vdots & \ddots & \vdots & \vdots \\ 0 & 0 & 0 & \cdots & 0 & 1 \\ 0 & 0 & 0 & \cdots & 1 & 1 \end{pmatrix} \in \mathrm{GL}_{n-1}(\mathbb{F}_2).$$

This also equals the companion matrix of the polynomial

$$f(x) = x^{n-1} + x^{n-2} + \cdots + x + 1.)$$

B1. (126, 17, 17, 0, 0, 0, 0, 0, 6, 0, 23, 16)

Right triangle ABC has right angle at C and $\angle BAC = \theta$; the point D is chosen on AB so that $|AC| = |AD| = 1$; the point E is chosen on BC so that $\angle CDE = \theta$. The perpendicular to BC at E meets AB at F. Evaluate $\lim_{\theta \to 0} |EF|$. [Here $|PQ|$ denotes the length of the line segment PQ.][†]

Answer. The limit of $|EF|$ as $\theta \to 0$ is $1/3$.

Solution. (See Figure 39.) The triangles BEF and BCA are similar. Since $AC = 1$, we have $EF = BE/BC$. Also, since $\triangle ACD$ is isosceles, $\angle ACD = \angle ADC = \pi/2 - \theta/2$. We then compute $\angle BCD = \pi/2 - \angle DCA = \theta/2$ and $\angle BDE = \pi - \angle CDE - \angle CDA = \pi/2 - \theta/2$.

[†] The figure is omitted here, since it is contained in Figure 39 used in the solution.

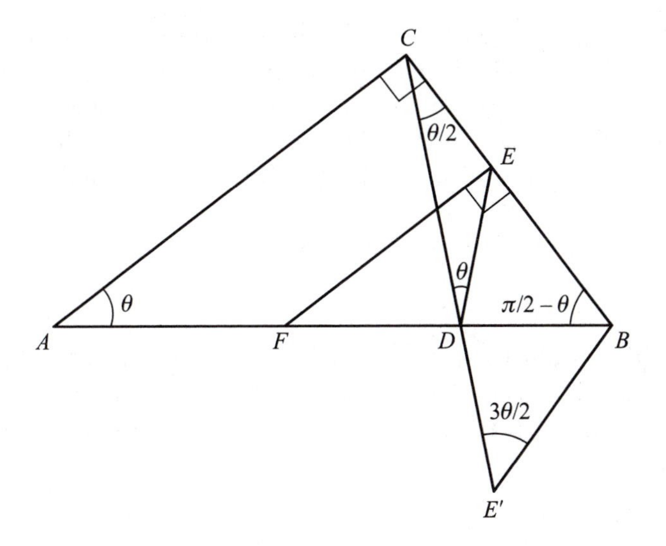

FIGURE 39.

The reflection E' of E across AB lies on CD, so by the Law of Sines,

$$\frac{BE}{BC} = \frac{BE'}{BC} = \frac{\sin \angle BCD}{\sin \angle CE'B}.$$

Since

$$\angle CBD = \frac{\pi}{2} - \angle BAC = \frac{\pi}{2} - \theta$$

and

$$\angle CE'B = \angle DEB = \frac{\pi}{2} - \angle CBD - \angle BDE = \frac{3\theta}{2},$$

we obtain

$$EF = \frac{BE}{BC} = \frac{\sin(\theta/2)}{\sin(3\theta/2)}.$$

By L'Hôpital's Rule,

$$\lim_{\theta \to 0} \frac{\sin(\theta/2)}{\sin(3\theta/2)} = \lim_{\theta \to 0} \frac{\cos(\theta/2)}{3\cos(3\theta/2)} = \frac{1}{3}. \qquad \blacksquare$$

Remark. One can avoid using the reflection by writing

$$\frac{BE}{BC} = \frac{BE}{BD} \cdot \frac{BD}{BC}$$

and using the Law of Sines in triangles BDE and BCD.

B2. (38, 27, 9, 0, 0, 0, 0, 0, 9, 0, 74, 48)

Let $P(x)$ be a polynomial of degree n such that $P(x) = Q(x)P''(x)$, where $Q(x)$ is a quadratic polynomial and $P''(x)$ is the second derivative of $P(x)$. Show that if $P(x)$ has at least two distinct roots then it must have n distinct roots. [The roots may be either real or complex.]

Solution. Suppose that P does not have n distinct zeros; then it has a zero of multiplicity $k \geq 2$, which we may assume without loss of generality is $x = 0$. Differentiating P term by term shows that the highest power of x dividing $P''(x)$ is x^{k-2}. But $P(x) = Q(x)P''(x)$, so x^2 divides $Q(x)$. Since Q is quadratic, $Q(x) = Cx^2$ for some constant C. Comparing the leading coefficients of $P(x)$ and $Q(x)P''(x)$ yields $C = \frac{1}{n(n-1)}$.

Write $P(x) = \sum_{j=0}^{n} a_j x^j$; equating coefficients in $P(x) = Cx^2P''(x)$ implies that $a_j = Cj(j-1)a_j$ for all j. Hence $a_j = 0$ for $j \leq n-1$, and $P(x) = a_n x^n$, which has all identical zeros. ∎

Remark. If Q has distinct zeros, then after replacing x by $ax+b$ for suitable $a, b \in \mathbb{C}$ we may assume that $Q = c(1-x^2)$ for some $c \in \mathbb{C}$. Then $P''(x)$ is a scalar multiple of $R_{n-2}(x)$, where $R_n(x)$ is the unique monic polynomial of degree n satisfying the differential equation

$$\frac{d^2}{dx^2}[(1-x^2)R_n(x)] = -(n+1)(n+2)R_n(x).$$

(This equation implies a recursion which uniquely determines the coefficients of R_n.) Integration by parts yields

$$\int_{-1}^{1} (1-x^2)R_m(x)R_n(x)\,dx = \frac{1}{(n+1)(n+2)} \int_{-1}^{1} [(1-x^2)R_m(x)]'[(1-x^2)R_n(x)]'\,dx;$$

interchanging m and n and comparing, we find

$$\int_{-1}^{1} (1-x^2)R_m(x)R_n(x)\,dx = 0$$

for $m \neq n$. That is, the $R_m(x)$ form a family of orthogonal polynomials with respect to the measure $(1-x^2)\,dx$. (In other terminology, the R_m are eigenvectors of the operator $f \mapsto \frac{d^2}{dx^2}[(1-x^2)f]$. This operator is self-adjoint with respect to the measure $(1-x^2)\,dx$, so the eigenvectors are orthogonal with respect to that measure, because eigenspaces corresponding to different eigenvalues are orthogonal.)

The solution above implies that R_n has distinct zeros. A result from the theory of orthogonal polynomials implies that the zeros of R_n are real numbers in $[-1, 1]$. This can also be deduced by applying Lucas' Theorem (1991A3) twice. Namely, if the convex hull of the zeros of the polynomial $P(x) = (1-x^2)R_n(x)$ has a vertex z other than ± 1, then Lucas' Theorem implies that z lies outside of the convex hull of the zeros of P' (because z is not a multiple zero of P) and then implies that z lies outside of the convex hull of the zeros of $P''(x) = R_n(x)$, a contradiction.

Noam Elkies points out that according to [GR], $R_n(x)$ is a scalar multiple of the nth Gegenbauer polynomial with parameter $\lambda = 3/2$, also called the nth *ultraspherical polynomial.* Also, if we put $P_n(x) = (1-x^2)R_n(x)$, then $P_n'(x)$ is (a scalar multiple) of the $(n+1)$st Legendre polynomial.

Related question. Problem 6 from the final round of the 1978 Swedish Mathematical Olympiad [SMO] is related:

The polynomials

$$P(x) = cx^n + a_{n-1}x^{n-1} + \cdots + a_1x + a_0$$
$$Q(x) = cx^m + b_{m-1}x^{m-1} + \cdots + b_1x + b_0$$

with $c \neq 0$ satisfy the identity

$$P(x)^2 = (x^2 - 1)Q(x)^2 + 1.$$

Show that $P'(x) = nQ(x)$.

B3. (44, 2, 2, 0, 0, 0, 0, 0, 13, 14, 56, 74)
Let $A = \{(x,y) : 0 \le x, y < 1\}$. For $(x,y) \in A$, let

$$S(x,y) = \sum_{\frac{1}{2} \le \frac{m}{n} \le 2} x^m y^n,$$

where the sum ranges over all pairs (m,n) of positive integers satisfying the indicated inequalities. Evaluate

$$\lim_{\substack{(x,y)\to(1,1) \\ (x,y)\in A}} (1 - xy^2)(1 - x^2y)S(x,y).$$

Answer. The limit is equal to 3.

Solution 1. For $(x,y) \in A$,

$$\sum_{m,n>0} x^m y^n = \frac{xy}{(1-x)(1-y)}.$$

Subtracting S from this gives two sums, one of which is

$$\sum_{m \ge 2n+1} x^m y^n = \sum_n y^n \frac{x^{2n+1}}{1-x} = \frac{x^3 y}{(1-x)(1-x^2y)}$$

and the other of which analogously sums to

$$\frac{xy^3}{(1-y)(1-xy^2)}.$$

Therefore

$$\begin{aligned} S(x,y) &= \frac{xy}{(1-x)(1-y)} - \frac{x^3y}{(1-x)(1-x^2y)} - \frac{xy^3}{(1-y)(1-xy^2)} \\ &= \frac{xy(1 + x + y + xy - x^2y^2)}{(1-x^2y)(1-xy^2)} \end{aligned}$$

and the desired limit is $\lim_{(x,y)\to(1,1)} xy(1 + x + y + xy - x^2y^2) = 3$. ■

Solution 2. A pair (m,n) of positive integers satisfies $1/2 \le m/n \le 2$ if and only if $(m,n) = a(1,2) + b(2,1)$ for some nonnegative rational numbers a, b. The pairs (a,b) with $0 \le a, b < 1$ which yield integral values of m and n are $(0,0), (1/3,1/3), (2/3,2/3)$, whose corresponding pairs are $(0,0), (1,1), (2,2)$. Thus every pair (m,n) with $1/2 \le m/n \le 2$ can be written as one of these three pairs plus

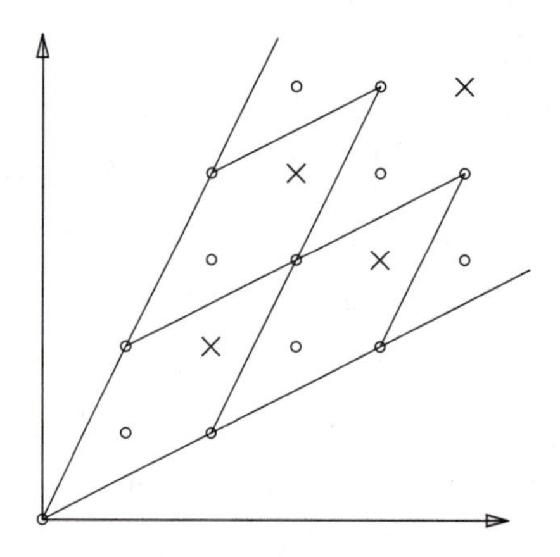

FIGURE 40.
Points of the form $a(1,2)+b(1,2)$, where a and b are nonnegative integers; two translates of this semigroup are also indicated.

an integral linear combination of $(1,2)$ and $(2,1)$. (See Figure 40 for those points of the form $a(1,2)+b(1,2)$ where a and b are nonnegative integers.) In particular,

$$\begin{aligned} S(x,y)+1 &= (1+xy+x^2y^2)\sum_{a=0}^{\infty} x^a y^{2a} \sum_{b=0}^{\infty} x^{2b}y^b \\ &= \frac{1+xy+x^2y^2}{(1-xy^2)(1-x^2y)} \end{aligned}$$

and the desired limit is 3. ■

Stronger result. Let p/q and r/s be positive rational numbers with $p/q < r/s$, and define $T(x,y) = \sum_{(m,n)} x^m y^n$, the sum taken over all pairs (m,n) of positive integers with $p/q \le m/n \le r/s$. Then

$$\lim_{\substack{(x,y)\to(1,1) \\ (x,y)\in A}} (1-x^p y^q)(1-x^r y^s)T(x,y) = qr - ps.$$

This can be proved by either of the above methods.

B4. (0, 0, 0, 0, 0, 0, 0, 0, 1, 0, 31, 173)

Let f be a real function with a continuous third derivative such that $f(x)$, $f'(x)$, $f''(x)$, $f'''(x)$ are positive for all x. Suppose that $f'''(x) \le f(x)$ for all x. Show that $f'(x) < 2f(x)$ for all x.

Remark. The *total* score of the top 205 writers in 1999 on this problem was 2. The same is true of the next problem, 1999B5. By this measure, these two are the hardest Putnam problems in the years covered in this volume. Nevertheless, we feel that there are other problems in this volume that are at least as difficult as these two. At the other extreme, the problems having the highest average score of the top 200 or so participants were 1988A1 and then 1988B1.

Solution 1. For simplicity, we will only show $f'(0) < 2f(0)$. Applying this result to $f(x+c)$ shows that $f'(c) < 2f(c)$ for all c.

Since f' is positive, f is an increasing function. Thus for $x \le 0$, $f'''(x) \le f(x) \le f(0)$. Integrating $f'''(x) \le f(0)$ from x to 0 gives $f''(x) \ge f''(0) + f(0)x$ for $x \le 0$, and integrating again gives the second inequality in $0 < f'(x) \le f'(0) + f''(0)x + f(0)x^2/2$. Thus the polynomial $f'(0) + f''(0)x + f(0)x^2/2$ has no negative zeros. Since its coefficients are positive, it also has no nonnegative zeros. Therefore its discriminant $f''(0)^2 - 2f(0)f'(0)$ must be negative.

In a similar vein, since f''' is positive, f'' is increasing. Thus for $x \le 0$, $f''(x) \le f''(0)$, so $f'(x) \ge f'(0) + f''(0)x$, and $0 < f(x) \le f(0) + f'(0)x + f''(0)x^2/2$. Again, the discriminant of the quadratic must be negative: $f'(0)^2 - 2f(0)f''(0) < 0$.

Combining the conclusions of the last two paragraphs, we obtain

$$f'(0)^4 < 4f(0)^2 f''(0)^2 < 8f(0)^3 f'(0),$$

which implies $f'(0) < 2f(0)$. ∎

Solution 2. We prove that $f'(x) < (9/2)^{1/3} f(x) < 1.651 f(x)$. Since $f(x)$ is bounded below (by 0) and increasing, it has an infimum m, and $\lim_{x\to-\infty} f(x) = m$. Then $f'(x)$ tends to 0 as $x \to -\infty$ since it is positive and its integral from $-\infty$ to 0 converges. Similarly $f''(x)$ and $f'''(x)$ tend to 0 as $x \to -\infty$.

By Taylor's Formula with remainder (or integration by parts), for any function g,

$$g(x) - g(x-s) = \int_0^s g'(x-t)\,dt,$$

$$g(x) - g(x-s) = sg'(x-s) + \int_0^s tg''(x-t)\,dt, \tag{2}$$

$$g(x) - g(x-s) = sg'(x-s) + \frac{1}{2}s^2 g''(x-s) + \int_0^s \frac{1}{2}t^2 g'''(x-t)\,dt, \tag{3}$$

as long as the specified derivatives exist and are continuous. For $g = f$, as $s \to +\infty$ for fixed x, the left side of (2) is bounded above, the term $sf'(x-s)$ is nonnegative and the integrand $tf''(x-t)$ is everywhere nonnegative; consequently, the integrand tends to 0 as $t \to \infty$. Thus in (2) with $g = f'$, letting s tend to ∞ yields

$$f'(x) = \int_0^\infty tf'''(x-t)\,dt.$$

We cannot *a priori* take limits in any of the equations with $g = f$, but from (3) we have

$$f(x) - \int_0^s \frac{1}{2}t^2 f'''(x-t)\,dt = f(x-s) + sf'(x-s) + \frac{1}{2}s^2 f''(x-s) \ge 0$$

for all s, so

$$f(x) \ge \int_0^\infty \frac{1}{2}t^2 f'''(x-t)\,dt.$$

Thus for $c \ge 0$, we have

$$cf(x) - f'(x) \ge \int_0^\infty \left(\frac{1}{2}ct^2 - t\right) f'''(x-t)\,dt.$$

The integrand is positive precisely when $t > 2/c$. Moreover, $f'''(x-t) \le f(x-t)$ by hypothesis and $f(x-t) \le f(x)$ for $t \ge 0$, since f' is positive. We conclude

$$cf(x) - f'(x) > \int_0^{2/c} \left(\frac{1}{2}ct^2 - t\right) f(x)\, dt = -\frac{2}{3c^2} f(x),$$

or equivalently

$$f'(x) < f(x)\left(c + \frac{2}{3}c^{-2}\right).$$

The minimum value of the expression in parentheses is $(9/2)^{1/3}$, achieved when $c = (4/3)^{1/3}$: this can be proved by calculus or by the $n = 3$ case of the AM-GM Inequality applied to $c/2$, $c/2$, and $\frac{2}{3}c^{-2}$. (The AM-GM Inequality is given at the end of 1985A2.) ■

Solution 3. We prove that $f'(x) < 2^{1/6} f(x) < 1.123 f(x)$. As in the previous solution, $f^{(i)}(x) \to 0$ as $x \to -\infty$ for $i = 1, 2, 3$. For similar reasons, $\lim_{x\to-\infty} f(x)f'(x) = 0$.

First notice that

$$2f(x)f''(x) + 2f'(x)^2 - 2f''(x)f'''(x) > 0$$

because the first term plus the third term is nonnegative and the second term is positive. Consequently,

$$2f(x)f'(x) - f''(x)^2 > 0$$

since this expression has positive derivative and tends to 0 as $x \to -\infty$.

Combining this with $f'''(x) \le f(x)$ yields

$$2f(x)^2 f'(x) - f''(x)^2 f'''(x) > 0,$$

from which we obtain

$$2f(x)^3 - f''(x)^3 > 2m^3 \ge 0$$

by similar reasoning, where $m = \lim_{x\to-\infty} f(x)$. This inequality implies

$$2^{1/3} f(x)f'(x) - f'(x)f''(x) > 0,$$

and integrating yields

$$2^{1/3} f(x)^2 - f'(x)^2 > 2^{1/3} m^2 \ge 0,$$

which implies the desired result. ■

Remark. Let C be the infimum of the set of $\gamma > 0$ such that the conditions of the problem imply $f'(x) \le \gamma f(x)$ for all x. Solution 3 shows that $C \le 2^{1/6}$, and the example $f(x) = e^x$ shows that $C \ge 1$. One might guess from this that $C = 1$, but we now give an example to show $C > 1$. (We do not know the exact value of C.)

We first construct an example where f''' has a discontinuity at 0, then show that f''' can be modified slightly while retaining the property that $f'(x) > f(x)$ at some point. Our strategy is to consider

$$H_3(x) = \begin{cases} e^x, & \text{if } x \le 0 \\ e^x - cg(x), & \text{if } x > 0, \end{cases}$$

where $g(x)$ is a solution of the differential equation $g''' = g$, to set $H_j(x) = \int_{-\infty}^{x} H_{j+1}(t)\,dt$ for $j = 2, 1, 0$, then to take $f(x) = H_0(x)$. For $x > 0$, we have

$$\begin{aligned} H_3(x) &= e^x - cg'''(x) \\ H_2(x) &= e^x - cg''(x) + cg''(0) \\ H_1(x) &= e^x - cg'(x) + c[g'(0) + g''(0)x] \\ H_0(x) &= e^x - cg(x) + c\left[g(0) + g'(0)x + \frac{1}{2}g''(0)x^2\right]. \end{aligned}$$

All solutions of $g''' = g$ are $O(e^x)$ as $x \to \infty$, so for c sufficiently small, $H_3(x) > 0$ for all $x > 0$. Hence $H_i(x) > 0$ for $x > 0$, $i = 2, 1, 0$ as well. Now

$$H_0(x) - H_3(x) = c\left(g(0) + g'(0)x + \frac{1}{2}g''(0)x^2\right).$$

We make the right-hand side positive by taking g to be the solution to $g''' = g$ whose Taylor polynomial of degree 2 equals $(x - d)^2$ for some $d > 0$, For sufficiently small c, the function $f(x) = H_0(x)$ now satisfies the conditions of the problem, except for continuity of the third derivative.

Also,

$$f'(d) - f(d) = H_1(d) - H_0(d) = c[g(d) - g'(d)].$$

If there exists $d > 0$ such that the corresponding g satisfies $g(d) > g'(d)$, we have $f'(d) > f(d)$ as desired. A numerical computation shows that this occurs for $d = 8$.

To modify this example so that f''' is continuous, choose $\delta > 0$ and define h_3 to be the function that is equal to H_3 outside $(0, \delta)$ and is linear between 0 and δ: see Figure 41. Define $h_i(x) = \int_{-\infty}^{x} h_{i+1}(t)\,dt$ for $i = 2, 1, 0$ and take $f(x) = h_0(x)$. Since

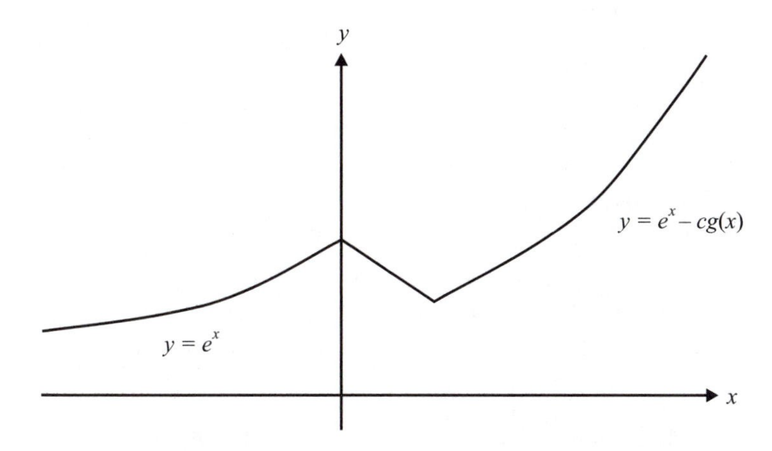

FIGURE 41.
The graph of $y = h_3(x)$.

$h_3(x) \geq H_3(x)$ everywhere (for δ sufficiently small), we also have $f^{(i)}(x) = h_i(x) \geq H_i(x) > 0$ for $i = 2, 1, 0$. For $x > \delta$, we have $f(x) \geq H_0(x) \geq H_3(x) = f'''(x)$; for $0 < x < \delta$, we have $f(x) \geq f(0) = 1 \geq h_3(x) = f'''(x)$, provided that δ is small enough that $H_3(\delta) < 1$. Thus we have $f'''(x) \leq f(x)$ for all x.

B5. (0, 0, 0, 0, 0, 0, 0, 0, 1, 0, 43, 161)

For an integer $n \geq 3$, let $\theta = 2\pi/n$. Evaluate the determinant of the $n \times n$ matrix $I + A$, where I is the $n \times n$ identity matrix and $A = (a_{jk})$ has entries $a_{jk} = \cos(j\theta + k\theta)$ for all j, k.

Remark. The *total* score of the top 205 writers in 1999 on this problem was 2. See the comment at the beginning of 1999B4.

Answer. The determinant of $I + A$ is $1 - n^2/4$.

Solution 1. We compute the determinant of $I + A$ by computing its eigenvalues. The eigenvalues of $I + A$ are obtained by adding 1 to each of the eigenvalues of A, so it suffices to compute the latter.

We claim that the eigenvalues of A are $n/2, -n/2, 0, \ldots, 0$, where 0 occurs with multiplicity $n - 2$. To prove this claim, define vectors $v^{(m)}$, $0 \leq m \leq n - 1$, componentwise by $(v^{(m)})_k = e^{ikm\theta}$ (where $\theta = 2\pi/n$). If we form a matrix from the $v^{(m)}$, its determinant is a Vandermonde product and hence is nonzero. (See 1986A6 for a short explanation of the Vandermonde determinant.) Thus the $v^{(m)}$ form a basis for $\mathbb{C}^n$. Since $\cos z = (e^{iz} + e^{-iz})/2$ for any z,

$$\begin{aligned}
(Av^{(m)})_j &= \sum_{k=1}^{n} \cos(j\theta + k\theta) e^{ikm\theta} \\
&= \frac{e^{ij\theta}}{2} \sum_{k=1}^{n} e^{ik(m+1)\theta} + \frac{e^{-ij\theta}}{2} \sum_{k=1}^{n} e^{ik(m-1)\theta}.
\end{aligned}$$

Since $\sum_{k=1}^{n} e^{ik\ell\theta} = 0$ for integer ℓ unless $n \mid \ell$, we conclude that $Av^{(m)} = 0$ for $m = 0$ and for $2 \leq m \leq n - 1$. In addition, we find that $(Av^{(1)})_j = \frac{n}{2} e^{-ij\theta} = \frac{n}{2}(v^{(n-1)})_j$ and $(Av^{(n-1)})_j = \frac{n}{2} e^{ij\theta} = \frac{n}{2}(v^{(1)})_j$, so $A(v^{(1)} \pm v^{(n-1)}) = \pm\frac{n}{2}(v^{(1)} \pm v^{(n-1)})$. Thus $\{v^{(0)}, v^{(2)}, v^{(3)}, \ldots, v^{(n-2)}, v^{(1)} + v^{(n-1)}, v^{(1)} - v^{(n-1)}\}$ is a basis for $\mathbb{C}^n$ of eigenvectors of A with the claimed eigenvalues.

Finally, the determinant of $I + A$ is the product of $(1 + \lambda)$ over all eigenvalues λ of A, namely,

$$\det(I + A) = (1 + n/2)(1 - n/2) = 1 - n^2/4. \qquad \blacksquare$$

Motivation. An $n \times n$ matrix A with the property that A_{jk} depends only on $j - k$ (mod n) is called a *circulant matrix*. Such a matrix always has $v^{(0)}, \ldots, v^{(n-1)}$ as eigenvectors. In this problem, A is not circulant but A^2 is. See Solution 3 to 1988B5 for more on circulant matrices.

Solution 2. As in Solution 1, to compute $\det(I + A)$, it suffices to compute the eigenvalues of A. Let C be the vector with components $(\cos\theta, \cos 2\theta, \ldots, \cos n\theta)$, and let D be the vector with components $(\sin\theta, \sin 2\theta, \ldots, \sin n\theta)$. If we identify C and D

with the corresponding $n \times 1$ matrices, then the addition formula

$$\cos(j+k)\theta = \cos j\theta \cos k\theta - \sin j\theta \sin k\theta$$

implies the matrix identity $A = CC^T - DD^T$. Since CC^T and DD^T are matrices of rank 1, A has rank at most 2. In fact, the image of A (as a linear transformation on column vectors) is spanned by C and D. Thus the zero eigenspace of A has dimension at least $n-2$, and the remaining eigenspaces lie in the span of C and D.

Now

$$n = \sum_{j=1}^{n} \left(\cos^2 j\theta + \sin^2 j\theta\right) = C \cdot C + D \cdot D$$

and

$$\begin{aligned} 0 &= \sum_{j=1}^{n} \left(e^{2\pi i j} n\right)^2 \\ &= \sum_{j=1}^{n} \left(\cos^2 j\theta - \sin^2 j\theta + 2i \cos j\theta \sin j\theta\right) \\ &= C \cdot C - D \cdot D + 2iC \cdot D. \end{aligned}$$

Therefore, $C \cdot C = D \cdot D = n/2$ and $C \cdot D = 0$. We now compute

$$\begin{aligned} AC &= \left(CC^T - DD^T\right) C = C(C \cdot C) - D(D \cdot C) = (n/2)C \\ AD &= \left(CC^T - DD^T\right) D = C(C \cdot D) - D(D \cdot D) = -(n/2)D \end{aligned}$$

and conclude that the two remaining eigenvalues of A are $n/2$ and $-n/2$. ■

B6. (3, 5, 5, 0, 0, 0, 0, 0, 3, 0, 80, 109)

Let S be a finite set of integers, each greater than 1. Suppose that for each integer n there is some $s \in S$ such that $\gcd(s,n) = 1$ or $\gcd(s,n) = s$. Show that there exist $s, t \in S$ such that $\gcd(s,t)$ is prime.

Solution. Let n be the smallest positive integer such that $\gcd(s,n) > 1$ for all s in n; note that n has no repeated prime factors. By the condition on S, there exists $s \in S$ which divides n.

On the other hand, if p is a prime divisor of s, then by the minimality of n, n/p is relatively prime to some element t of S. Since n cannot be relatively prime to t, t is divisible by p, but not by any other prime divisor of s (any such prime divides n/p). Thus $\gcd(s,t) = p$, as desired. ■

Remark. The problem fails if S is allowed to be infinite. For example, let $p_1, p_2, \ldots$ be distinct primes and let $S = \{p_1p_2, p_3p_4, p_5p_6, \ldots\}$.

Remark. One can rephrase the problem in combinatorial terms, by replacing each integer n with the sequence $(n_2, n_3, n_5, \ldots)$, where n_p is the exponent of p in the factorization of n.

The Sixty-First William Lowell Putnam Mathematical Competition
December 2, 2000

A1. (61, 9, 5, 0, 0, 0, 0, 0, 14, 54, 41, 11)

Let A be a positive real number. What are the possible values of $\sum_{j=0}^{\infty} x_j^2$, given that $x_0, x_1, \ldots$ are positive numbers for which $\sum_{j=0}^{\infty} x_j = A$?

Answer. The possible values comprise the interval $(0, A^2)$.

Solution 1. Since all terms in the series are positive, we may rearrange terms to deduce

$$0 < \sum x_i^2 < \sum x_i^2 + \sum_{i<j} 2x_i x_j = \left(\sum x_i\right)^2 = A^2.$$

Thus it remains to show that each number in $(0, A^2)$ is a possible value of $\sum x_i^2$.

We use geometric series. Given $0 < r < 1$, there is a geometric series $\sum x_i$ with common ratio r and sum A: it has $x_0/(1-r) = A$ so $x_0 = (1-r)A$ and $x_j = r^j(1-r)A$. Then

$$\sum_{j=0}^{\infty} x_j^2 = \frac{x_0^2}{1-r^2} = \frac{1-r}{1+r} A^2.$$

To make this equal a given number $B \in (0, A^2)$, take $r = (A^2 - B)/(A^2 + B)$. ■

Solution 2. As in Solution 1, we prove $0 < \sum x_i^2 < A^2$, and it remains to show that each number in $(0, A^2)$ is a possible value of $\sum x_i^2$. There exists a series of positive numbers $\sum x_i$ with sum A. Then $x_0/2$, $x_0/2$, $x_1/2$, $x_1/2$, $x_2/2$, ... also sums to A but its squares sum to half the previous sum of squares. Iterating shows that the sum of squares can be arbitrarily small.

Given any series $\sum x_i$ of positive terms with sum A, form a weighted average of it with the series $A+0+0+\cdots$; in other words, choose $t \in (0,1)$ and define a new series of positive terms $\sum y_i$ by setting $y_0 = tx_0 + (1-t)A$ and $y_i = tx_i$ for $i \geq 1$. Then $\sum y_i = A$, and

$$\sum y_i^2 = t^2 \sum x_i^2 + 2t(1-t)x_0 A + (1-t)^2 A^2.$$

As t runs from 0 to 1, the Intermediate Value Theorem shows that this quadratic polynomial in t takes on all values strictly between $\sum x_i^2$ and A^2. Since $\sum x_i^2$ can be made arbitarily small, any number in $(0, A^2)$ can occur as $\sum y_i^2$. ■

A2. (150, 1, 0, 0, 0, 0, 0, 0, 1, 0, 23, 20)

Prove that there exist infinitely many integers n such that n, $n+1$, $n+2$ are each the sum of two squares of integers. [Example: $0 = 0^2+0^2$, $1 = 0^2+1^2$, and $2 = 1^2+1^2$.]

In all of the following solutions, we take $n = x^2 - 1$. Then $n + 1 = x^2 + 0^2$, $n + 2 = x^2 + 1^2$ and it suffices to exhibit infinitely many x so that $x^2 - 1$ is the sum of two squares.

Solution 1. Let a be an even integer such that a^2+1 is not prime. (For example, choose $a = 10k+2$ for some integer $k \geq 1$, so that $a^2+1 = 100k^2+40k+5$ is divisible by 5.) Then we can write a^2+1 as a difference of squares $x^2 - b^2$, by factoring a^2+1

as rs with $r \geq s > 1$, and setting $x = (r+s)/2, b = (r-s)/2$. (These are integers because r and s must both be odd.) It follows that $x^2 - 1$ is the sum of two squares $a^2 + b^2$, as desired. ■

Solution 2. The equation $u^2 - 2v^2 = 1$ is an example of Pell's equation [NZM, Section 7.8], so it has infinitely many solutions, and we can take $x = u$. ■

Remark. The positive integer solutions to $u^2 - 2v^2 = 1$ are the pairs (u, v) satisfying $u + v\sqrt{2} = (1+\sqrt{2})^{2n}$ for some $n \geq 1$. Incidentally, the positive possibilities for v are the same as the odd terms in the sequence $(a_n)_{n\geq 0}$ of Problem 1999A3.

Solution 3. Take $x = 2a^2 + 1$; then $x^2 - 1 = (2a)^2 + (2a^2)^2$. ■

Solution 4 (Abhinav Kumar). If $x^2 - 1$ is the sum of two squares, then so is $x^4 - 1$, using

$$x^4 - 1 = (x^2 - 1)(x^2 + 1)$$

and the identity

$$(a^2 + b^2)(c^2 + d^2) = (ac - bd)^2 + (ad + bc)^2 \tag{1}$$

(obtained by computing the norm of $(a+bi)(c+di)$ in two ways). Hence by induction on n, if $x^2 - 1$ is the sum of two squares (for instance, if $x = 3$) then so is $(x^{2^n})^2 - 1$ for all nonnegative integers n. ■

Related question. Let $S = \{\, a^2 + b^2 : a, b \in \mathbb{Z} \,\}$. Show that S does not contain four consecutive integers. But show that S does contain a (nonconstant) 4-term arithmetic progression.

Related question. Does S contain an arithmetic progression of length k for every integer $k \geq 1$? This is currently an unsolved problem! A positive answer would follow from either of the following conjectures:

- *Dickson's Conjecture.* Given linear polynomials $a_1 n + b_1, \ldots, a_k n + b_k$ in n, with $a_i, b_i \in \mathbb{Z}$ and $a_i > 0$ for all i, such that no prime p divides $(a_1 n + b_1) \cdots (a_k n + b_k)$ for all $n \in \mathbb{Z}$, there exist infinitely many integers $n \geq 1$ such that the values $a_1 n + b_1, \ldots, a_k n + b_k$ are simultaneously prime. (The hypothesis that p not divide $(a_1 n + b_1) \cdots (a_k n + b_k)$ is automatic if $p > k$ and $p \nmid \gcd(a_1, b_1) \cdots \gcd(a_k, b_k)$, so checking it for all p is a finite computation.)

  The special case $a_1 = \cdots = a_k = 1$ of Dickson's Conjecture is the qualitative form of the *Hardy-Littlewood Prime k-tuple Conjecture*, which itself is a generalization of the *Twin Prime Conjecture*, which is the statement that there exist infinitely many $n \geq 1$ such that n and $n+2$ are both prime. On the other hand, Dickson's Conjecture is a special case of "Hypothesis H" of Schinzel and Sierpiński, in which the linear polynomials are replaced by distinct irreducible polynomials $f_1(n), \ldots, f_k(n)$ with positive leading coefficients. Moreover, for each of these conjectures, there is a heuristic that predicts that the number of n less than or equal to x satisfying the conclusion is $(c + o(1))x/(\ln x)^k$ as $x \to \infty$, where c is a constant given by an explicit formula in terms of the k polynomials. (See 1988A3 for the definition of $o(1)$.) The quantitative form of Hypothesis H is known as the Bateman-Horn Conjecture. For more on all of these conjectures, see Chapter 6 of [Ri], especially pages 372, 391, and 409.

- *A conjecture of P. Erdős.* If T is a set of positive integers such that $\sum_{n \in T} 1/n$ diverges, then T contains arbitrarily long finite arithmetic progressions. Erdős, who frequently offered cash rewards for the solution to problems, offered his highest-valued reward of $3000 for a proof or disproof of this statement. See [Gr1, p. 24] or [Guy, p. 16].

Let us explain why either of the two conjectures above would imply that our set S contains arithmetic progressions of arbitrary length. We will use the theorem mentioned in 1991B5, that any prime congruent to 1 modulo 4 is in S.

Fix $k \geq 4$, and let $\ell_i(n) = 4n + 1 + i(k!)$ for $i = 1, 2, \ldots, k$. Let $P(n) = \ell_1(n)\ell_2(n)\ldots\ell_k(n)$. If p is prime and $p \leq k$, then p does not divide $P(0)$. If p is prime and $p > k$, then for each i, the set $\{n \in \mathbb{Z} : p \mid \ell_i(n)\}$ is a residue class modulo p, so there remains at least one residue class modulo p consisting of n such that $p \nmid P(n)$. Hence Dickson's Conjecture predicts that there are infinitely many n such that $\ell_1(n), \ldots, \ell_k(n)$ are simultaneously prime. Each of these k primes would be congruent to 1 modulo 4, and hence would be in S.

Now we show that the conjecture of Erdős implies that S contains arithmetic progressions of arbitrary length, and even better, that the subset T of primes congruent to 1 mod 4 contains such progressions. It suffices to prove that $\sum_{p \in T} 1/p$ diverges, or equivalently that $\lim_{s \to 1^+} \sum_{p \in T} 1/p^s = \infty$. The proof of Dirichlet's Theorem on primes in arithmetic progressions gives a precise form of this, namely:

$$\lim_{s \to 1^+} \left(\sum_{p \in T} \frac{1}{p^s} \right) \Big/ \left(\ln \frac{1}{s-1} \right) = \frac{1}{2}.$$

Since $\lim_{s \to 1^+} \ln \frac{1}{s-1} = \infty$, this implies $\lim_{s \to 1^+} \sum_{p \in T} 1/p^s = \infty$.

More generally, one says that a subset P of the set of prime numbers has *Dirichlet density* α if

$$\lim_{s \to 1^+} \left(\sum_{p \in P} \frac{1}{p^s} \right) \Big/ \left(\ln \frac{1}{s-1} \right) = \alpha.$$

Dirichlet's Theorem states that if a and m are relatively prime positive integers, then the Dirichlet density of the set of primes congruent to a modulo m equals $1/\phi(m)$, where $\phi(m)$ is the Euler ϕ-function defined in 1985A4. See Theorem 2 in Chapter VI, §4 of [Se1] for details.

A3. (64, 20, 20, 0, 0, 0, 0, 0, 23, 9, 27, 32)

The octagon $P_1P_2P_3P_4P_5P_6P_7P_8$ is inscribed in a circle, with the vertices around the circumference in the given order. Given that the polygon $P_1P_3P_5P_7$ is a square of area 5 and the polygon $P_2P_4P_6P_8$ is a rectangle of area 4, find the maximum possible area of the octagon.

Answer. The maximum possible area is $3\sqrt{5}$.

Solution. (See Figure 42.) We deduce from the area of $P_1P_3P_5P_7$ that the radius of the circle is $\sqrt{5/2}$. If $s > t$ are the sides of the rectangle $P_2P_4P_6P_8$, then $s^2 + t^2 = 10$ and $st = 4$, so $(s+t)^2 = 18$ and $(s-t)^2 = 2$. Therefore $s + t = 3\sqrt{2}$ and $s - t = \sqrt{2}$,

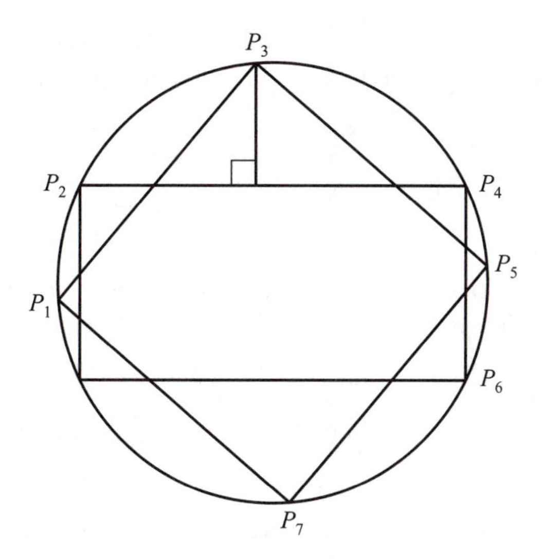

FIGURE 42.

yielding $s = 2\sqrt{2}$ and $t = \sqrt{2}$. Without loss of generality, assume that $P_2P_4 = 2\sqrt{2}$ and $P_4P_6 = \sqrt{2}$.

For notational ease, let $[Q_1Q_2\cdots Q_n]$ denote the area of the polygon $Q_1Q_2\cdots Q_n$. By symmetry, the area of the octagon can be expressed as

$$[P_2P_4P_6P_8] + 2[P_2P_3P_4] + 2[P_4P_5P_6].$$

Note that $[P_2P_3P_4]$ is $\sqrt{2}$ times the distance from P_3 to P_2P_4, which is maximized when P_3 lies on the midpoint of arc P_2P_4; similarly, $[P_4P_5P_6]$ is $\sqrt{2}/2$ times the distance from P_5 to P_4P_6, which is maximized when P_5 lies on the midpoint of arc P_4P_6. Thus the area of the octagon is maximized when P_3 is the midpoint of arc P_2P_4 and P_5 is the midpoint of arc P_4P_6 (in which case $P_1P_3P_5P_7$ is indeed a square). In this case, the distance from P_3 to P_2P_4 equals the radius of the circle minus half of P_4P_6, so $[P_2P_3P_4] = \sqrt{5} - 1$. Similarly $[P_4P_5P_6] = \sqrt{5}/2 - 1$, so the area of the octagon is $3\sqrt{5}$. ■

A4. (10, 2, 2, 0, 0, 0, 0, 0, 6, 2, 43, 130)

Show that the improper integral

$$\lim_{B\to\infty} \int_0^B \sin(x)\sin(x^2)\,dx$$

converges.

Solution 1. We may shift the lower limit to 1 without affecting convergence. That done, we use integration by parts:

$$\begin{aligned}\int_1^B \sin x \sin x^2\,dx &= \int_1^B \frac{\sin x}{2x}\sin x^2(2x\,dx)\\ &= -\frac{\sin x}{2x}\cos x^2\Big|_1^B + \int_1^B \left(\frac{\cos x}{2x} - \frac{\sin x}{2x^2}\right)\cos x^2\,dx.\end{aligned}$$

Now $\frac{\sin x}{2x} \cos x^2$ tends to 0 as $x \to \infty$, and the integral of $\frac{\sin x}{2x^2} \cos x^2$ converges absolutely as $B \to \infty$ by comparison to $1/x^2$. It remains to consider

$$\begin{aligned}\int_1^B \frac{\cos x}{2x} \cos x^2 \, dx &= \int_1^B \frac{\cos x}{4x^2} \cos x^2 (2x \, dx) \\ &= \frac{\cos x}{4x^2} \sin x^2 \Big|_0^B - \int_1^B \frac{-2\cos x - x \sin x}{4x^3} \sin x^2 \, dx.\end{aligned}$$

Now $\frac{\cos B}{4B^2} \sin B^2 \to 0$ as $B \to \infty$, and the final integral converges absolutely as $B \to \infty$ by comparison to the integral of $1/x^2$. ■

Solution 2. The addition formula for cosine implies that

$$\sin x \sin x^2 = \frac{1}{2}\big(\cos(x^2 - x) - \cos(x^2 + x)\big).$$

The substitution $x = y + 1$ transforms $x^2 - x$ into $y^2 + y$, so it suffices to show that $\int_0^\infty \cos(x^2 + x)\, dx$ converges. Now substitute $u = x^2 + x$; then $x = -1/2 + \sqrt{u + 1/4}$ and

$$\int_0^\infty \cos(x^2 + x)\, dx = \int_0^\infty \frac{\cos u}{2\sqrt{u + 1/4}}\, du. \tag{1}$$

The latter integrand is bounded, so we may replace the lower limit of integration with $\pi/2$ without affecting convergence. Moreover, the integrand tends to zero as $u \to \infty$, so the error introduced by replacing the upper limit of integration in

$$\int_{\pi/2}^B \frac{\cos u}{2\sqrt{u + 1/4}}\, du$$

by the nearest odd integer multiple of $\pi/2$ tends to zero as $B \to \infty$. Therefore (1) converges if and only if $\sum_{n=1}^\infty a_n$ converges, where

$$a_n = \int_{(n-\frac{1}{2})\pi}^{(n+\frac{1}{2})\pi} \frac{\cos u}{2\sqrt{u + 1/4}}\, du = (-1)^n \int_{-\pi/2}^{\pi/2} \frac{\cos t}{2\sqrt{t + n\pi + 1/4}}\, dt.$$

The integrand in the last expression decreases to 0 uniformly as $n \to \infty$. Therefore a_n alternate in sign, tend to 0 as $n \to \infty$ and satisfy $|a_n| \ge |a_{n+1}|$. By the alternating series test, $\sum_{n=1}^\infty a_n$ converges, so the original integral converges. ■

Remark. One can also combine the first two solutions, by rewriting in terms of $\cos(x^2 + x)$ and then integrating by parts.

Solution 3. The integrand is the imaginary part of $(\sin x)e^{ix^2}$, and $\sin x = \frac{e^{ix} - e^{-ix}}{2i}$, so it suffices to show that

$$\int_0^\infty e^{ix^2 + ix}\, dx \qquad \text{and} \qquad \int_0^\infty e^{ix^2 - ix}\, dx$$

converge. The functions $e^{ix^2 \pm ix}$ are entire, so for any real $B > 0$, Cauchy's Theorem [Ah, p. 141] applied to the counterclockwise triangular path with vertices at 0, B, and $B + Bi$ in the complex plane (Figure 43) yields

$$\int_0^B e^{ix^2 \pm ix}\, dx = \int_0^B e^{i(t+ti)^2 \pm i(t+ti)}(1+i)\, dt - \int_0^B e^{i(B+ui)^2 \pm i(B+ui)} i\, du. \tag{2}$$

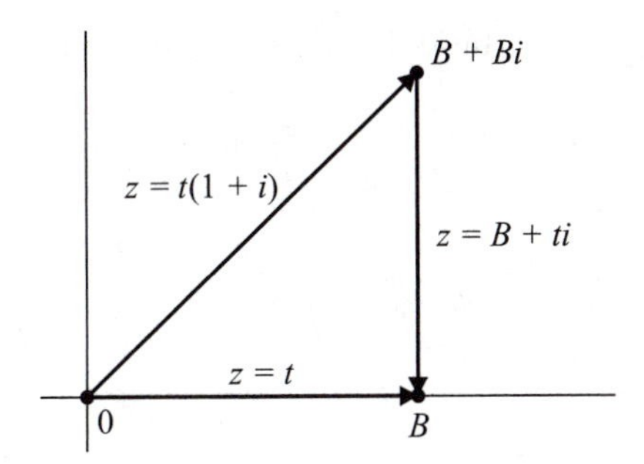

FIGURE 43.

Since

$$|e^{i(t+ti)^2 \pm i(t+ti)}| = e^{-2t^2 \mp t} \le e^{-t^2}$$

for $t \ge 1$, and

$$|e^{i(B+ui)^2 \pm i(B+ui)}| = e^{-2Bu \mp u} \le e^{-u}$$

once $B \ge 1$, the two integrals on the right of (2) have finite limits as $B \to \infty$, as desired. ■

Remark. By extending the previous analysis, we can evaluate

$$I(t) = \int_0^\infty (\sin tx)e^{ix^2}\,dx$$

for $t \in \mathbb{R}$ in terms of generalized hypergeometric functions, and hence we can evaluate the integral of the problem, which is $\operatorname{Im}(I(1))$. (A definition of the generalized hypergeometric functions is given below; for more on these functions, see [O, p. 168].) Since $\int_0^\infty e^{-2Bu+u}\,du \to 0$ as $B \to \infty$, the Cauchy's Theorem argument implies that the path of integration in

$$\int_0^\infty (\sin x)e^{ix^2}\,dx$$

can be changed to the diagonal ray $x = e^{i\pi/4}y$ with y ranging from 0 to ∞, without changing its value. More generally, this argument shows that for any $t \in \mathbb{R}$,

$$I(t) = \int_0^\infty \sin(e^{i\pi/4}ty)e^{-y^2}e^{i\pi/4}\,dy.$$

The substitution $z = y^2$ transforms this into

$$\frac{1}{2}\int_0^\infty e^{i\pi/4}z^{-1/2}\sin(e^{i\pi/4}t\sqrt{z})e^{-z}\,dz.$$

We now expand in a power series in t and integrate term-by-term: this is justified by the Dominated Convergence Theorem [Ru, p. 321], since the series of integrals of absolute values of the terms,

$$\sum_{n=0}^\infty \int_0^\infty z^{-1/2}\frac{|t\sqrt{z}|^{2n+1}}{(2n+1)!}e^{-z}\,dz = \sum_{n=0}^\infty \frac{n!\,|t|^{2n+1}}{(2n+1)!},$$

converges for any real t, by the Ratio Test. (We used the identity

$$\int_0^\infty z^n e^{-z}\,dz = \Gamma(n+1) = n!.)$$

Thus

$$I(t) = \frac{1}{2}\sum_{n=0}^{\infty} \frac{e^{i(2n+2)\pi/4}(-1)^n t^{2n+1}}{(2n+1)!} \int_0^\infty z^n e^{-z}\,dz = \frac{1}{2}\sum_{n=0}^{\infty} \frac{(-1)^n i^{n+1} n!\, t^{2n+1}}{(2n+1)!}$$

$$\operatorname{Im}(I(t)) = \frac{t}{2}\sum_{m=0}^{\infty} \frac{(-1)^m (2m)!\, t^{4m}}{(4m+1)!}.$$

Pairing the factor j in $(2m)!$ with the factor $2j$ of $(4m+1)!$, and dividing numerator and denominator by $2m$ factors of 4, which we distribute among the odd factors of $(4m+1)!$ in the denominator, we obtain

$$\operatorname{Im}(I(t)) = \left(\frac{t}{2}\right) {}_1F_2\left(1; \frac{3}{4}, \frac{5}{4}; -\frac{t^4}{64}\right),$$

where the generalized hypergeometric function is defined by

$${}_pF_q\left(a_1, \ldots, a_p; b_1, \ldots, b_q; z\right) = \sum_{m=0}^{\infty} \frac{(a_1)_m \cdots (a_p)_m}{(b_1)_m \cdots (b_q)_m} \cdot \frac{z^m}{m!},$$

using the Pochhammer symbol

$$(a)_n = a(a+1)\cdots(a+n-1) = \frac{\Gamma(a+n)}{\Gamma(a)}.$$

(One can similarly evaluate $\operatorname{Re}(I(t))$.) In particular, the integral in the original problem converges to

$$\operatorname{Im}(I(1)) = \left(\frac{1}{2}\right) {}_1F_2\left(1; \frac{3}{4}, \frac{5}{4}; -\frac{1}{64}\right).$$

Remark. One can also show that

$$\int_0^\infty \sin(bx)\sin\left(ax^2\right)dx = \sqrt{\frac{\pi}{2a}}\left(\cos\left(b^2/4a\right)C\left(b^2/4a\right) + \sin\left(b^2/4a\right)S\left(b^2/4a\right)\right)$$

where the two *Fresnel integrals* are

$$C(x) = \frac{1}{\sqrt{2\pi}}\int_0^x \cos t \frac{dt}{\sqrt{t}}$$

and

$$S(x) = \frac{1}{\sqrt{2\pi}}\int_0^x \sin t \frac{dt}{\sqrt{t}}.$$

Similarly,

$$\int_0^\infty \sin(bx)\cos\left(ax^2\right)dx = \sqrt{\frac{\pi}{2a}}\left(\sin\left(b^2/4a\right)C\left(b^2/4a\right) - \cos\left(b^2/4a\right)S\left(b^2/4a\right)\right).$$

If $\sin(bx)$ is replaced by $\cos(bx)$ in the above two integrals, then they can be evaluated in terms of elementary functions:

$$\int_0^\infty \cos(bx)\sin(ax^2)\,dx = \frac{1}{2}\sqrt{\frac{\pi}{2a}}\left(\cos(b^2/4a) - \sin(b^2/4a)\right)$$

$$\int_0^\infty \cos(bx)\cos(ax^2)\,dx = \frac{1}{2}\sqrt{\frac{\pi}{2a}}\left(\cos(b^2/4a) + \sin(b^2/4a)\right).$$

All of these integrals are originally due to Cauchy [Tal].

Remark. Oscillatory integrals like the one in this problem occur frequently in analysis and mathematical physics. They can be bounded or evaluated by the methods given here in some cases, and often can be estimated by a technique known as the stationary phase approximation [CKP, Section 6.4]. This method was first used by Stokes, to obtain an asymptotic representation for the function

$$f(x) = \int_0^\infty \cos\left(x(\omega^3 - \omega)\right)d\omega$$

valid for large positive real values of x.

A5. (30, 11, 4, 0, 0, 0, 0, 0, 2, 0, 19, 129)

Three distinct points with integer coordinates lie in the plane on a circle of radius $r > 0$. Show that two of these points are separated by a distance of at least $r^{1/3}$.

Solution 1. We will prove a stronger result, with $r^{1/3}$ replaced by $(4r)^{1/3}$. Let A, B, C be the three points. Examining small triangles with integer coordinates shows that either $\triangle ABC$ has sides 1, 1, $\sqrt{2}$, or else some side has length at least 2 and hence $r \geq 1$. In the first case, $r = \sqrt{2}/2$, so $(4r)^{1/3} = \sqrt{2}$ and we are done. So assume $r \geq 1$. If the *minor* arcs AB, BC, CA cover the circle, then one has measure at least $2\pi/3$, and the length of the corresponding chord is at least

$$2r\sin(\pi/3) = r\sqrt{3} > (4r)^{1/3},$$

where the last equality follows from $r \geq 1$. Thus we may assume, without loss of generality, that C lies in the *interior* of minor arc AB.

Since A, B, C have integer coordinates, the area K of $\triangle ABC$ is half an integer (see the first remark in 1990A3). Thus $K \geq 1/2$. Let 2θ be the measure of minor arc AB. Then the base AB of $\triangle ABC$ is $2r\sin\theta$, and the height is at most $r - r\cos\theta$, with equality if and only if C is the midpoint of arc AB (see Figure 44). Thus

$$\begin{aligned}
\frac{1}{2} &\leq K \\
&\leq \frac{1}{2}(2r\sin\theta)(r - r\cos\theta) \\
&= \frac{r^2\sin\theta(1-\cos^2\theta)}{1+\cos\theta} \\
&= \frac{r^2\sin^3\theta}{1+\cos\theta} \\
&\leq r^2\sin^3\theta \qquad \text{(since } 0 < \theta \leq \pi/2\text{)},
\end{aligned}$$

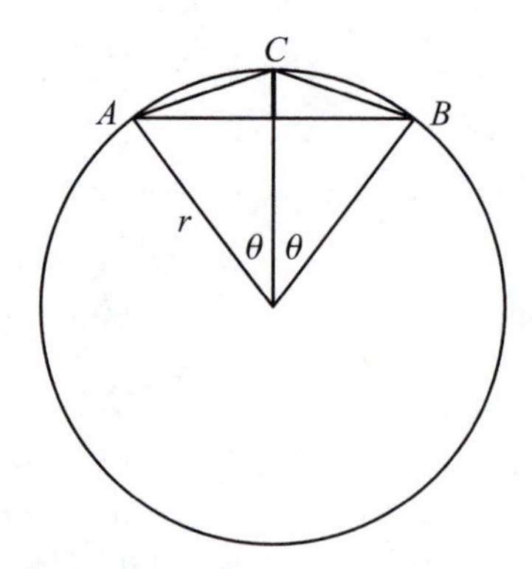

FIGURE 44.
The point C on arc AB maximizing the height of $\triangle ABC$ in Solution 1 to 2000A5.

so $\sin\theta \geq 1/(2r^2)^{1/3}$ and

$$AB = 2r\sin\theta \geq 2r/(2r^2)^{1/3} = (4r)^{1/3}.$$

■

Related question. Examine Solution 1 to prove that if the longest side of $\triangle ABC$ has length *equal* to $(4r)^{1/3}$, it is an isosceles right triangle with sides 1, 1, $\sqrt{2}$. (We will prove a stronger result soon.)

Solution 2. We will prove the result with $r^{1/3}$ replaced by $(2r)^{1/3}$, using:

Lemma. *If a, b, c are the lengths of sides of a triangle with area K and circumradius r, then $K = abc/(4r)$.*

Proof. Let A, B, C be the vertices opposite the sides of length a, b, c, respectively. If we view BC as base, the height is $b\sin C$, so $K = \frac{1}{2}ab\sin C$. Now use the Extended Law of Sines,

$$\frac{\sin A}{a} = \frac{\sin B}{b} = \frac{\sin C}{c} = \frac{1}{2r},$$

to replace $\sin C$ by $c/(2r)$. □

Let a, b, c be the distances between the points. By the lemma, the area of the triangle with the three points as vertices is $abc/(4r)$. On the other hand, the area is half an integer (see the first remark in 1990A3). Thus $abc/(4r) \geq 1/2$, and

$$\max\{a, b, c\} \geq (abc)^{1/3} \geq (2r)^{1/3}.$$

■

Stronger result. In the notation of Solution 2, we prove that if r is sufficiently large, then $\max\{a, b, c\} \geq 2r^{1/3}$.

If $K \geq 1$, then the lower bound $(4r)^{1/3}$ of Solution 1 is improved to $2r^{1/3}$, and we are done. Therefore assume $K = 1/2$. Without loss of generality, assume $a \leq b \leq c$. Suppose that $c < 2r^{1/3}$. Then $c^3 < 8r = 8(abc/4K) = 4abc$, so $4ab > c^2$. If $a \leq \frac{1}{2}r^{1/3}$, then $abc < \frac{1}{2}r^{1/3}(2r^{1/3})(2r^{1/3}) = 2r$, contradicting $abc/(4r) = 1/2$ (the lemma of Solution 2). Thus $\frac{1}{2}r^{1/3} < a \leq b \leq c < 2r^{1/3}$.

Viewing AB as base of $\triangle ABC$, the height h equals $1/c$, since $K = 1/2$. Let a' and b' be the lengths of the projections of the sides of lengths a and b onto AB (see

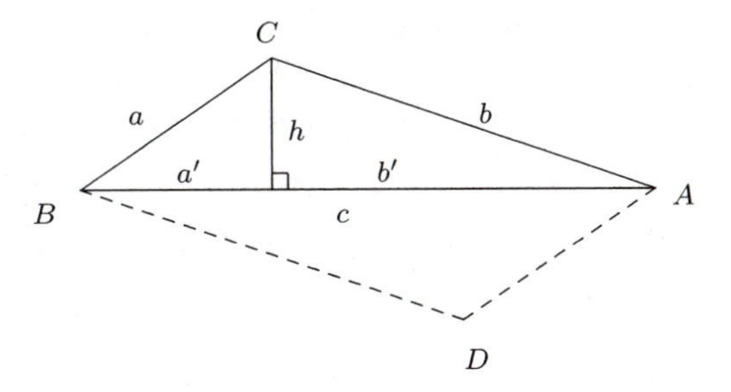

FIGURE 45.
Defining a' and b', and constructing the parallelogram $ACBD$.

Figure 45). Since AB is the longest side of $\triangle ABC$, point C lies directly above the segment AB, so $a' + b' = c$. By the Pythagorean Theorem,

$$\begin{aligned} a' &= \sqrt{a^2 - h^2} = \sqrt{a^2\left(1 - \frac{1}{a^2c^2}\right)} \\ &= a\left(1 - O\left(\frac{1}{a^2c^2}\right)\right) = a - O\left(\frac{1}{ac^2}\right) \\ &= a - O\left(\frac{1}{c^3}\right), \end{aligned}$$

and similarly $b' = b - O(1/c^3)$.

If we form a parallelogram $ACBD$ as in Figure 45, then D also has integer coordinates, so

$$1 \le CD^2 = (a' - b')^2 + (2h)^2 = (a' - b')^2 + O(1/c^2),$$

by the Pythagorean Theorem.

Now

$$\begin{aligned} 4ab &= 4\big(a' + O(1/c^3)\big)\big(b' + O(1/c^3)\big) \\ &= 4a'b' + O(\max\{a', b'\}/c^3) \\ &= 4a'b' + O(1/c^2) \\ &= (a' + b')^2 - (a' - b')^2 + O(1/c^2) \\ &\le c^2 - \big(1 - O(1/c^2)\big) + O(1/c^2) \\ &= c^2 - 1 + O(1/c^2). \end{aligned}$$

If r is sufficiently large, then $c \ge (4r)^{1/3}$ is also large, so this contradicts $4ab > c^2$.

Stronger result. With a little more work, we can prove that $\max\{a, b, c\} \ge 2r^{1/3}$ *except* when a, b, c are 1, 1, $\sqrt{2}$ or 1, $\sqrt{2}$, $\sqrt{5}$ (in some order). We continue with the notation and results introduced in the previous proof. Let $d = a' - b'$, so $d^2 + 4h^2 = CD^2 \ge 1$. Thus $1 - 4h^2 \le d^2 \le c^2$. Also $a' = (c + d)/2$ and $b' = (c - d)/2$. Recall that in a counterexample, $K = 1/2$ and $4ab > c^2$. Thus

$$c^4 < (4ab)^2 = 16(a'^2 + h^2)(b'^2 + h^2) = \big((c + d)^2 + 4h^2\big)\big((c - d)^2 + 4h^2\big).$$

For fixed c, the right-hand side is a monic quadratic in d^2 so its maximum value for $d^2 \in [1-4h^2, c^2]$ is attained at an endpoint. In other words, the inequality remains true either when d^2 is replaced by $1-4h^2$ or when d^2 is replaced by c^2. At $d^2 = 1-4h^2$, the inequality becomes

$$c^4 < (c^2+2cd+1)(c^2-2cd+1) = (c^4+2c^2+1) - 4c^2(1-4h^2),$$

which, since $h = 1/c$, is equivalent to $c^2 < 17/2$. At $d^2 = c^2$, the inequality becomes

$$c^4 < (4c^2+4h^2)(4h^2) = 16(1+c^{-4}),$$

which implies $c^4 < 17$. Thus in either case, $c^2 < 17/2$. But c^2 is a sum of integer squares, so $c^2 \in \{1,2,4,5,8\}$. The only lattice triangles of area $1/2$ with such a value of c^2 are those with side lengths 1, 1, $\sqrt{2}$ and 1, $\sqrt{2}$, $\sqrt{5}$.

Remark. The result just proved is nearly best possible. We now construct examples with $r \to \infty$ such that $\max\{a,b,c\} = 2r^{1/3} + O(r^{-1/3})$.

Let n be a large positive integer, and take the circle passing through $(0,0)$, $(n,1)$, $(2n+1,2)$. Then the three sides of the triangle are $\sqrt{n^2+1}$, $\sqrt{(n+1)^2+1}$, $\sqrt{(2n+1)^2+4}$, and $K = 1/2$, so the lemma of Solution 2 implies

$$r = \frac{1}{2}\sqrt{(n^2+1)(n^2+2n+2)(4n^2+4n+5)}.$$

Thus

$$\begin{aligned} 2r^{1/3} &= 2\left((n^2+1)(n^2+2n+2)(n^2+n+5/4)\right)^{1/6} \\ &= 2n\left((1+n^{-2})(1+2n^{-1}+2n^{-2})(1+n^{-1}+(5/4)n^{-2})\right)^{1/6} \\ &= 2n+1+\frac{5}{6}n^{-1}+O(n^{-2}), \end{aligned}$$

and similarly

$$\max\{a,b,c\} = \sqrt{(2n+1)^2+4} = 2n+1+n^{-1}+O(n^{-2}),$$

so

$$\max\{a,b,c\} = 2r^{1/3} + \frac{1}{6}n^{-1} + O(n^{-2}) = 2r^{1/3} + \frac{1}{6}r^{-1/3} + O(r^{-2/3}).$$

A6. (1, 2, 2, 0, 0, 0, 0, 0, 3, 19, 57, 111)

Let $f(x)$ be a polynomial with integer coefficients. Define a sequence $a_0, a_1, \ldots$ of integers such that $a_0 = 0$ and $a_{n+1} = f(a_n)$ for all $n \geq 0$. Prove that if there exists a positive integer m for which $a_m = 0$ then either $a_1 = 0$ or $a_2 = 0$.

Solution 1. Recall that if $f(x)$ is a polynomial with integer coefficients, then $m-n$ divides $f(m)-f(n)$ for any integers m and n. In particular, if we put $b_n = a_{n+1} - a_n$, then b_n divides b_{n+1} for all n. On the other hand, we are given that $a_0 = a_m = 0$, which implies that $a_1 = a_{m+1}$ and so $b_0 = b_n$. If $b_0 = 0$, then $a_0 = a_1 = \cdots = a_m$ and we are done. Otherwise, $|b_0| = |b_1| = |b_2| = \cdots$, so $b_n = \pm b_0$ for all n.

Now $b_0 + \cdots + b_{m-1} = a_m - a_0 = 0$, so half of the integers $b_0, \ldots, b_{m-1}$ are positive and half are negative. In particular, there exists an integer $0 < k < m$ such that

$b_{k-1} = -b_k$, which is to say, $a_{k-1} = a_{k+1}$. From this it follows that $a_n = a_{n+2}$ for all $n \geq k-1$; in particular, for $n = m$, we have

$$0 = a_m = a_{m+2} = f(f(a_m)) = f(f(a_0)) = a_2.$$

■

Solution 2. Choose $m \geq 1$ minimal such that $a_m = 0$. If $a_i = a_j$ for some $0 \leq i < j \leq m-1$, then $m-j$ applications of f lead to $a_{m-(j-i)} = a_m = 0$, contradicting the minimality of m. Hence, for arbitrary i and j, we have $a_i = a_j$ if and only if $i - j$ is divisible by m.

If $m = 1$, we are done. Otherwise let $a_i > a_j$ be the maximum and minimum terms of $a_0, \ldots, a_{m-1}$. Then $a_i - a_j$ divides $f(a_i) - f(a_j) = a_{i+1} - a_{j+1}$, which is nonzero since $(i+1) - (j+1) = i - j$ is not divisible by m. On the other hand, $|a_{i+1} - a_{j+1}| \leq a_i - a_j$ because a_i and a_j are the maximum and minimum of all terms in the sequence. Therefore equality must occur, which implies that a_{i+1} and a_{j+1} equal a_i and a_j in some order. If $a_{i+1} = a_i$, then $m = 1$; otherwise $a_{i+1} = a_j$ implies $a_{i+2} = a_{j+1}$, which with $a_{j+1} = a_i$ yields $a_{i+2} = a_i$, so m divides 2. Hence m is 1 or 2. ■

Remark. A special case of this problem (the fact that $a_3 = 0$ implies $a_1 = 0$ or $a_2 = 0$) was Problem 1 on the 1974 USA Mathematical Olympiad; the solution given in [USAMO7286] is similar to Solution 1 above.

Literature note. By a shift of variable, it follows that for $f(x) \in \mathbb{Z}[x]$, if $a \in \mathbb{Z}$ is such that a, $f(a)$, $f(f(a))$, $\ldots$ is periodic, then the minimal period is at most 2. See Chapter XII of [Na] for many other results of this type.

If one replaces $\mathbb{Z}$ everywhere with $\mathbb{Q}$, then the period can be arbitrarily long: given distinct $a_1, \ldots, a_n \in \mathbb{Q}$, Lagrange interpolation (see 1998B6) lets one construct a polynomial $f(x) \in \mathbb{Q}[x]$ such that $f(a_i) = a_{i+1}$ for $1 \leq i \leq n-1$ and $f(a_n) = a_1$. On the other hand, there might be a bound on the period in terms of the degree of the polynomial $f(x) \in \mathbb{Q}[x]$. Whether such a bound exists is unknown even in the case of quadratic polynomials. For quadratic polynomials, Lagrange interpolation shows that periods 1, 2, and 3 are possible; it is also true (but much harder to prove) that periods 4 and 5 are not possible. See [Mort], [MS], [FPS], and [P2] for more details.

B1. (126, 14, 4, 0, 0, 0, 0, 0, 8, 14, 14, 15)

Let a_j, b_j, c_j be integers for $1 \leq j \leq N$. Assume, for each j, at least one of a_j, b_j, c_j is odd. Show that there exist integers r, s, t such that $ra_j + sb_j + tc_j$ is odd for at least $4N/7$ values of j, $1 \leq j \leq N$.

Solution. Consider the seven triples (r, s, t) with $r, s, t \in \{0, 1\}$ not all zero. If a_j, b_j, c_j are not all even, then four of the sums $ra_j + bs_j + tc_j$ with $r, s, t \in \{0, 1\}$ are even and four are odd. (For example, if a_j is odd, then flipping r changes the sum between even and odd, so half of the sums must be even and half odd.) Of course the sum with $r = s = t = 0$ is even, so at least four of the seven triples with r, s, t not all zero yield an odd sum. In other words, at least $4N$ of the quadruples (r, s, t, j) yield odd sums. By the Pigeonhole Principle, there is a triple (r, s, t) for which at least $4N/7$ of the sums are odd. ■

Reinterpretation. If (r,s,t) is chosen randomly from the seven triples with $(r,s,t) \in \{0,1\}$ not all zero, then for each j, the probability that $ra_j + sb_j + tc_j$ is odd is $4/7$, so the expected number of j for which this sum is odd equals $4N/7$. For some particular choice of (r,s,t), the number of such j must be greater than or equal to the expectation.

B2. (114, 7, 2, 0, 0, 0, 0, 0, 2, 6, 35, 29)

Prove that the expression

$$\frac{\gcd(m,n)}{n}\binom{n}{m}$$

is an integer for all pairs of integers $n \geq m \geq 1$.

Solution 1. Let $a = \frac{m}{\gcd(m,n)}$ and $b = \frac{n}{\gcd(m,n)}$. Then

$$\frac{a}{b}\binom{n}{m} = \frac{m}{n}\binom{n}{m} = \binom{n-1}{m-1}$$

is an integer, so $b \mid a\binom{n}{m}$. But $\gcd(a,b) = 1$, so $b \mid \binom{n}{m}$. Hence

$$\frac{\gcd(m,n)}{n}\binom{n}{m} = \frac{1}{b}\binom{n}{m}$$

is an integer. ■

Solution 2. Since $\gcd(m,n)$ is an integer linear combination of m and n, it follows that

$$\frac{\gcd(m,n)}{n}\binom{n}{m}$$

is an integer linear combination of the integers

$$\frac{m}{n}\binom{n}{m} = \binom{n-1}{m-1} \text{ and } \frac{n}{n}\binom{n}{m} = \binom{n}{m}$$

and hence is itself an integer. ■

Solution 3. To show that a nonzero rational number is an integer, it suffices to check that the exponent of each prime in its factorization is nonnegative. So let p be a prime, and suppose that the highest powers of p dividing n and m are p^a and p^b, respectively. If $a \leq b$, then p has a nonnegative exponent in both $\gcd(m,n)/n$ and in $\binom{n}{m}$. If $a > b$, it suffices to show that $\binom{n}{m}$ is divisible by p^{a-b}, but this follows from Kummer's Theorem, described in Solution 2 to 1997A5. ■

B3. (5, 3, 3, 1, 0, 0, 0, 16, 9, 26, 3, 129)

Let $f(t) = \sum_{j=1}^{N} a_j \sin(2\pi jt)$, where each a_j is real and a_N is not equal to 0. Let N_k denote the number of zeros† (including multiplicities) of $\frac{d^k f}{dt^k}$. Prove that

$$N_0 \leq N_1 \leq N_2 \leq \cdots \quad \textbf{and} \quad \lim_{k\to\infty} N_k = 2N.$$

---

† The proposers intended for N_k to count only the zeros in the interval $[0,1)$.

Solution. We first show $N_k \le 2N$ for all $k \ge 0$. Write $f^{(k)}(t)$ for $\frac{df^k}{dt^k}$. If we use the identity $\sin x = (e^{ix} - e^{-ix})/(2i)$ and set $z = e^{2\pi it}$, then

$$f^{(k)}(t) = \frac{1}{2i}\sum_{j=1}^{N}(2\pi ij)^k a_j(z^j - z^{-j})$$

is rewritten as z^{-N} times a polynomial of degree $2N$ in z. Hence as a function of z, it has at most $2N$ zeros. Therefore $f_k(t)$ has at most $2N$ zeros in $[0,1)$; that is, $N_k \le 2N$. In particular, $f^{(k)}$ has at most finitely many zeros in $[0,1)$.

To prove $N_k \le N_{k+1}$, we use Rolle's Theorem, and the fact that at every zero of f, f' has a zero of multiplicity one less. If $0 \le t_1 < t_2 < \cdots < t_r < 1$ are the zeros of $f^{(k)}$ in $[0,1)$, occurring with respective multiplicities $m_1, \ldots, m_r$, then $f^{(k+1)}$ has at least one zero in each of the open intervals $(t_1,t_2), (t_2,t_3), \ldots, (t_{r-1},t_r), (t_r, t_1+1)$; we may translate the part $[1, t_1+1)$ of the last interval to $[0,t_1)$, on which $f^{(k+1)}$ takes the same values. This gives r zeros of $f^{(k+1)}$. Adding to these the multiplicities $m_1 - 1, \ldots, m_r - 1$ at $t_1, \ldots, t_r$, we find that $f^{(k+1)}$ has at least $m_1 + \cdots + m_r = N_k$ zeros. Thus $N_{k+1} \ge N_k$.

To establish that $N_k \to 2N$, it suffices to prove $N_{4k} \ge 2N$ for sufficiently large k. This we do by making precise the assertion that

$$f^{(4k)}(t) = \sum_{j=1}^{N}(2\pi j)^{4k} a_j \sin(2\pi jt)$$

is dominated by the term with $j = N$ at each point $t = t_m = (2m+1)/(4N)$ for $m = 0, 1, \ldots, 2N-1$. At $t = t_m$, the $j = N$ term is

$$(2\pi N)^{4k} a_N \sin(2\pi N t_m) = (-1)^m (2\pi N)^{4k} a_N.$$

The absolute value of the sum of the other terms at $t = t_m$, divided by the $j = N$ term, is bounded by

$$|a_1|\left(\frac{1}{N}\right)^{4k} + \cdots + |a_{N-1}|\left(\frac{N-1}{N}\right)^{4k},$$

which tends to 0 as $k \to \infty$; in particular, this ratio is less than 1 for sufficiently large k. Then $f^{(4k)}(t_m)$ has the same sign as $(-1)^m(2\pi N)^{4k}a_N$. In particular, the sequence

$$f^{(4k)}(t_0), f^{(4k)}(t_1), \ldots, f^{(4k)}(t_{2N-1})$$

alternates in sign, when k is sufficiently large. Between these points (again including a final "wraparound" interval) we find $2N$ sign changes of $f^{(4k)}$. By the Intermediate Value Theorem , this implies $N_{4k} \ge 2N$ for large k, and we are done. ■

Remark. A similar argument was used in Solution 1 to 1989A3.

Remark. A more analytic proof that $N_k \to 2N$ involves the observation that $(2\pi N)^{-4k} f^{(4k)}(x)$ and its derivative converge uniformly to $a_N \sin(2\pi Nt)$ and $2\pi N a_N \cos(2\pi Nt)$, respectively, on $[0,1]$. That implies that for large k, we can divide $[0,1]$ into intervals on which $f^{(4k)}(x)$ does not change sign, and intervals on which $f^{(4k+1)}(x)$ does not change sign while $f^{(4k)}(x)$ has a sign change.

B4. (14, 5, 1, 0, 0, 0, 0, 0, 3, 6, 106, 60)

Let $f(x)$ be a continuous function such that $f(2x^2 - 1) = 2xf(x)$ for all x. Show that $f(x) = 0$ for $-1 \le x \le 1$.

Solution 1. Since $x = -1/2$ and $x = 1$ satisfy $2x^2 - 1 = x$, the given equation implies $f(-1/2) = -f(-1/2)$ and $f(1) = 2f(1)$, so $f(-1/2)$ and $f(1)$ both equal 0. The former equation can be rewritten

$$f\big(\cos(2\pi/3 + 2\pi n)\big) = 0$$

for any $n \in \mathbb{Z}$. Taking $x = \cos\theta$ in the functional equation shows that if $f(\cos 2\theta) = 0$ and $\cos\theta \neq 0$, then $f(\cos\theta) = 0$ as well. Therefore

$$f\Big(\cos\left(2^{-k}(2\pi/3 + 2\pi n)\right)\Big) = 0$$

for all $k, n \in \mathbb{Z}$ with $k \ge 0$. The numbers $2^{-k}(2\pi/3 + 2\pi n)$ are dense in $\mathbb{R}$, so continuity of $f(\cos x)$ yields $f(\cos r) = 0$ for all $r \in \mathbb{R}$. Thus f is zero on $[-1, 1]$. ■

Remark. We started with $f(-1/2) = 0$ instead of $f(1) = 0$ to avoid complications arising from the condition $\cos\theta \neq 0$ in the iteration.

Solution 2. For t real and not a multiple of π, write $g(t) = \frac{f(\cos t)}{\sin t}$. Then $g(t + 2\pi) = g(t)$; furthermore, the given equation implies that for t not a multiple of $\pi/2$,

$$g(2t) = \frac{f(2\cos^2 t - 1)}{2\sin t\cos t} = \frac{2(\cos t)f(\cos t)}{2\sin t\cos t} = g(t).$$

In particular, for any integers n and k, we have

$$g\left(1 + n\pi/2^{k-1}\right) = g\left(2^k + 2n\pi\right) = g\left(2^k\right) = g(1).$$

Since f is continuous, g is continuous where it is defined; but the set $\{1 + n\pi/2^{k-1} : n, k \in \mathbb{Z}\}$ is dense in the reals (because π is irrational), and so g must be constant on its domain. Since $g(-t) = -g(t)$ for any t not a multiple of π, this constant must be zero. Hence $f(x) = 0$ for $x \in (-1, 1)$. Finally, setting $x = 0$ and $x = 1$ in the given equation yields $f(-1) = f(1) = 0$ (which can also be deduced by continuity). ■

Remark. A variation of the above solution starts by noting that for $x = -1/2$, we have $f(-1/2) = -f(-1/2) = 0$, that is, $f(\cos(2\pi/3)) = 0$. As above, we deduce that $f(\cos\theta) = 0$ for $\theta = (2\pi/3 + 2n)/2^k$. These θ are dense in the interval $[0, 2\pi]$.

Remark. We describe all functions f that satisfy the conditions of the problem: in particular, we see that they need not be constant outside $[-1, 1]$. As shown above, f must be zero on $[-1, 1]$. From the functional equation,

$$2xf(x) = f(2x^2 - 1) = 2(-x)f(-x),$$

so $f(x) = -f(-x)$; so it suffices to classify continuous functions f on $[1, \infty)$ with $f(1) = 0$ that satisfy the functional equation.

Given such a function, the function $g(t) = \frac{f(\cosh t)}{\sinh t}$ on $(0, \infty)$ is continuous and satisfies $g(2t) = g(t)$, so the function $h(u) = g(e^u)$ is continuous on $\mathbb{R}$ and periodic with period $\ln 2$. Conversely, given $h : \mathbb{R} \to \mathbb{R}$ continuous of period $\ln 2$, set $g(t) = h(\ln t)$ for

$t > 0$; then $f(x) = g(\cosh^{-1} x)\sqrt{x^2 - 1}$ for $x > 1$ and $f(1) = 0$ define an f satisfying the conditions of the problem. (Continuity of f at 1 follows because h is bounded.)

B5. (29, 16, 14, 0, 0, 0, 0, 0, 13, 1, 11, 111)

Let S_0 be a finite set of positive integers. We define finite sets $S_1, S_2, \ldots$ of positive integers as follows: the integer a is in S_{n+1} if and only if exactly one of $a - 1$ or a is in S_n. Show that there exist infinitely many integers N for which $S_N = S_0 \cup \{ N + a : a \in S_0 \}$.

Solution. We claim that all integers N of the form 2^k, with k a positive integer and $N > \max S_0$, satisfy the desired condition.

It follows from the definition of S_n that

$$\sum_{j \in S_n} x^j \equiv (1 + x) \sum_{j \in S_{n-1}} x^j \pmod{2}.$$

(When we write that two polynomials $\sum a_i x^i$ and $\sum b_i x^i$ in $\mathbb{Z}[x]$ are congruent modulo 2, we mean that $a_i \equiv b_i \pmod{2}$ for all $i \geq 0$, or equivalently, that the two polynomials are equal when considered in $\mathbb{F}_2[x]$. Thus for instance, $x^2 \not\equiv x \pmod{2}$ as polynomials, even though $m^2 \equiv m \pmod{2}$ holds for any particular integer m.) By induction on n,

$$\sum_{j \in S_n} x^j \equiv (1 + x)^n \sum_{j \in S_0} x^j \pmod{2}.$$

From the identity $(x+y)^2 \equiv x^2 + y^2 \pmod{2}$ and induction on n, we have $(x+y)^{2^k} \equiv x^{2^k} + y^{2^k} \pmod{2}$. (This also follows from Kummer's Theorem, described in Solution 2 to 1997A5.) Hence if $N = 2^k$ for some $k \geq 1$, then

$$\sum_{j \in S_n} x^j \equiv (1 + x^N) \sum_{j \in S_0} x^j.$$

If moreover $N > \max S_0$, then $S_N = S_0 \cup \{ N + a : a \in S_0 \}$, as desired. ■

Remark. This solution can also be written in terms of binomial coefficients, without reference to polynomials with mod 2 coefficients.

Remark. See 1989A5 for another problem involving generating functions modulo 2.

Related question. Problem 6 of the 1993 International Mathematical Olympiad [IMO93] is related:

> There are n lamps $L_0, \ldots, L_{n-1}$ in a circle ($n > 1$), where we denote $L_{n+k} = L_k$. (A lamp at all times is either on or off.) Perform steps s_0, $s_1, \ldots$ as follows: at step s_i, if L_{i-1} is lit, switch L_i from on to off or vice versa; otherwise do nothing. Initially, all lamps are on. Show that:
>
> (a) There is a positive integer $M(n)$ such that after $M(n)$ steps all the lamps are on again;
>
> (b) If $n = 2^k$, we can take $M(n) = n^2 - 1$;
>
> (c) If $n = 2^k + 1$, we can take $M(n) = n^2 - n + 1$.

B6. (8, 1, 0, 0, 0, 0, 0, 0, 0, 0, 41, 145)

Let B be a set of more than $2^{n+1}/n$ distinct points with coordinates of the form $(\pm 1, \pm 1, \ldots, \pm 1)$ in n-dimensional space with $n \geq 3$. Show that there are three distinct points in B which are the vertices of an equilateral triangle.

Solution. Let S be the set of points with all coordinates equal to ± 1. For each $P \in B$, let S_P be the set of points in S which differ from P in exactly one coordinate. Since there are more than $2^{n+1}/n$ points in B, and each S_P has n elements, the cardinalities of the sets S_P sum to more than 2^{n+1}, which is to say, more than twice the number of points in S. By the Pigeonhole Principle, there must be a point of S in at least three of the sets, say in S_P, S_Q, S_R. But then any two of P, Q, R differ in exactly two coordinates, so PQR is an equilateral triangle of side length $2\sqrt{2}$, by the Pythagorean Theorem. ■

Remark. Noam Elkies points out that when $n = 2^k$, there exist sets of $2^{n+1}/n = 2^{n-k+1}$ points having no subset of three points, each pair of which differ in two coordinates. For example, for $i = 1, \ldots, k-1$, let T_i be the set of numbers in $\{0, 1, \ldots, n-1\}$ that, when expanded in base 2 as $\sum a_j 2^j$, have $a_i = 1$. Then take B to be the set of points $(e_1, \ldots, e_n) \in S$ with

$$\prod_{j \in T_1} e_j = \prod_{j \in T_2} e_j = \cdots = \prod_{j \in T_{k-1}} e_j = 1.$$

Now $|B| = 2^{n-k+1}$, since we can choose e_j for all j except $2, 2^2, \ldots, 2^{k-1}$, and then the condition $\prod_{j \in T_i} e_j = 1$ determines e_{t^i} in terms of the e_j already chosen. To see that there cannot be three points in B, any pair of which differ in two coordinates, suppose that such points exist; then there exist coordinates l, m, n at each of which two of the three points are the same and the third is different. At least two of l, m, n must have the same parity; suppose l and m have the same parity. Then there exists $i \in \{1, \ldots, k-1\}$ such that T_i contains one of l and m but not the other. Hence at the two points that only differ in coordinates l and m, the values of $\prod_{j \in T_i} e_i$ are different. This contradicts the definition of B, which says that both values should equal 1.

Remark. The construction of patterns such as in the previous remark is known as *design theory*. This topic is the subject of [BJL].

Results

Individual Results

The following is a list of the highest-ranking individual contestants (Putnam Fellows), in alphabetical order within each year. The list also mentions the winners of the Elizabeth Lowell Putnam Prize, which was first awarded in 1992. Educational and employment history up to 2001 is given, when known to the authors. If degrees are in subjects other than math, the subject is given when known. We use the following abbreviations: cs (computer science), I (Instructor), L (Lecturer), AP (Assistant Professor), AsP (Associate Professor), CCR (Center for Communications Research).

Forty-sixth Competition — 1985

Martin V. Hildebrand, Williams College
B.A. ('86), Ph.D. (Harvard, '90), AP (SUNY Albany)
Everett W. Howe, California Institute of Technology
B.S. ('86), Ph.D. (Berkeley, '93), AP (Michigan), researcher at CCR (La Jolla)
Douglas S. Jungreis, Harvard University
A.B. ('87), Ph.D. (Berkeley, '92), researcher at CCR (La Jolla)
Bjorn M. Poonen, Harvard University
A.B. (math, physics, '89), Ph.D. (Berkeley, '94), I (Princeton), AsP (Berkeley)
Keith A. Ramsay, University of Chicago
B.A. ('86), Ph.D. (Harvard, '91), computer programmer at Maptek

Forty-seventh Competition — 1986

David J. Grabiner, Princeton University
A.B. ('90), Ph.D. (Harvard, '95), Visiting AP (Arizona State)
Waldemar P. Horwat, Massachusetts Institute of Technology
B.S. (math, elec. eng., '88), M.S. (cs, MIT, '88), Ph.D. (cs, MIT, '94), Netscape
Douglas S. Jungreis, Harvard University
(see 1985 results)
David J. Moews, Harvard University
A.B. ('89), Ph.D. (Berkeley, '93), researcher at CCR (La Jolla)
Bjorn M. Poonen, Harvard University
(see 1985 results)
David I. Zuckerman, Harvard University
A.B. ('87), Ph.D. (cs, Berkeley, '91), AsP (UT Austin)

Forty-eighth Competition — 1987

David J. Grabiner, Princeton University
(see 1986 results)
David J. Moews, Harvard University
(see 1986 results)
Bjorn M. Poonen, Harvard University
(see 1985 results)
Michael Reid, Harvard University
A.B. ('88), Ph.D. (Brown, '00), Visiting AP (U. Mass.)
Constantin S. Teleman, Harvard University
A.B./A.M. ('91), Ph.D. (Harvard, '94), AP (Stanford), AP (UT Austin), L (Cambridge)
John S. Tillinghast, University of California, Davis
B.A. ('89), M.A. (Harvard, '92), Ph.D. (Stanford, '97), Director of Technical Business Development, Pathmetrics (bioinformatics)

Forty-ninth Competition — 1988

David Grabiner, Princeton University
(see 1986 results)
Jeremy A. Kahn, Harvard University
A.B. ('91), Ph.D. (Berkeley, '95)
David J. Moews, Harvard University
(see 1986 results)
Bjorn M. Poonen, Harvard University
(see 1985 results)
Ravi D. Vakil, University of Toronto
B.Sc. ('92), Ph.D. (Harvard, '97), I (Princeton), I (MIT), AP (Stanford)

Fiftieth Competition — 1989

Christo Athanasiadis, Massachusetts Institute of Technology
B.Sc. ('92), Ph.D. (MIT, '96), KTH Stockholm, compulsory military service
William P. Cross, California Institute of Technology
B.A. ('90), M.Sc. (Chicago), Ph.D. (Michigan, Industrial and Operations Engineering, '95), actuary
Andrew H. Kresch, Yale University
B.S. ('93), M.S. (Yale, '93), Ph.D. (Chicago, '98), L (Penn)
Colin M. Springer, University of Waterloo
Ravi D. Vakil, University of Toronto
(see 1988 results)
Sihao Wu, Yale University

Fifty-first Competition — 1990

Jordan S. Ellenberg, Harvard University
A.B. ('93), M.A. ("The Writing Seminars", Johns Hopkins, '94), Ph.D. (Harvard, '98), I (Princeton), AP (Princeton)

Jordan Lampe, University of California, Berkeley
B.A. ('91), computer programmer
Raymond M. Sidney, Harvard University
A.B. ('91), Ph.D. (MIT, '95), software engineer at Google
Ravi D. Vakil, University of Toronto
(see 1988 results)
Eric K. Wepsic, Harvard University
A.B. ('92), A.M. (Harvard, '94), Managing Director, Proprietary Trading, D.E. Shaw & Co.

Fifty-second Competition — 1991

Xi Chen, University of Missouri, Rolla
B.A ('92), Ph.D. (Harvard, '97), AP (UC Santa Barbara)
Joshua B. Fischman, Princeton University
A.B. ('94), J.D. (Yale, '99), quantitative analyst at KBC Financial Products
Samuel A. Kutin, Harvard University
A.B. ('93), M.A. (Chicago, '94), Ph.D. (Chicago, expected)
Ravi D. Vakil, University of Toronto
(see 1988 results)
Eric K. Wepsic, Harvard University
(see 1990 results)

Fifty-third Competition — 1992

Jordan S. Ellenberg, Harvard University
(see 1990 results)
Samuel A. Kutin, Harvard University
(see 1991 results)
Adam M. Logan, Princeton University
A.B. ('95), Ph.D. (Harvard, '99), AP (Berkeley)
Şerban M. Nacu, Harvard University
A.B. ('96), D.E. Shaw & Co., Ph.D. (statistics, Berkeley, expected)
Jeffrey M. VanderKam, Duke University
B.S. (math, physics, '94), Ph.D. ('98), researcher at CCR (Princeton)
Elizabeth Lowell Putnam Prize: Dana Pascovici, Dartmouth College
B.A. ('95), Ph.D. (MIT, '00), I (Purdue), Enterprise Speech Recognition

Fifty-fourth Competition — 1993

Craig B. Gentry, Duke University
B.S. ('95), J.D. (Harvard, '98), intellectual property lawyer, cryptography researcher at NTT DoCoMo Communications Labs
J.P. Grossman, University of Toronto
B.Sc. ('96), M.Sc. (comp. graphics, MIT, '98), Ph.D. (comp. eng., MIT, expected)
Wei-Hwa Huang, California Institute of Technology
B.S. ('98), Armillaire Technologies
Kiran S. Kedlaya, Harvard University
A.B. (math and physics, '96), M.A. (Princeton, '97), Ph.D. (MIT, '00), AP (Berkeley)

Adam M. Logan, Princeton University
(see 1992 results)
Lenhard L. Ng, Harvard University
A.B. (math and physics, '96), Ph.D. (MIT, '01)

Fifty-fifth Competition — 1994

Jeremy L. Bem, Cornell University
B.A. ('98), Ph.D. (Berkeley, expected)
J.P. Grossman, University of Toronto
(see 1993 results)
Kiran S. Kedlaya, Harvard University
(see 1993 results)
William R. Mann, Princeton University
A.B. ('95), C.P.G.S. (Cambridge, '96), A.M. (Harvard, '97), Ph.D. (Harvard, '01)
Lenhard L. Ng, Harvard University
(see 1993 results)
Elizabeth Lowell Putnam Prize: Ruth A. Britto-Pacumio, Massachusetts Institute of Technology
B.S. ('96), Ph.D. (Harvard, physics, expected)

Fifty-sixth Competition — 1995

Yevgeniy Dodis, New York University
B.A. (math, cs '96), Ph.D. (cs, '00), Visitor (IBM T.J. Watson Research Center), AP (cs, NYU)
J.P. Grossman, University of Toronto
(see 1993 results)
Kiran S. Kedlaya, Harvard University
(see 1993 results)
Sergey V. Levin, Harvard University
A.B. ('96), quantitative trader for Credit Suisse First Boston
Lenhard L. Ng, Harvard University
(see 1993 results)
Elizabeth Lowell Putnam Prize: Ioana Dumitriu, New York University
B.A. ('99), Ph.D. (MIT, expected)

Fifty-seventh Competition — 1996

Jeremy L. Bem, Cornell University
(see 1994 results)
Ioana Dumitriu, New York University
(see 1995 results)
Robert D. Kleinberg, Cornell University
B.A. ('97), Akamai Technologies
Dragos N. Oprea, Harvard University
A.B. ('00), Ph.D. (MIT, expected)
Daniel K. Schepler, Washington University, St. Louis
B.A. ('98), Ph.D. (Berkeley, expected)

Stephen S. Wang, Harvard University
A.B. ('98), A.M. (Harvard, '98), programmer, Ph.D. (Chicago, expected)
Elizabeth Lowell Putnam Prize: Ioana Dumitriu, New York University
(see 1995 results)

Fifty-eighth Competition — 1997

Patrick K. Corn, Harvard University
A.B. ('98), Ph.D. (Berkeley, expected)
Michael L. Develin, Harvard University
A.B. ('00), Ph.D. (Berkeley, expected)
Samuel Grushevsky, Harvard University
A.B. (math and physics, '98), Ph.D. (Harvard, expected)
Ciprian Manolescu, Harvard University
A.B. ('01), Ph.D. (Harvard, expected)
Ovidiu Savin, University of Pittsburgh
B.A., Ph.D. (U.T. Austin, expected)
Daniel K. Schepler, Washington University, St. Louis
(see 1996 results)
Elizabeth Lowell Putnam Prize: Ioana Dumitriu, New York University
(see 1995 results)

Fifty-ninth Competition — 1998

Nathan G. Curtis, Duke University
B.A. (music, math minor, '02 expected)
Michael L. Develin, Harvard University
(see 1997 results)
Kevin D. Lacker, Duke University
B.A. (math, cs, '02 expected)
Ciprian Manolescu, Harvard University
(see 1997 results)
Ari M. Turner, Princeton University
A.B. ('00), Ph.D. (physics, Harvard, expected)

Sixtieth Competition — 1999

Sabin Cautis, University of Waterloo
B.Math. ('01), Ph.D. (Harvard, expected)
Derek I.E. Kisman, University of Waterloo
B.Math. ('00), M.Sc. (Toronto, '01)
Abhinav Kumar, Massachusetts Institute of Technology
S.B. (math, physics, cs, '02 expected)
Davesh Maulik, Harvard University
A.B. ('01), Marshall Scholar (Cambridge)
Christopher C. Mihelich, Harvard University
A.B. ('02, expected)
Colin A. Percival, Simon Fraser University
B.Sc. ('00)

Elizabeth Lowell Putnam Prize: Wai Ling Yee, University of Waterloo
B.Math. (math, cs, '00), Ph.D. (MIT, expected)

Sixty-first Competition — 2000

Gabriel D. Carroll, University of California, Berkeley
A.B. (Harvard, '05, expected)
Abhinav Kumar, Massachusetts Institute of Technology
(see 1999 results)
Ciprian Manolescu, Harvard University
(see 1997 results)
Pavlo Pylyavskyy, Massachusetts Institute of Technology
B.A. ('03, expected)
Alexander B. Schwartz, Harvard University
A.B. ('04, expected)

Team Results

The following is a list of the top teams, in order of rank. Schools choose teams in advance of the competition. School ranks are obtained by summing the individual ranks of the team members.

Forty-sixth Competition — 1985

Harvard University
(Glenn D. Ellison, Douglas S. Jungreis, Michael Reid)
Princeton University
(Michael A. Abramson, Douglas R. Davidson, James C. Yeh)
University of California, Berkeley
(Michael J. McGrath, David P. Moulton, Jonathan E. Shapiro)
Rice University
(Charles R. Ferenbaugh, Thomas M. Hyer, Thomas M. Zavist)
University of Waterloo
(David W. Ash, Yong Yao Du, Kenneth W. Shirriff)

Forty-seventh Competition — 1986

Harvard University
(Douglas S. Jungreis, Bjorn M. Poonen, David I. Zuckerman)
Washington University, St. Louis
(Daniel N. Ropp, Dougin A. Walker, Japheth L.M. Wood)
University of California, Berkeley
(Michael J. McGrath, David P. Moulton, Christopher S. Welty)
Yale University
(Thomas O. Andrews, Kamal F. Khuri-Makdisi, David R. Steinsaltz)
Massachusetts Institute of Technology
(David Blackston, James P. Ferry, Waldemar P. Horwat)

Forty-eighth Competition — 1987

Harvard University
(David J. Moews, Bjorn M. Poonen, Michael Reid)
Princeton University
(Daniel J. Bernstein, David J. Grabiner, Matthew D. Mullin)
Carnegie Mellon University
(Petros I. Hadjicostas, Joseph G. Keane, Karl M. Westerberg)
University of California, Berkeley
(David P. Moulton, Jonathan E. Shapiro, Christopher S. Welty)
Massachusetts Institute of Technology
(David T. Blackston, James P. Ferry, Waldemar P. Horwat)

Forty-ninth Competition — 1988

Harvard University
(David J. Moews, Bjorn M. Poonen, Constantin S. Teleman)
Princeton University
(Daniel J. Bernstein, David J. Grabiner, Matthew D. Mullin)

Rice University
(Hubert L. Bray, Thomas M. Hyer, John W. McIntosh)
University of Waterloo
(Frank M. D'Ippolito, Colin M. Springer, Minh-Tue Vo)
California Institute of Technology
(William P. Cross, Robert G. Southworth, Glenn P. Tesler)

Fiftieth Competition — 1989

Harvard University
(Jeremy A. Kahn, Raymond M. Sidney, Eric K. Wepsic)
Princeton University
(David J. Grabiner, Matthew D. Mullin, Rahul V. Pandharipande)
University of Waterloo
(Grayden Hazenberg, Stephen M. Smith, Colin M. Springer)
Yale University
(Bruce E. Kaskel, Andrew H. Kresch, Sihao Wu)
Rice University
(Hubert L. Bray, John W. McIntosh, David S. Metzler)

Fifty-first Competition — 1990

Harvard University
(Jordan S. Ellenberg, Raymond M. Sidney, Eric K. Wepsic)
Duke University
(Jeanne A. Nielsen, Will A. Schneeberger, Jeffrey M. VanderKam)
University of Waterloo
(Dorian Birsan, Daniel R.L. Brown, Colin M. Springer)
Yale University
(Zuwei Thomas Feng, Andrew H. Kresch, Zhaoliang Zhu)
Washington University, St. Louis
(William Chen, Adam M. Costello, Jordan A. Samuels)

Fifty-second Competition — 1991

Harvard University
(Jordan S. Ellenberg, Samuel A. Kutin, Eric K. Wepsic)
University of Waterloo
(Daniel R.L. Brown, Ian A. Goldberg, Colin M. Springer)
Harvey Mudd College
(Timothy P. Kokesh, Jon H. Leonard, Guy D. Moore)
Stanford University
(Gregory G. Martin, Garrett R. Vargas, András Vasy)
Yale University
(Zuwei Thomas Feng, Evan M. Gilbert, Andrew H. Kresch)

Fifty-third Competition — 1992

Harvard University
(Jordan S. Ellenberg, Samuel A. Kutin, Royce Y. Peng)

University of Toronto
(J.P. Grossman, Jeff T. Higham, Hugh R. Thomas)
University of Waterloo
(Dorian Birsan, Daniel R.L. Brown, Ian A. Goldberg)
Princeton University
(Joshua B. Fischman, Adam M. Logan, Joel E. Rosenberg)
Cornell University
(Jon M. Kleinberg, Mark Krosky, Demetrio A. Muñoz)

Fifty-fourth Competition — 1993

Duke University
(Andrew O. Dittmer, Craig B. Gentry, Jeffrey M. VanderKam)
Harvard University
(Kiran S. Kedlaya, Şerban M. Nacu, Royce Y. Peng)
Miami University, Ohio
(John D. Davenport, Jason A. Howald, Matthew D. Wolf)
Massachusetts Institute of Technology
(Henry L. Cohn, Alexandru D. Ionescu, Andrew Przeworski)
University of Michigan, Ann Arbor
(Philip L. Beineke, Brian D. Ewald, Kannan Soundararajan)

Fifty-fifth Competition — 1994

Harvard University
(Kiran S. Kedlaya, Lenhard L. Ng, Dylan P. Thurston)
Cornell University
(Jeremy L. Bem, Robert D. Kleinberg, Mark Krosky)
Massachusetts Institute of Technology
(Henry L. Cohn, Adam W. Meyerson, Thomas A. Weston)
Princeton University
(William R. Mann, Joel E. Rosenberg, Michail Sunitsky)
University of Waterloo
(Ian A. Goldberg, Peter L. Milley, Kevin Purbhoo)

Fifty-sixth Competition — 1995

Harvard University
(Kiran S. Kedlaya, Lenhard L. Ng, Hong Zhou)
Cornell University
(Jeremy L. Bem, Robert D. Kleinberg, Mark Krosky)
Massachusetts Institute of Technology
(Ruth A. Britto-Pacumio, Sergey M. Ioffe, Thomas A. Weston)
University of Toronto
(Edward Goldstein, J.P. Grossman, Naoki Sato)
Princeton University
(Michael J. Goldberg, Alex Heneveld, Jacob A. Rasmussen)

Fifty-seventh Competition — 1996

Duke University
(Andrew O. Dittmer, Robert R. Schneck, Noam M. Shazeer)
Princeton University
(Michael J. Goldberg, Craig R. Helfgott, Jacob A. Rasmussen)
Harvard University
(Chung-chieh Shan, Stephen S. Wang, Hong Zhou)
Washington University, St. Louis
(Mathew B. Crawford, Daniel K. Schepler, Jade P. Vinson)
California Institute of Technology
(Christopher C. Chang, Hui Jin, Hanhui Yuan)

Fifty-eighth Competition — 1997

Harvard University
(Samuel Grushevsky, Dragos N. Oprea, Stephen S. Wang)
Duke University
(Jonathan G. Curtis, Andrew O. Dittmer, Noam M. Shazeer)
Princeton University
(Craig R. Helfgott, Michael R. Korn, Alexandru-Anton A.M. Popa)
Massachusetts Institute of Technology
(Federico Ardila, Constantin Chiscanu, Amit Khetan)
Washington University, St. Louis
(Daniel B. Johnston, Daniel K. Schepler, Arun K. Sharma)

Fifty-ninth Competition — 1998

Harvard University
(Michael L. Develin, Ciprian Manolescu, Dragos N. Oprea)
Massachusetts Institute of Technology
(Amit Khetan, Eric H. Kuo, Edward D. Lee)
Princeton University
(Craig R. Helfgott, Michael R. Korn, Yuliy V. Sannikov)
California Institute of Technology
(Christopher C. Chang, Christopher M. Hirata, Hanhui Yuan)
University of Waterloo
(Sabin Cautis, Derek I.E. Kisman, Soroosh Yazdani)

Sixtieth Competition — 1999

University of Waterloo
(Sabin Cautis, Donny C. Cheung, Derek I.E. Kisman)
Harvard University
(Michael L. Develin, Ciprian Manolescu, Alexander H. Saltman)
Duke University
(Kevin D. Lacker, Carl A. Miller, Melanie E. Wood)
University of Michigan, Ann Arbor
(Chetan T. Balwe, Rishi Raj, Dapeng Zhu)
University of Chicago

(Matthew T. Gealy, Christopher D. Malon, Sergey Vasseliev)

Sixty-first Competition — 2000

Duke University
(John J. Clyde, Jonathan G. Curtis, Kevin D. Lacker)
Massachusetts Institute of Technology
(Aram W. Harrow, Abhinav Kumar, Ivan Petrakiev)
Harvard University
(Lukasz Fidkowski, Davesh Maulik, Christopher C. Mihelich)
California Institute of Technology
(Kevin P. Costello, Christopher M. Hirata, Michael Shulman)
University of Toronto
(Jimmy Chui, Pavel T. Gyrya, Pompiliu Manuel Zamfir)

Putnam Trivia for the Nineties

Joseph A. Gallian[†]
University of Minnesota, Duluth, Duluth, MN 55812, jgallian@d.umn.edu

The annual Putnam competition has a long and glorious history of identifying extraordinary mathematical talent. Indeed, three Putnam Fellows (top 5 finishers) have won the Fields Medal and two have won a Nobel prize in Physics. In fact, the 1954 Harvard Putnam team included a future Fields Medalist and a future Nobel Laureate! And of course many Putnam Fellows have had distinguished careers starting with the very first winner of the Putnam Fellowship to Harvard — Irving Kaplansky. In the early days (1930s) only a few hundred students competed in the competition whereas by the 1990s over 2000 per year took part. In a 1989 issue of the *Monthly* I gave a list of trivia questions based on the first fifty years of the competition [G1]. In this article I offer trivia questions based on the competitions of the 1990s. Answers are given on page 321.

1. Among the five winners of the AMS-MAA-SIAM Morgan Prize for undergraduate research in the 90s, how many have been Putnam Fellows?
   A. 0 B. 1 C. 2 D. 3

2. Among the five people named honorable mention for the Morgan Prize in the 90s, how many have been Putnam Fellows?
   A. 0 B. 1 C. 2 D. 3

3. How many times during the 90s was an individual from a school that does not grant a Ph.D. degree in mathematics a Putnam Fellow?
   A. 0 B. 1 C. 2 D. more than 2

4. How many times during the 90s did a team from a school that does not grant a Ph.D. degree in mathematics finish in the top 5?
   A. 0 B. 1 C. 2 D. more than 2

[†] This article first appeared in the *American Mathematical Monthly* [G2].

5. How many times during the 90s did a team from a U.S. state university finish in the top 5?
   A. 0 B. 1 C. 2 D. more than 2

6. How many times during the 90s was an individual from a U. S. state university a Putnam Fellow?
   A. 0 B. 1 C. 2 D. more than 2

7. How many times during the 90s did the University of Waterloo finish in the top 5?
   A. 2 B. 4 C. 6 D. more than 6

8. How many times during the 90s has a Putnam Fellow been from a Canadian school?
   A. 4 B. 6 C. 8 D. more than 8

9. How many women were Putnam Fellows in the 90s?
   A. 0 B. 1 C. 2 D. more than 2

10. Which schools have won the team competition in the 90s?
    A. Harvard B. Princeton C. Duke D. University of Waterloo

11. Rank the following schools according to the greatest number of times they have placed in the top 5 in the team competition in the 90s.
    A. Duke B. MIT C. University of Waterloo D. Princeton

12. Rank the following schools according to the greatest number of Putnam Fellows in the 90s.
    A. Duke B. Cornell C. University of Waterloo D. Princeton

13. How many times during the 90s has the winning team had all three team members finish in the top 5 that year?
    A. 0 B. 1 C. 2 D. 3

14. How many times during the 90s has the winning team had no Putnam Fellows?
    A. 0 B. 1 C. 2 D. 3

15. How many times during the 90s has the same person been a Putnam Fellow four times?
    A. 0 B. 1 C. 2 D. 3

16. How many times during the 90s has the same person been a Putnam Fellow three

times?
A. 0 B. 1 C. 2 D. 3

17. What is the lowest ranking during the 90s by the Harvard team?
A. 2nd B. 3rd C. 4th D. 5th

18. What is the only school to have both its Putnam team and its football team finish in the top 5 in the 90s?
A. Michigan B. Stanford C. Wisconsin D. Notre Dame

19. What is the only school to have both its Putnam team and its men's basketball team (in the NCAA tournament) finish first in the 90s?
A. North Carolina B. Michigan C. Duke D. UCLA

20. What is the only school to have its Putnam team finish in the top 5 and its men's hockey team finish first in the NCAA tournament in the 90s?
A. Michigan B. Harvard C. Cornell D. Wisconsin

21. How many years in the 90s was every Putnam Fellow from a different school?
A. 0 B. 1 C. 2 D. more than 2

22. What is the highest team finish with a woman team member in the 90s?
A. 1st B. 2nd C. 3rd D. not in top 5

23. How many schools had two or more Putnam Fellows in the same year in the 90s?
A. 0 B. 1 C. 3 D. more than 3

24. Which school(s) had two Putnam Fellows in the same year but did not finish in the top 5 of the team competition?
A. Cornell B. Duke C. MIT D. Princeton

25. Who is the Fields Medalist whose son was on the winning team during the 90s?
A. William Thurston B. Vaughan Jones C. Ed Witten D. Charles Fefferman

26. Rank the following schools according to the most second place finishes in the 90s.
A. Harvard B. Duke C. MIT D. Cornell

27. What was the monetary prize given to the winning team in 1990?
A. $1000 B. $2000 C. $5000 D. $10000

28. What was the monetary prize given to the winning team in 1999?

A. \$5000 B. \$10000 C. \$20000 D. \$25000

29. How many times during the 90s did Harvard have three or more Putnam Fellows in the same year?
A. 1 B. 3 C. 4 D. more than 4

30. How many times during the 90s has someone from Harvard been a Putnam Fellow?
A. 10 B. 15 C. 20 D. 25

Answers are given on page 321.

Some Thoughts on Writing for the Putnam

Bruce Reznick†
University of Illinois at Urbana-Champaign

What is the Putnam?

This chapter describes the process of composing problems for the William Lowell Putnam Mathematical Competition. Inevitably, this leads to the more general issue of mathematical problem writing. I shall be anecdotal and probably idiosyncratic, and do not purport that my opinions are definitive or comprehensive. The reader should not look for "How to Pose It," but rather sit back and enjoy the heart-warming tale of a boy and his problems.

The following is a quotation from the official brochure (Putnam, 1992). The William Lowell Putnam Mathematical Competition "began in 1939 and is designed to stimulate a healthful rivalry in mathematical studies in the colleges and universities of the United States and Canada. It exists because Mr. William Lowell Putnam had a profound conviction in the value of organized team competition in regular college studies." The Putnam, as it is universally called, is administered by the Mathematical Association of America. It is offered annually (since 1962, on the first Saturday in December) to students who have not yet received a college degree. In 1991, 2,375 students at 383 colleges and universities took the exam. (For more on a history of the Putnam, see [PutnamI, PutnamII].) The problems and solutions for Competitions through 1984 are in [PutnamI, PutnamII]. The problems, solutions, and winners' names are also published in the *American Mathematical Monthly*, usually about a year after the exam.

The Putnam consists of two independent 3-hour sessions, each consisting of six problems arranged roughly in order of increasing difficulty. The exam is administered by proctors who cannot comment on its content. Contestants work alone and without notes, books, calculators, or other external resources. They are ranked by their scores, except that the top five are officially reported en bloc. Teams are preselected by their coach, and team rankings are determined by the sum of the ranks (not the sum of the scores).

---

† This chapter originally appeared in [Re4]. The author was supported in part by the National Science Foundation.

> The examination will be constructed to test originality as well as technical competence. It is expected that the contestant will be familiar with the formal theories embodied in undergraduate mathematics. It is assumed that such training, designed for mathematics and physical science majors, will include somewhat more sophisticated mathematical concepts than is the case in minimal courses.
> Questions will be included that cut across the bounds of various disciplines and self-contained questions which do not fit into any of the usual categories may be included. It will be assumed that the contestant has acquired a familiarity with the body of mathematical lore commonly discussed in mathematics clubs or in courses with such titles as "Survey of the foundations of mathematics." It is also expected that self-contained questions involving elementary concepts from group theory, set theory, graph theory, lattice theory, number theory, and cardinal arithmetic will not be entirely foreign to the contestant's experience. (from official brochure)

Between 1969 and 1985, I participated in all but three Putnams. I competed four times, the last two wearing the silks of the Caltech team, which placed first. As a graduate student and faculty member, I coached or assisted at Stanford, Duke, Berkeley, and Urbana. I was a grader in 1982 and a member of the Problems Subcommittee for the 1983, 1984, and 1985 Competitions. (I had been living quietly in Putnam retirement when Alan Schoenfeld invited me to this conference.†)

Who writes the Putnam?

The Problems Subcommittee of the MAA Putnam Committee consists of three question writers, who serve staggered 3-year terms. The most senior member chairs the Subcommittee. During my service, the Problems Subcommittee consisted of successive blocks of three consecutive people from the following list: Doug Hensley, Mel Hochster, myself, Richard Stanley, and Harold Stark. Outgoing members are invited to suggest possible replacements, but these are not acted on immediately. This is an old-boy network, but in practice one often suggests the names of strangers whose problems one has admired. (It turned out that the 1985 group inadvertently consisted of three Caltech alumni, and, as chair, I was relieved when the Harvard team won.)

The rest of the Putnam Committee consisted of three permanent members, two of whom (Jerry Alexanderson and Leonard Klosinski) arranged the massive logistics of the Competition and a liaison with the Problems Subcommittee (Abe Hillman, then Loren Larson). They do an incredible amount of work, which is not germane to this essay, but should not go unappreciated. They also provide a valuable final check to the Subcommittee's work.

---

† This article was originally written for the conference "Mathematical Thinking and Problem Solving", held in Berkeley on December 14–16, 1989.

How was the 1985 Putnam Written?

The following is an overview of how the 1985 Putnam was written. In November 1984, I wrote a welcoming instructional letter to Richard and Harold, describing our timetable and goals. In early December, each of us circulated about a dozen problems to the other two. After a decent interval, we circulated solutions to our own problems and comments about the others and added some more into the pot. A few more rounds of correspondence ensued. In March, we met for a weekend with the rest of the Committee to construct the Competition. After a small flurry of additional correspondence, the material was handed over to the logistics team.

The greeting letter of the chair of the Problems Subcommittee is a quilt to which each chair adds (or rips up) patches. The following excerpts, lightly rewritten for style, are thus simultaneously traditional and fully my own responsibility.

> It used to be said that a Broadway musical was a success if the audience left the theater whistling the tunes. I want to see contestants leave the Putnam whistling the problems. They should be vivid and striking enough to be shared with roommates and teachers.
>
> Security should be a major concern ... the problems should be handwritten or typed by ourselves, and our files should either be unmarked or kept home. Putnam problems ought to be pretty enough that you want to tell your friends about them. Do your best to resist this temptation.
>
> Lemmas in research papers are fair game, but material from well-known textbooks or problem collections are not. (If you submit a problem from a known source, please include this information.) Seminar material is OK unless undergraduates were present, and anything you have taught to an undergraduate honors class ought to dry out for a few years before gaining eligibility. In general, problems from other people are not reliably secure unless your source can vouch for their originality.
>
> As for the problems themselves, my feeling is that any problem solved by only one or two contestants is a failure, no matter how beautiful it might be. In the last couple of years, we have sought to turn A-1 and B-1 into "hello, welcome to the exam" problems, and their relative tractability has been appreciated. It is better to require one major insight than several minor ones (partial credit is undesirable). It is better to write a streamlined problem without many cases, so that we test perceptiveness, rather than stamina. Proofs by contradiction are, in general, unsuitable, both because they are ugly, and because they are harder to grade.
>
> Although concern for the graders is not our primary consideration, we should keep them in mind. There is no reason to exclude a problem such as 1983 B-2, merely because there are many different legitimate proofs. On the other hand, we must be at pains to write unambiguous questions even at the expense of simplicity in the phrasing. Answers in a particular numerical form are often desirable so that students won't puzzle over the phrase "simplest terms"; this is one reason that the current year stands in for "n". Answers that turn out to be 0, 1, π, $\sqrt{2}$, etc., should be avoided to eliminate the lucky

guess, and we should not present problems in which the solution is easy to guess but hard to prove. (The reverse is preferable.) I confess to a predilection for "garden variety" mathematical objects, such as powers of 2, binomial coefficients, pentagons, sines, cosines, and so on. I dislike problems with an elaborate notation, whose unraveling is a major portion of the solution.

What does the Putnam mean?

The previous three sections have given a theoretical description of how the Putnam works, and what it is intended to accomplish. I turn now to the Putnam in reality.

There is some evidence that the Putnam achieves its intended goals. Many schools run training sessions for contestants, in which interesting mathematics and useful techniques of problem solving are presented. A successful individual performance on the Putnam leads to fame and glory and an increased probability of a fellowship. (However, the results are announced in March, very late for seniors applying to graduate school.) There is also money; I was entertained for years by the William Lowell Putnam Stereo System. More importantly, a contestant can properly be satisfied in solving any Putnam problem, though this is tempered by the (larger number of) problems one does not solve. Putnam problems have occasionally led to research, and a problem may stick in a contestant's mind for years. The ultimate source of [Re2] was 1971 A-1.

The phrase "Putnam problem" has achieved a certain cachet among those mathematicians of the problem-solving temperament and is applied to suitably attractive problems which never appeared on the exam. One motivation for my joining the Problems Subcommittee was the aesthetic challenge of presenting to the mathematical community a worthy set of problems. In fact, the opportunity to maintain this "brand name of quality" was more enticing to me than the mere continuation of an undergraduate competition. Of course, the primary audience for the Putnam must always be the students, not one's colleagues.

At the same time, the Putnam causes a few negative effects, mainly because of its difficulty. Math contests are supposed to be hard, and the Putnam is the hardest one of all. In 1972, I scored less than 50% and finished seventh. In most years, the median Putnam paper has fewer than two largely correct solutions. For this reason, the first problem in each session is designed to require an "insightlet", though not a totally trivial one. We on the Committee tried to keep in mind that median Putnam contestants, willing to devote one of the last Saturdays before final exams to a math test, are likely to receive an advanced degree in the sciences. It is counterproductive on many levels to leave them feeling like total idiots.

Success on the Competition requires mathematical ability and problem-solving experience, but these are not sufficient; a "Putnam" temperament is also necessary. A contestant must be able to work quickly, independently, and without references and be willing to consider problems out of context. I have been saddened by reports of students who were discouraged in their academic careers by a poor performance on the Putnam. Fortunately for the mathematical community, there are many excellent, influential, and successful mathematicians who also did badly on the Putnam. As a result, the absence of Putnam kudos has a negligible effect on one's career. (At the same time, I confess to enjoying the squirmy defensiveness that the term "Putnam" evokes in some otherwise arrogant colleagues of the "wrong" temperament. They loudly deny an importance to the Competition that nobody else asserts.)

Among those who do very well on Putnam problems, there is little hard evidence that doing extremely well is significant. The best papers usually average about twice as many correct solutions as the thirtieth. My impression is that the likely future mathematical outputs of the writers are comparable.

In sum, the Putnam plays a valuable, but ultimately inessential, role in undergraduate mathematics. This is a test; this is only a test.

What makes a good Putnam problem?

Other considerations besides pure problem aesthetics afflict the Putnam writers. We wish to have a balance of questions in various subject areas and solving styles. We need an "easy" question for A-1 and B-1. As college math teachers, we are often astonished at how poorly we know what it is that our own students do and do not understand. This is magnified on the Putnam, in which contestants come from hundreds of different programs. The Committee tries to be sensible. It's unreasonable to have the trace of a matrix in one of the easier problems, but we used it in 1985 B-6, on the grounds that a contestant who had not heard of a trace would probably be unable to do the problem anyway.

We want to test, if possible, abstract problem-solving ability, rather than classroom knowledge; maturity "yes," facts "no." We try to avoid the traditional corpus of problem-solving courses to minimize the reward in "studying" for the Putnam. This leads to a tradeoff between familiarity and quality. We occasionally receive a complaint that a problem is not new or has even appeared in material used to train Putnam competitors at a particular school. This is unfortunate, but probably inevitable. It would be easy to write an exam with twelve highly convoluted, certifiably original, and thoroughly uninteresting problems.

Otherwise acceptable questions have been rejected on the grounds that they "sound" familiar or "must have appeared somewhere," even when no member of the Committee can cite a reference. Here's an example: A projectile is to be fired up a hill which makes an angle α with the horizontal. At what angle should the projectile be launched in order to maximize the distance it travels?

One April, the day after the Committee completed its deliberations, someone discovered that our A-1 was a problem posed in the most recent *Two-Year College Mathematics Journal.* Fortunately, we had bequeathed an easy problem to the following year's exam, and a few phone calls resolved the crisis. I do not know what we would have done if this had happened in November.

What follows are some representative comments evoked by the first round of proposed problems in the Subcommittee:

> "Seems routine, too easy."
>
> "I found the computations too messy, and it was easy to head off in the wrong direction."

"Can't use, it's been around for years."

"Hard (or did I miss something). A good problem."

"The trouble with this one is that an intuitive guess ... is incorrect. OK, not inspired."

"I was stumped, but it's a nice hard problem."

"I couldn't do this one either."

Every Putnam I helped write contained at least one problem I could not do on first sight. More comments follow:

"This looks very messy. I saw nothing that motivated me to take up pencil and scribble."

"I wasn't lit up by waves of excitement."

"We must use this one, I love it."

"Yes, yes, yes."

(The last block suggests the sensuality of a good problem to the discriminating solver. For more on this subject, see [Re1].)

How are Putnam problems polished?

The Committee acts by consensus; I do not ever recall voting on a problem. Most problems have one primary author, although the full group polishes the final version. Often, though, this version is a special case of the original problem. We tried to have at least one problem from calculus, geometry, and number theory on the exam. I was always amazed at the ability of the Committee to find unfamiliar problems in such fully excavated fields. It is hard to write serious algebra problems that are not basically manipulative, rather than conceptual, because we can assume so little knowledge. In analysis, any integrals must be innocent of measure theory.

We do not accept a problem until we have seen a solution written out in full; sometimes we produce more than one solution. It is not unusual for contestants to find new (and better) solutions. One problem (1983 B-2) evolved from the remarkable fact, familiar to our silicon friends, that a nonnegative integer n has a unique binary representation. Let $f(n)$ denote the number of ways that n can be written in the form $\sum a_i 2^i$, if the a_i's are allowed to take the values 0, 1, 2, or 3. It turns out that $f(n) = \lfloor n/2 \rfloor + 1$. (Here, and below, $\lfloor x \rfloor$ represents the largest integer $\leq x$.) This problem has at least three different solutions:

1. by generating functions — $\sum f(n)x^n = \{(1-x)(1-x^2)\}^{-1}$,
2. by induction — use the recurrence $f(2n) = f(2n+1) = f(n) + f(n-1)$, or
3. by direct manipulation — writing $a_i = 2b_i + c_i$, where b_i and c_i are 0 or 1, gives a bijection onto sums $n = 2k + m$, where $k = \sum b_i 2^i$ and $m = \sum c_i 2^i$.

I have explored this topic more extensively elsewhere [Re3].

Several times, there was true collaboration. Doug Hensley called me to say that he wanted a problem in which an algorithm terminated because a certain nonnegative integral parameter decremented by 1 after each iteration. This reminded me of a situation I was playing with, in which n was replaced by $n + \lfloor\sqrt{n}\rfloor$; the relevant parameter is $(n - \lfloor\sqrt{n}\rfloor^2) \pmod{\lfloor\sqrt{n}\rfloor}$, which decrements by 1, usually after two iterations. It follows that one eventually reaches a perfect square (see 1983 B-4).

Another time, Mel Hochster had been playing with a problem using the tips of the hands of an accurate clock (as we ultimately phrased it — we received complaints from students who were only familiar with digital clocks!). This was a nice "trapdoor" situation. A reasonably competent student could parametrize the positions of the tips, and after a half-hour of calculation, solve the question. The intended insight was to make a rotating set of coordinates in which the minute hand is fixed, so only the hour hand is rotating. We would have a problem, if only we could find the right question. It occurred to me to look at the distance between the tips when that distance was changing most rapidly. Solving it the long way, I uninspiredly computed an answer that shouted out, "You idiot, use the Pythagorean theorem!" In fact, the derivative of the difference vector from one tip to the other has constant magnitude and is normal to the hour vector, and the distance is changing most rapidly when the difference vector and its derivative are parallel (see 1983 A-2).

The length of service as a Putnam writer seems optimal. In my first year, I was bursting with problems I had saved for the exam and discovered that some were unsuitable. In the second year, I tried to rework the leftovers and develop some techniques for consciously writing other problems (these will be discussed below). (A few Putnam rejects have appeared in the *Monthly* problem section, where the lack of time constraints relieves concern over messy or evasive algebra.) By the third year, I felt drained of inspiration; in fact, my impression all three years was that the chair placed the fewest problems on the Putnam. Service on the Subcommittee was also beginning to have an adverse effect on my research. Ordinarily, a mathematician tries to nurture a neat idea in hopes that it will grow into a theorem or a paper. I found that I was trying to prune my ideas so that they would fit on the exam. Bonsai mathematics may be hazardous to your professional health!

Did Archimedes use δ's and ϵ's?

The story of one of my favorite problems (1984 B-6) serves as an object lesson in theft. In the course of researching the history of the Stern sequence and Minkowski's ?-function, I had run across a beautiful example of Georges de Rham. Let P_0 be a polygon with n sides, trisect each side, and snip off the corners, creating P_1, a polygon with $2n$ sides. Iterate. The boundary of the limiting figure, P_∞, has many interesting counterintuitive properties (see the last paragraph of the discussion of this problem in [PutnamII]). For example, it is a smooth convex curve which is flat almost everywhere. I was rather pleased with myself for having noticed a property of P_∞ itself. Suppose one corner, snipped from P_i, has area A. Then each of the two new adjacent corners snipped from P_{i+1} has one-third the altitude and one-third the base of the original corner, and so has area $A/9$. Further, if P_0 is a triangle, then P_1 is a hexagon whose area is two-thirds the area of P_0. This information can be combined with the formula for the sum of a geometric series with $r = 2/9$ to show that the area of P_∞ is four-sevenths the area of P_0. A sneaky new problem that requires only precalculus — Putnam heaven! I had mentioned this result in a seminar a few years before, but no undergraduates attended, and I was confident of security.

The Tuesday before the competition, I attended a Pi Mu Epsilon lecture on the approaches of Archimedes to calculus, given by Igor Kluvanik, an Australian mathematician visiting Urbana. To my horror, I learned that Archimedes had stolen my method in order to compute the area under a parabola! At least one colleague noticed that I lost my color, and I told her that I could explain the circumstances in about a week. Fortunately, our students did not do unusually well on that problem.

By the way, we stated this problem so that P_0 was an equilateral triangle of side 1, and asked for the area in the form $a\sqrt{c}$, where a is rational, and c is integral. The majority of solvers assumed, incorrectly, that P_∞ had to be the circle inscribed in the triangle and derived an answer involving π. (When I mentioned this problem at a colloquium, a famous mathematical physicist in the audience audibly made the same guess.) The reader might find it amusing to consider the following variation, in which the resulting figure is not a circle, but is piecewise algebraic: Suppose that, rather than a trisection, each side is split in ratio $1:2:1$ before the corners are snipped off. Describe P_∞.

What makes a good Putnam problem?

The instructions sent to composers do not include a description of the characteristics of a good Putnam problem. The aesthetic seems to be fairly universal among dedicated problem solvers and can be applied more generally to describing good mathematical problems. (Perhaps this reluctance to be specific also reflects the mathematician's cowboy taciturnity on such woolly subjects.)

I will hazard some definitions. A mathematical problem is simply a mathematical situation in which some information is implied by other information. The principal characteristics of a good problem are simplicity, surprise, and inevitability. By *inevitability*, I mean two things: once you've solved the problem, you cannot look at it without also seeing its solution; once you see the problem, you feel you *must* solve it. (A tacit rite of passage for the mathematician is the first sleepless night caused by an unsolved problem.)

As an undergraduate competitor, I told my housemates that doing well on the Putnam reflected one's ability to do very quickly, other people's tricky, solved problems. I'll stand by that today. Three of the most important preliminary questions a problem solver must face are: (a) Is there a solution? (b) What do I need to know to find the solution? and (c) What does the solution look like? These questions are all answered in advance for the Putnam competitor. You know that there *is* a solution, which is probably short and clever and does not require a great deal of knowledge. You know that you will recognize the solution when you see it. Tables and computer data and other references are irrelevant, and inaccessible in any case. You cannot collaborate, or even describe the problem to someone else in hopes of understanding it in the retelling. It is for these reasons that I am extremely unhappy when I hear that some problem-solving courses use the Putnam as a final exam. Chocolate decadence cake à la mode is a delicious dessert, but makes an unfilling main course.

This discussion begs the larger question of the role of problems within mathematics. Simplicity and surprise may be enjoyable, but they do not accurately characterize much of the mathematician's world, in which correct insights are often wrested after much reflection from a rich contextual matrix and are as snappy as a tension headache.

Okay, so how do you sit down and create a Putnam problem?

Okay, so how do you sit down and create a Putnam problem? One way is to keep your eyes open for anything in your own work that looks like a Putnam problem. You can make a votive offering to Ineedalemma, the tutelary goddess of mathematical inspiration.

You can also be somewhat more systematic. I was fortunate to have a father who addressed very similar questions in his own work: comedy writing. He wrote a book about writing jokes, from which I take the following quotation [ReS, p. 15].

> Very few writers can pound out a huge batch of jokes week after week relying solely on sheer inspiration. They need the help of some mechanical process. When a writer "has to be funny by Tuesday" he's not going to wait for hot flashes of genius, especially if he doesn't happen to be feeling too "geniusy." ... Switching is the gag writer's alchemy by which he takes the essence of old jokes from old settings and dresses them up in new clothes so they appear fresh.

(Putnam punks such as myself are often inveterate punsters. Punning requires the rapid formal combinatorial manipulation of strings of symbols, without much concern for content. This skill is also very helpful on the Putnam. More serious connections between humor and creativity are discussed in [Koe].)

The details of switching problems and switching jokes are substantially different, but the principle is the same. The following is one practical illustration of problem switching. I heard a seminar speaker refer to a beautiful result of Mills: There is a positive number α with the property that $p_n = \lfloor \alpha^{3^n} \rfloor$ is a prime for every $n \geq 1$. The construction is recursive, based on the observation that α must lie between $(p_n)^{3^{-n}}$ and $(p_n + 1)^{3^{-n}}$, and there is always a prime between any two consecutive cubes. It occurred to me that something similar might be wrought out of the simpler expression $\lfloor \alpha^n \rfloor$. One of the most familiar properties of α^n is that it is always even if α is an even integer (and $n \geq 1$) and always odd if α is odd. The most counterintuitive behavior for $\lfloor \alpha^n \rfloor$ would thus be for it to alternate between even and odd. If you start with $\lfloor \alpha \rfloor \geq 3$, then Mills's interval argument will work, and this is how 1983 A-5 came to life.

Although this problem was solved by fewer contestants than we had hoped, perhaps some of them later realized that the alternation of even and odd is basically irrelevant to the problem, and that any pattern of parities (mod 2) can be achieved using the same proof, as well as any pattern (mod m). Later on, William Waterhouse found an explicit algebraic integer with the property that $\lfloor \alpha^n \rfloor$ alternates in parity and submitted this version of A-5 to the *Monthly* Problems Section.† I refereed it, and we received dozens of correct solutions; see [Mon1, Mon2].

---

† Editors' note: one possible α is the positive root of the quadratic $x^2 - ax - b$, where a is odd, b is even, and $b < a$.

Another way to create Putnam problems is via Fowler's method. Gene Fowler once explained that it's very easy to write. All you have to do is sit at a typewriter and stare at a sheet of blank paper until blood begins to appear on your forehead. I often applied this technique at less exciting seminars and colloquia, when my neighbors thought I was doodling. I'd take a combination of simple mathematical objects and stare at them until I could see a Putnam problem. Sometimes it worked. For example, 1984 A-4 asks for the maximal possible area of a pentagon inscribed in a unit circle with the property that two of its chords intersect at right angles.

Contrary to popular opinion, it's unhelpful to read through old problem books very much for inspiration, because subconscious plagiarism is a great danger, and our larger audience is very alert. (It might be more useful to look through the back of books, because switches based on solutions are less transparent.)

So how do you sit down and create a Putnam problem? Let's apply Pólya's rules and generalize the question. How do you sit down and create? This is a very difficult and personal question. (It might not even have an answer; our romantic culture tends to identify the results of algorithmic thinking as mechanical, rather than creative.)

In the end, you can do everything you can, rely on the rest of the Committee for inspiration, and visualize two thousand fresh minds on the first Saturday in December, eager to be challenged.

Answers to Putnam trivia questions appearing on pages 307–310:

1. A
2. C (Kiran Kedlaya and Lenny Ng)
3. A
4. C (Miami University and Harvey Mudd College)
5. C (Michigan 1993 and 1999)
6. C (Xi Chen from U. Missouri–Rolla and Jordan Lampe from Berkeley. Ovidiu Savin was a Putnam Fellow from the state-related U. Pittsburgh)
7. C
8. C
9. B (Ioana Dumitriu)
10. A, C, D
11. C and D six times each; A and B five times each
12. D (5), A (4), B (3), C (2)
13. B (Harvard 1990)
14. B (Duke 1996)
15. A
16. D (Kiran Kedlaya, Lenny Ng, J.P. Grossman; they were Putnam Fellows in the same three years.)
17. B (1996)
18. A
19. C
20. A
21. A
22. B (Duke 1990)
23. D (Harvard, Cornell, Duke and Waterloo)
24. A (1996), B (1998)
25. A (Dylan Thurston)
26. A, B and D each finished second twice; C once
27. C
28. D
29. C (1990, 1992, 1995 and 1997)
30. D

Bibliography

[ACGH] Arbarello, E., M. Cornalba, P. A. Griffiths, and J. Harris, *Geometry of algebraic curves* vol. 1, Springer-Verlag, New York, 1985. (1991B3)

[Ah] Ahlfors, L., *Complex analysis*, third ed., McGraw-Hill, 1979. (1989B3)

[An] Andrews, P., Where not to find the critical points of a polynomial — Variation on a Putnam theme, *Amer. Math. Monthly* **102** (1995), 155–158. (1991A3)

[Ap1] Apostol, T., *Calculus* vol. 1, second ed., John Wiley & Sons, New York, 1967. (1992A4, 1993B4)

[Ap2] ——, *Calculus* vol. 2, second ed., John Wiley & Sons, New York, 1969. (1985B6, 1987A5, 1990A5, 1995B2, 1996A1, 1996B2)

[APMO] The Asian-Pacific Mathematical Olympiad; problems available at Problems Corner section of American Mathematics Competition website `http://www.unl.edu/amc`. (1988A5, 1992B1)

[Ar] Artin, E., *Galois theory,* Univ. of Notre Dame Press, Notre Dame, 1944. (1992B6)

[AS] Allouche, J.-P. and J. Shallit, "The ubiquitous Prouhet-Thue-Morse sequence," in C. Ding, T. Helleseth, and H. Niederreiter, eds., *Sequences and their applications: Proceedings of SETA '98*, Springer-Verlag, 1999, pp. 1–16. (1992A5)

[At] Atkinson, K. E., *An introduction to numerical analysis,* second ed., John Wiley & Sons, New York, 1989. (1996B2)

[AZ] Almkvist, G. and D. Zeilberger, The method of differentiating under the integral sign, *J. Symbolic Computation* **10** (1990), 571–591. (1997A3)

[Bak1] Baker, A., *A concise introduction to the theory of numbers,* Cambridge Univ. Press, Cambridge, 1984. (1991B4)

[Bak2] ——, *Transcendental number theory,* Cambridge Univ. Press, Cambridge, 1975. (1998B6)

[Bar] Barnett, S., *Matrices: Methods and applications,* Oxford Univ. Press, New York, 1990. (1988B5)

[BB] Borwein, J. M. and P. B. Borwein, Strange series and high precision fraud, *Amer. Math. Monthly* **99** (1992), no. 7, 622–640. (1987A6)

[BC] Borwein, J. and K.-K. S. Choi, On the representations of $xy + yz + zx$, *Experiment. Math.* **9** (2000), no. 1, 153–158. (1988B1)

[BCG] Berlekamp, E. R., J. H. Conway and R. K. Guy, *Winning ways for your mathematical plays,* Academic Press, New York, 1982. (1993A6, 1995B5, 1997A1)

[BD] Boyce, W. E. and R. C. DiPrima, *Elementary differential equations and boundary value problems,* seventh ed., John Wiley & Sons, New York, 2001. (1988A2, 1995A5, 1997A3)

[Bel] Bell, E. T., Class numbers and the form $xy + yz + zx$, *Tôhoku Math. J.* **19** (1921), 105–116. (1988B1)

[Berg] Berger, M., *Geometry I,* Springer-Verlag, New York, 1994. (1990B6)

[Bernau] Bernau, S. J., The evaluation of a Putnam integral, *Amer. Math. Monthly* **95** (1988), 935. (1985B5)

[Berndt] Berndt, B. C., *Ramanujan's notebooks, part I,* Springer-Verlag, New York, 1985. (1986A3)

[Bi] Birkhoff, G., Extensions of Jentzsch's theorem, *Trans. Amer. Math. Soc.* **85** (1957), 219–227. (1997A6)

[BJL] Beth, T., D. Jungnickel and H. Lenz, *Design theory,* vols. I and II, second ed., Cambridge Univ. Press, Cambridge, 1999. (2000B6)

[Bl] Blumenthal, L. M., *Theory and applications of distance geometry,* second ed., Chelsea, New York, 1970. (1993B5)

[Br] Brauer, A., On a problem of partitions, *Amer. J. Math.* **64** (1942), 299–312. (1991B3)

[BX] Brown, J. E. and G. Xiang, Proof of the Sendov conjecture for polynomials of degree at most eight, *J. Math. Anal. Appl.* **232** (1999), no. 2, 272–292. (1991A3)

[C] Christol, G., Ensembles presque périodiques k-reconnaissables, *Theoret. Comput. Sci.* **9** (1979), no. 1, 141–145. (1989A6)

[CF] Cusick, T. W. and M. E. Flahive, *The Markoff and Lagrange spectra,* Amer. Math. Soc., Providence, RI, 1989.(1986B5)

[CG] Coxeter, H. S. M. and S. Greitzer, *Geometry revisited,* New Mathematical Library **19**, Math. Association of America, Washington, DC, 1967. (1997A1, 1998B2)

[CKMR] Christol, G., T. Kamae, M. Mendès France, and G. Rauzy, Suites algébriques, automates et substitutions, *Bull. Soc. Math. France* **108** (1980), no. 4, 401–419. (1989A6)

[CKP] Carrier, G. F., M. Kroork, C. E. Pearson, *Functions of a complex variable: Theory and technique,* McGraw-Hill Book Co., New York, 1966. (2000A4)

[Co] Conrad, K., The origin of representation theory, *Enseign. Math.* **44** (1998), 361–392. (1988B5)

[Da] Davis, P., *Circulant matrices,* John Wiley & Sons, New York, 1979. (1988B5).

[De] Devaney, R., *An introduction to chaotic dynamical systems,* second ed., Addison-Wesley, Redwood City, 1989. (1992B3)

[DMc] Dym, H. and H. P. McKean, *Fourier series and integrals,* Academic Press, New York, 1972. (1995A6)

[EFT] Ebbinghaus, H.-D., J. Flum, and W. Thomas, *Mathematical logic,* second ed., translated from the German by M. Meßmer, Springer-Verlag, New York, 1994. (1988B2)

[Ei] Eisenbud, D., *Commutative algebra with a view towards algebraic geometry,* Graduate Texts in Math. **150**, Springer-Verlag, New York, 1995. (1986B3)

[En] Enderton, H. B., *Elements of set theory*, Academic Press, New York-London, 1977. (1989B4)

[Er] Erdős, P., On the set of distances of n points, *Amer. Math. Monthly* **53** (1946), 248–250. (1990A4)

[Fe] Feynman, R. P., *"Surely you're joking, Mr. Feynman!": Adventures of a curious character*, Bantam Books, New York, 1985. (1997A3)

[FN] Fel'dman, N. I. and Yu. V. Nesterenko, "Number theory, IV, transcendental numbers," in A. N. Parshin and I. R. Shafarevich eds., *Encyclopaedia Math. Sci.* **44**, Springer-Verlag, Berlin, 1998. (1993A5)

[Fo] Folland, G. B., *Real analysis*, John Wiley & Sons, New York, 1984. (1993B3)

[FPS] Flynn, E. V., B. Poonen, and E. Schaefer, Cycles of quadratic polynomials and rational points on a genus 2 curve, *Duke Math. J.* **90** (1997), no. 3, 435–463. (2000A6)

[Fr] Fraleigh, J. B., *A first course in abstract algebra*, sixth ed., Addison-Wesley, Reading, MA, 1999. (1990B4)

[Fu] Fulton, W., *Introduction to toric varieties*, Princeton Univ. Press, Princeton, 1993. (1996B6)

[FW] Frankl, P. and R. M. Wilson, Intersection theorems with geometric consequences, *Combinatorica* **1** (1981), no. 4, 357–368. (1988A4)

[G1] Gallian, J. A., Fifty years of Putnam trivia, *Amer. Math. Monthly* **96** (1989), 711–713.

[G2] ——, Putnam trivia for the 90s, *Amer. Math. Monthly* **107** (2000), 733–735, 766.

[GK] Glasser M. L. and M. S. Klamkin, On some inverse tangent summations, *Fibonacci Quart.* **14** (1976), no. 5, 385–388. (1986A3)

[GKP] Graham, R. L., D. E. Knuth and O. Patashnik, *Concrete mathematics: A foundation for computer science*, second ed., Addison-Wesley, Reading, MA, 1994. (1995A4)

[Gl] Glaisher, J. W. L., A theorem in trigonometry, *Quart. J. Math.* **15** (1878), 151–157. (1986A3)

[Göd] Gödel, K., Über formal unentscheidbare Sätze der Principia Mathematica und verwandter System I, *Monatshefte für Math. und Physik* **38** (1931), 173–198. English translation by Elliot Mendelson: "On formally undecidable propositions of Principia Mathematica and related systems I" in M. Davis, ed., *The undecidable*, Raven Press, 1965. (1988B2)

[Gor] Gorenstein, D., *Finite groups*, second ed., Chelsea Publishing Co., New York, 1980. (1992B6)

[GR] Gradshteyn, I. S. and I. M. Ryzhik, *Table of integrals, series and products* (translated from the Russian), Academic Press, Boston, 1994. (1999B2)

[Gr1] Graham, R. L., "Rudiments of Ramsey theory," in *CBMS Regional Conference Series in Mathematics* **45**, Amer. Math. Soc., Providence, R.I., 1981. (1996A3, 2000A2)

[Gr2] ——, "Old and new Euclidean Ramsey theorems," in *Discrete geometry and convexity (New York 1982)*, Ann. New York Acad. Sci. **440**, New York Acad. Sci., New York, 1985, pp. 20–30. (1988A4)

[Guy] Guy, R. K., *Unsolved problems in number theory*, second ed., Springer-Verlag, New York, 1994. (1986B5, 2000A2)

[Had] Hadwiger, H., Ein Überdeckungssatz für den Euklidischen Raum, *Portugaliae Math.* **4** (1944), 140–144. (1988A4)

[Hal] Halmos, P., *Problems for mathematicians, young and old*, Math. Association of America, Washington, DC, 1991. (1988A2, 1989B4, 1992A5)

[Har] Hartshorne, R., *Algebraic geometry*, Graduate Texts in Math. **52**, Springer-Verlag, New York, 1977. (1991B5)

[He] Hensley, D., Lattice vertex polytopes with interior lattice points, *Pacific J. Math.* **105** (1983), 183–191. (1990A3)

[HiS] Hindry, M. and J. H. Silverman, *Diophantine geometry: An introduction*, Graduate Texts in Math. **201**, Springer-Verlag, New York, 2000. (1998B6)

[HJ] Horn, R. A. and C. R. Johnson, *Matrix analysis*, Cambridge Univ. Press, Cambridge, 1985. (1997A6)

[HLP] Hardy, G. H., J. E. Littlewood and G. Pólya, *Inequalities*, reprint of the 1952 ed., Cambridge Univ. Press, Cambridge, 1988. (1985A2)

[Hof] Hoffman, K., *Analysis in Euclidean space*, Prentice-Hall, Englewood Cliffs, NJ, 1975. (1999A5)

[Hol] Hollis, S., Cones, k-cycles of linear operators, and problem B4 on the 1993 Putnam competition, *Math. Mag.* **72** (1999), no. 4, 299–303. (1993B4)

[Hon1] Honsberger, R., *Ingenuity in mathematics*, Math. Association of America, Washington, DC, 1970. (1992A6, 1993A6, 1993B1, 1995B6)

[Hon2] ——, *Mathematical gems II*, Math. Association of America, Washington, DC, 1976. (1990A4, 1991B3)

[Hon3] ——, *Mathematical gems III*, Math. Association of America, Washington, DC, 1985. (1993A6)

[HoS] Howard, R. and P. Sisson, Capturing the origin with random points: Generalizations of a Putnam problem, *College Math. J.* **27** (1996), no. 3, 186–192. (1992A6)

[HoU] Hopcroft, J. E. and J. D. Ullman, *Introduction to automata theory, languages, and computation.* Addison-Wesley, Reading, MA, 1979. (1990A5)

[Hu] Hume, A., A tale of two greps, *Software — Practice and experience* **18** (1988), no. 11, 1063–1072. (1990A5)

[HW] Hardy, G. H. and E. M. Wright, *An introduction to the theory of numbers*, fifth ed., Clarendon Press, Oxford, 1988. (1993B3, 1996B2)

[IMO59–77] Greitzer, S., *International Mathematical Olympiads 1959–1977*, Math. Assoc. of Amer., Washington, DC, 1978.

[IMO79–85] Klamkin, M. S., *International Mathematical Olympiads 1979–1985*, Math. Assoc. of Amer., 1986. (1987B6, 1991B3, 1993A6, 1996A5) Despite its title, this book contains the 1978 IMO as well.

[IMO86] The 1986 International Mathematical Olympiad; problems and solutions published in a pamphlet by American Mathematics Competition. (1993B6)

[IMO87] The 1987 International Mathematical Olympiad; problems and solutions published in a pamphlet by American Mathematics Competition. (1990A4)

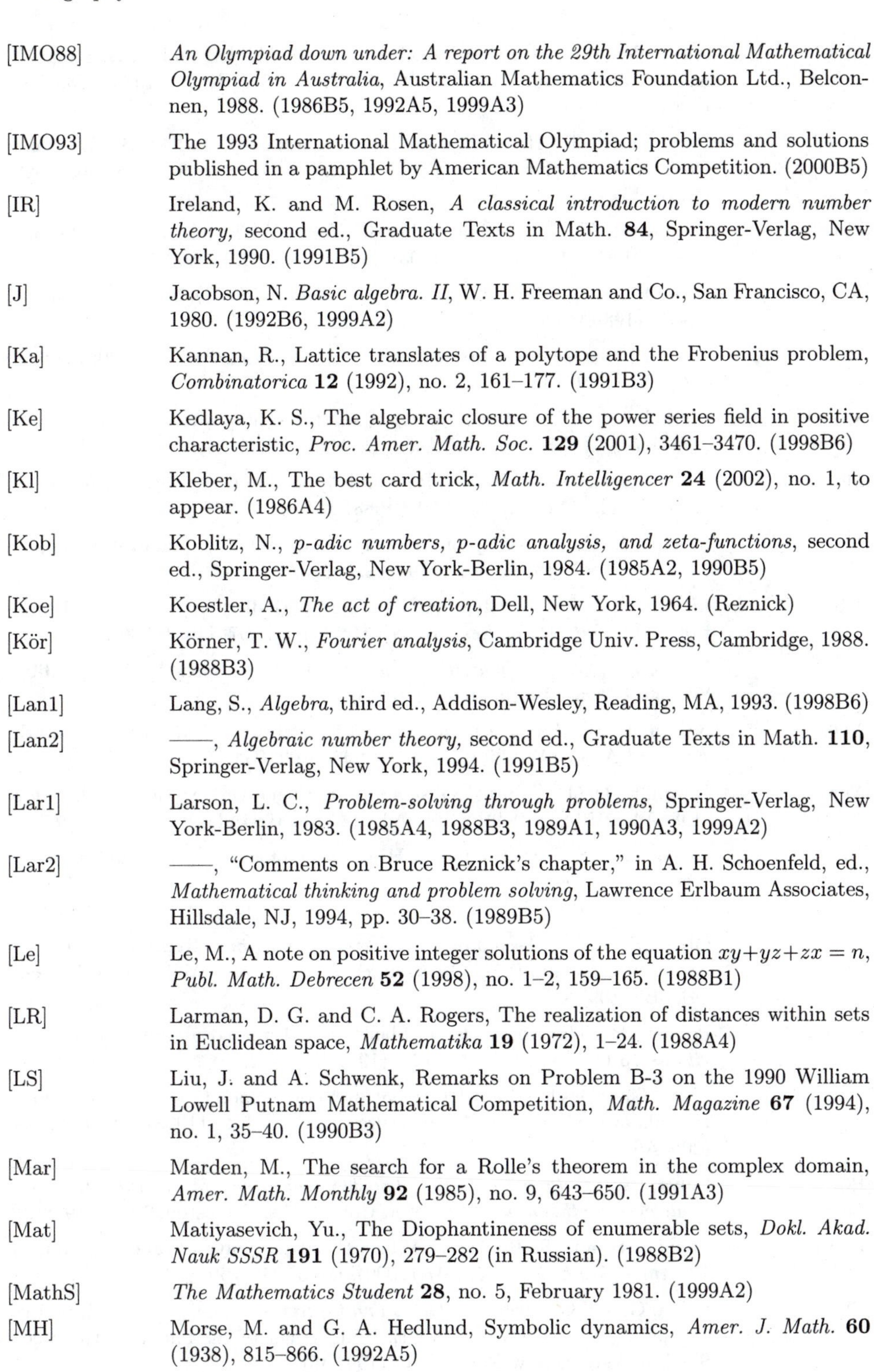

[IMO88] *An Olympiad down under: A report on the 29th International Mathematical Olympiad in Australia*, Australian Mathematics Foundation Ltd., Belconnen, 1988. (1986B5, 1992A5, 1999A3)

[IMO93] The 1993 International Mathematical Olympiad; problems and solutions published in a pamphlet by American Mathematics Competition. (2000B5)

[IR] Ireland, K. and M. Rosen, *A classical introduction to modern number theory,* second ed., Graduate Texts in Math. **84**, Springer-Verlag, New York, 1990. (1991B5)

[J] Jacobson, N. *Basic algebra. II*, W. H. Freeman and Co., San Francisco, CA, 1980. (1992B6, 1999A2)

[Ka] Kannan, R., Lattice translates of a polytope and the Frobenius problem, *Combinatorica* **12** (1992), no. 2, 161–177. (1991B3)

[Ke] Kedlaya, K. S., The algebraic closure of the power series field in positive characteristic, *Proc. Amer. Math. Soc.* **129** (2001), 3461–3470. (1998B6)

[Kl] Kleber, M., The best card trick, *Math. Intelligencer* **24** (2002), no. 1, to appear. (1986A4)

[Kob] Koblitz, N., *p-adic numbers, p-adic analysis, and zeta-functions*, second ed., Springer-Verlag, New York-Berlin, 1984. (1985A2, 1990B5)

[Koe] Koestler, A., *The act of creation*, Dell, New York, 1964. (Reznick)

[Kör] Körner, T. W., *Fourier analysis*, Cambridge Univ. Press, Cambridge, 1988. (1988B3)

[Lan1] Lang, S., *Algebra*, third ed., Addison-Wesley, Reading, MA, 1993. (1998B6)

[Lan2] ——, *Algebraic number theory,* second ed., Graduate Texts in Math. **110**, Springer-Verlag, New York, 1994. (1991B5)

[Lar1] Larson, L. C., *Problem-solving through problems*, Springer-Verlag, New York-Berlin, 1983. (1985A4, 1988B3, 1989A1, 1990A3, 1999A2)

[Lar2] ——, "Comments on Bruce Reznick's chapter," in A. H. Schoenfeld, ed., *Mathematical thinking and problem solving*, Lawrence Erlbaum Associates, Hillsdale, NJ, 1994, pp. 30–38. (1989B5)

[Le] Le, M., A note on positive integer solutions of the equation $xy+yz+zx = n$, *Publ. Math. Debrecen* **52** (1998), no. 1–2, 159–165. (1988B1)

[LR] Larman, D. G. and C. A. Rogers, The realization of distances within sets in Euclidean space, *Mathematika* **19** (1972), 1–24. (1988A4)

[LS] Liu, J. and A. Schwenk, Remarks on Problem B-3 on the 1990 William Lowell Putnam Mathematical Competition, *Math. Magazine* **67** (1994), no. 1, 35–40. (1990B3)

[Mar] Marden, M., The search for a Rolle's theorem in the complex domain, *Amer. Math. Monthly* **92** (1985), no. 9, 643–650. (1991A3)

[Mat] Matiyasevich, Yu., The Diophantineness of enumerable sets, *Dokl. Akad. Nauk SSSR* **191** (1970), 279–282 (in Russian). (1988B2)

[MathS] *The Mathematics Student* **28**, no. 5, February 1981. (1999A2)

[MH] Morse, M. and G. A. Hedlund, Symbolic dynamics, *Amer. J. Math.* **60** (1938), 815–866. (1992A5)

[Mi] Milnor, J., *Dynamics in one complex variable*, Vieweg, Wiesbaden, 1999. (1992B3)

[Mon1] Alexanderson, G. L., H. M. W. Edgar, D. H. Mugler, and K. B. Stolarsky eds., Problems and solutions, *Amer. Math. Monthly* **92** (1985), 735–736. (Reznick)

[Mon2] Bateman, P. T., H. G. Diamond, K. B. Stolarsky, and D. B. West eds., Problems and solutions, *Amer. Mathematical Monthly* **94** (1987) 691–692. (Reznick)

[Mon3] Bumby, R. T., F. Kochman, D. B. West eds., Problems and solutions, *Amer. Math. Monthly* **100** (1993), no. 3, 290–303. (1986A3)

[Mon4] ——, Problems and solutions, *Amer. Math. Monthly* **102** (1995), no. 1, 74–75. (1989A5)

[Mon5] ——, Problems and solutions, *Amer. Math. Monthly* **103** (1996), no. 3, 266–274. (1986A3)

[Mord1] Mordell, L. J., On the number of solutions in positive integers of the equation $yz + zx + xy = n$, *Amer. J. Math.* **45** (1923), 1–4. (1988B1)

[Mord2] ——, *Diophantine equations*, Pure and Applied Mathematics **30**, Academic Press, London-New York, 1969. (1988B1)

[Mort] Morton, P., Arithmetic properties of periodic points of quadratic maps II, *Acta Arith.* **87** (1998), no. 2, 89–102. (2000A6)

[MS] Morton P. and J. H. Silverman, Rational periodic points of rational functions, *Internat. Math. Res. Notices* **1994**, no. 2, 97–110. (2000A6)

[Na] Narkiewicz, W., *Polynomial mappings*, Lecture Notes in Math. **1600**, Springer-Verlag, Berlin, 1995. (2000A6)

[Nel] Nelsen, R. B., *Proofs without words: Exercises in visual thinking*, Math. Assoc. of Amer., Washington, DC, 1993. (1993B1)

[New] Newman, D. J., *A problem seminar*, Springer-Verlag, New York, 1982. (1988A4, 1989A4, 1989B4, 1990A4, 1993A6, 1994A1)

[NZM] Niven, I., H. S. Zuckerman, and H. L. Montgomery, *An introduction to the theory of numbers*, fifth ed., John Wiley & Sons, New York, 1991. (1986B3, 1987B3, 1987B6, 1988A5, 1991A6, 1991B3, 1993B3, 1998B6, 2000A2)

[O] Olver, F. W. J., *Asymptotics and special functions*, Reprint of the 1974 original (Academic Press, New York), A. K. Peters, Wellesley, MA, 1997. (1985B5, 1997A3)

[P1] Poonen, B., The worst case in Shellsort and related algorithms, *J. Algorithms* **15** (1993), no. 1, 101–124. (1991B3)

[P2] ——, The classification of rational preperiodic points of quadratic polynomials over **Q**: a refined conjecture, *Math. Z.* **228** (1998), no. 1, 11–29. (2000A6)

[Pó] Pólya, G., *Mathematics and plausible reasoning vol. I: Induction and analogy in mathematics*, Princeton Univ. Press, Princeton, 1954. (1998B2)

[PRV] Poonen, B. and F. Rodriguez-Villegas, Lattice polygons and the number 12, *Amer. Math. Monthly* **107** (2000), no. 3, 238–250. (1990A3)

[PS] Pólya, G. and G. Szegő, *Problems and theorems in analysis,* vol. II, revised and enlarged translation by C. E. Billigheimer of the fourth German ed., Springer-Verlag, New York, 1976. (1992B5)

[PutnamI] Gleason, A. M., R. E. Greenwood, and L. M. Kelly, *The William Lowell Putnam Mathematical Competition, Problems and solutions: 1938–1964*,

Math. Association of America, Washington, DC, 1980. (1986A6, 1992B3, 1993A6, 1994A3, Reznick)

[PutnamII] Alexanderson, G. L., L. F. Klosinski, and L. C. Larson eds., *The William Lowell Putnam Mathematical Competition, Problems and solutions: 1965–1984*, Math. Association of America, Washington, DC, 1985. (1990A1, 1990A3, 1991B3, 1992B5, 1993A5, 1994B2, 1995B6, 1998B6, Reznick)

[PZ] Pheidas, T. and K. Zahidi, "Undecidability of existential theories of rings and fields," in Denef et al. eds., *Hilbert's Tenth Problem: Relations with arithmetic and algebraic geometry*, Contemporary Math. **270**, Amer. Math. Soc., Providence, 2000, pp. 49–105. (1988B2)

[Re1] Reznick, B., Review of "A Problem Seminar" by D. J. Newman, *Bull. Amer. Math. Soc.* **11** 223–227. (Reznick)

[Re2] ——, Lattice point simplices, *Discrete Math.* **60** (1986), 219–242. (Reznick)

[Re3] ——, "Some binary partition functions," in *Analytic number theory (Allerton Park, IL, 1989)*, Progr. Math. **85**, Birkhäuser, Boston, MA, 1990, pp. 451–477. (Reznick)

[Re4] ——, "Some thoughts on writing for the Putnam," in A. H. Schoenfeld, ed., *Mathematical thinking and problem solving*, Lawrence Erlbaum Associates, Hillsdale, NJ, 1994, pp. 19–29. (Reznick)

[ReS] Reznick, S., *How to write jokes*, Townley, New York, 1954. (Reznick)

[Ri] Ribenboim, P., *The new book of prime number records*, Springer-Verlag, New York, 1996. (2000A2)

[Robe] Robert, A. M., *A course in p-adic analysis*, Graduate Texts in Math. **198**, Springer-Verlag, New York, 2000. (1991B4)

[Robi] Robinson, J., Definability and decision problems in arithmetic, *J. Symbolic Logic* **14** (1949), 98–114. (1988B2)

[Ros1] Rosen, K. H., *Elementary number theory and its applications*, fourth ed., Addison-Wesley, Reading, MA, 2000. (1985A4)

[Ros2] ——, *Discrete mathematics and its applications,* fourth ed., WCB/ McGraw-Hill, Boston, MA, 1999. (1988B5, 1990B4)

[Ru] Rudin, W., *Principles of mathematical analysis*, third ed., McGraw-Hill Book Co., New York-Auckland-Düsseldorf, 1976. (1990A4, 1991A5, 1993B3, 1997A3, 2000A4)

[Sa] Salomaa, A., *Computation and automata*, Encyclopedia of Mathematics and its Applications **25**, Cambridge Univ. Press, Cambridge, 1985. (1990A5)

[Sel] Selmer, E., On the linear Diophantine problem of Frobenius, *J. reine angew. Math.* **294** (1977), 1–17. (1991B3)

[Se1] Serre, J.-P., *A course in arithmetic*, translated from the French, Graduate Texts in Math. **7**, Springer-Verlag, New York-Heidelberg, 1973. (1991B5, 2000A2)

[Se2] ——, *Linear representations of finite groups*, translated from the second French ed. by L. L. Scott, Springer-Verlag, New York, 1977. (1985B6, 1992B6)

[Se3] ——, *Lectures on the Mordell-Weil theorem*, translated from the French and edited by M. Brown from notes by M. Waldschmidt, Aspects of Mathematics **E15**, Friedr. Vieweg & Sohn, Braunschweig, 1989. (1988B1)

[Shaf] Shafarevich, I. R., *Basic algebraic geometry 1*, second ed., translated from the 1988 Russian ed. and with notes by M. Reid, Springer-Verlag, Berlin, 1994. (1987B3)

[Shar] Sharp, W. J. C., Solution to Problem 7382 (Mathematics) proposed by J. J. Sylvester, *Ed. Times* **41** (1884), London. (1991B3)

[She] Shen, A., Three-dimensional solutions for two-dimensional problems, *Math. Intelligencer* **19** (1997), no. 3, 44–47. (1998B3)

[Sil] Silverman, J. H., *The arithmetic of elliptic curves*, Graduate Texts in Math. **106**, Springer-Verlag, New York, 1986. (1998B6)

[Sim] Simmons, G. F., *Calculus with analytic geometry*, second ed., McGraw-Hill, New York, 1996. (1993B3)

[SMO] The Swedish Mathematical Olympiad. (1999B2)

[So] Solomon, H., *Geometric probability*, SIAM, Philadelphia, PA, 1978, p. 124. (1992A6)

[Spt] Spitzer, F., *Principles of random walk*, second ed., Graduate Texts in Math. **34**, Springer-Verlag, New York, 1976. (1995A6)

[Spv] Spivak, M., *Calculus*, second ed., Publish or Perish, Inc., 1980. (1985A5, 1988B4, 1989A3, 1989A5, 1989B3, 1989B6, 1990A6, 1991B6, 1994B2)

[St] Stanley, R. P., *Enumerative combinatorics*, vol. 2, with a foreword by Gian-Carlo Rota and appendix 1 by Sergey Fomin, Cambridge Studies in Advanced Math. **62**, Cambridge Univ. Press, Cambridge, 1999. (1995A4)

[Tak] Takács, L., On cyclic permutations, *Math. Sci.* **23** (1998), no. 2, 91–94. (1995A4)

[Tal] Talvila, E., Some divergent trigonometric integrals, *Amer. Math. Monthly* **108** (2001), no. 5, 432–435. (2000A4)

[Tar] Tarski, A., *A decision method for elementary algebra and geometry*, second ed., Univ. of California Press, Berkeley and Los Angeles, CA, 1951. (1988B2)

[USAMO7286] Klamkin, M., *U.S.A. Mathematical Olympiads 1972–1986*, Math. Assoc. of Amer., Washington, DC, 1988. (2000A6)

[USAMO] The USA Mathematical Olympiad; problems and solutions from individual years published in pamphlets by American Mathematics Competitions. Problems from recent years are available at Problems Corner section of American Mathematics Competition website `http://www.unl.edu/amc`. (1995B6, 1997B5)

[Wag] Wagon, S., Fourteen proofs of a result about tiling a rectangle, *Amer. Math. Monthly* **97** (1987), no. 7, 601–617. (1991B3)

[War] Warner, F. W., *Foundations of differentiable manifolds and Lie groups*, Springer-Verlag, 1983. (1991B2)

[Weil1] Weil, A., *Sur les courbes algébriques et les variétés qui s'en déduisent*, Hermann, Paris, 1948. This volume was later combined with another volume of Weil and republished as *Courbes algébriques et variétés abéliennes*, Hermann, Paris, 1971. (1991B5)

[Weil2] ——, Numbers of solutions of equations in finite fields, *Bull. Amer. Math. Soc.* **55** (1949), 497–508. (1991B5)

[Wen] Wendel, J. G., A problem in geometric probability, *Math. Scand.* **11** (1962), 109–111. (1992A6)

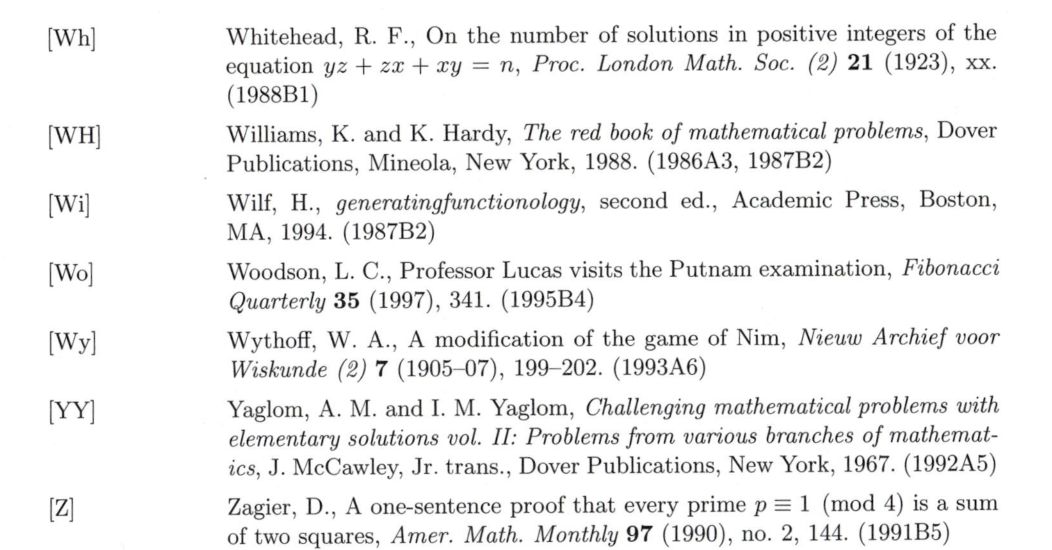

[Wh] Whitehead, R. F., On the number of solutions in positive integers of the equation $yz + zx + xy = n$, *Proc. London Math. Soc. (2)* **21** (1923), xx. (1988B1)

[WH] Williams, K. and K. Hardy, *The red book of mathematical problems*, Dover Publications, Mineola, New York, 1988. (1986A3, 1987B2)

[Wi] Wilf, H., *generatingfunctionology*, second ed., Academic Press, Boston, MA, 1994. (1987B2)

[Wo] Woodson, L. C., Professor Lucas visits the Putnam examination, *Fibonacci Quarterly* **35** (1997), 341. (1995B4)

[Wy] Wythoff, W. A., A modification of the game of Nim, *Nieuw Archief voor Wiskunde (2)* **7** (1905–07), 199–202. (1993A6)

[YY] Yaglom, A. M. and I. M. Yaglom, *Challenging mathematical problems with elementary solutions vol. II: Problems from various branches of mathematics*, J. McCawley, Jr. trans., Dover Publications, New York, 1967. (1992A5)

[Z] Zagier, D., A one-sentence proof that every prime $p \equiv 1 \pmod 4$ is a sum of two squares, *Amer. Math. Monthly* **97** (1990), no. 2, 144. (1991B5)

Index

Bold face page numbers indicate pages with background information about the entry, e.g., a detailed explanation or reference, while page numbers in normal type indicate a textual reference.

algebraic function, 259
algebraic geometry, 83, 84, 231
algebraic number, **121**
Arithmetic-Mean–Geometric-Mean (AM-GM) Inequality, **56**, 95, 153, 253, 254, 274
automata, 108, 122, 123
Axiom of Choice, **112**

Bateman-Horn Conjecture, 279
Beatty sequences, **179**, 180
Beatty's Theorem, **179**, 216
Bernoulli numbers, 184, **225**
Bessel functions, **63**, **236**
big-O notation, **89**, 90, 107, 111, 117, 118, 134, 204, 208, 258, 275
bijection, 21, 53, 82, 123, 124, 140, 141, 173, 174, 195, 223, 229
binary representation, 11, 17, 39, 43, 107, 108, 156, 157, 188
Birkhoff-von Neumann Theorem, **68**
block matrix, 75
Brauer group, 169

Carmichael's lambda function, **57**
Catalan numbers, **206**
Cauchy's Lemma, **143**
Cauchy's Theorem, **282**, 283
Cauchy-Schwarz Inequality, **55**
Cayley diagram, **127**
Cayley digraph, **127**
Cayley-Hamilton Theorem, 63, **122**, 199, 228
Cayley-Menger determinant, **188**
centroid, 40, 47, 129, 130, **130**, 131, 132, 233
Chain Rule, 139
change of variables, 183, 184
character, 64, 150, 151
Chebychev's Inequality, **55**
chess, 158
Chinese Remainder Theorem, 58, 202, 257
circuit, **126**
circulant matrix, **98**, 99, 276
class number, 94
combinatorial games, 212, 214
compactness, 44, 49, 104, 134, 195, 230, 266
companion matrix, 268
Comparison Test, **97**, 282
compass and straightedge, 218
complementary sequences, **179**, 180
complex numbers, 143
concavity of functions, 48, 55, 132–134, 197, 251
content (of a polynomial over $\mathbb{Z}$), **259**
continued fraction, **74**, 212
continuity, 4, 7, 12, 13, 20, 25, 29, 32, 34, 42, 49, 56, 68, 79, 80, 85, 111, 113, 124, 130, 132–134, 138, 139, 143, 155, 156, 166, 176, 184, 220–222, 242, 251, 272, 274, 275, 292, 293
contour integration, 59, 283
contraction mapping, 241
convergence, 10, 37, 41, 42, 61, 62, 72, 80, 89, 97, 110, 111, 125, 162, 183, 191, 194, 195, 204, 221, 222, 228, 234–236, 259, 265, 282–284, 291
convexity, 13, 14, 26, 55, 68, 118, 120, 128, 130, 131, 134, 159, 160, 230, 231, 270
cosine law, *see* Law of Cosines
countability, 12, 44, 111, 112, 120, 121, 195, 219
cycle, **126**

de Moivre's Theorem, 35, **58**
derangement, **99**
Descartes' Rule of Signs, 60, **60**, 198
design theory, 294
determinants, xi, 10, 18, 24, 32, 45, 70, 97–99, 164, 188, 193, 200, 211, 276
diagonalization, 227
Dickson's Conjecture, **279**
difference operator, 259

differential equations, 36–38, 63, 70, 78, 89, 93, 110, 111, 124, 207, 235, 239, 243, 270, 275
dimension, 134, 167, 187, 209, 215, 216, 266, 277
directed multigraph, **126**, 127
Dirichlet L-function, 95
Dirichlet density, **280**
Dirichlet's Theorem, **280**
discriminant, 74, 178, 198, 260, 261, 273
divisibility of binomial coefficients, 25, 147, 220, 238, 290, 293
division algorithm, 163
Dominated Convergence Theorem, 236, 283
doubly stochastic matrices, **68**
dynamical systems, 85, 159, **163**, 212, 222

eigenvalue, 49, **63**, 93, 98, 99, 167, 184, 199, 200, 207, 227–229, 241, 242, 276–277
eigenvector, 9, 38, 45, 49, **63**, 93, 99, 184, 207, 227, 241, 242, 270, 276
Eisenstein integers, **91**
elliptic curve, **258**
Euler ϕ-function, 35, **57**, 58, 247, 267, 280
Euler line, **233**
Euler's Theorem, **57**, 58
Euler-Maclaurin summation formula, **225**, 227
Eulerian circuit, **126**, 127
Eulerian path, **126**, 127
Extreme Value Theorem, **184**, 266

Farey series, **182**
Fermat's Little Theorem, **86**, 202
Fibonacci numbers, 40, **93**, 117, 123, 124, 139, 173, 222, 223, 252
field, xi, 5, 7, 8, 11, 16, 74, 83, 84, 86, 93–95, 107, 108, 128, 148, 158, 164, 167, 178, 261
field of complex numbers, **xi**
field of rational numbers, **xi**
field of real numbers, **xi**
finite directed multigraph, **126**
finite field, **xi**, 8, 11, 16, 86, 107, 108, 148, 151, 158, 203, 261
first order sentence, **95**, 96
Fourier analysis, 58, 184, 216
Fourier series, 58, 216
Fresnel integrals, 284
functional equation, 92, 110, 142, 154, 178, 220, 221, 292
Fundamental Theorem of Calculus, 138, 176

Gabriel's Horn, **138**
Gall-Peters projection, 255
games, 11, 20, 24, 27, 47, 102, 103, 144, 158, 180, 182, 212–214, 233, 234
Gauss sum, **150**, 151
Gauss's Lemma, 86, 258, **259**
Gauss-Lucas Theorem, *see* Lucas' Theorem
Gaussian elimination, 97
Gaussian integers, **150**
Gegenbauer polynomials, **270**
generalized hypergeometric functions, *see* hypergeometric functions
Generalized Riemann Hypothesis (GRH), vii, 94, 95
generating function, 47, 82, **83**, 93, 145, 158, 161, 239, 246, 293
generating functions, 266
geometric probability, 159, 160, 182
geometric transformations, 254
geometry, 39, 83
gradient, 46, **68**, 217
graph theory, 40, 126, 242
greedy algorithms, 36, 48, 252
grep, 123
group theory, 3, 11, 14, 27, 57, 63, 99, 109, 120, 126, 143, 168–170, 178, 195, 202, 237, 238

harmonic series, 61, 164, 244
height function, **258**
Hensel's Lemma, **72**, 261
Hilbert metric, 241
Hilbert's Seventeenth Problem, 264
Hilbert's Tenth Problem, 96
hypergeometric functions, **283**, 284
Hypothesis H, 279

Inclusion-Exclusion Inequality, 201
Inclusion-Exclusion Principle, 99, 126, 201
indegree, **126**
inequalities, 32, 49, 55, 56, 65, 88, 89, 95, 103, 104, 132, 139, 151, 153, 162, 180, 181, 201, 218, 219, 224, 252, 253, 271, 274
inequalities of integrals, 49, 85, 139, 225, 274
Integral Comparison Test, **89**
integral domain, **107**
integration by parts, 39, 47, 50, **110**, 132, 234, 235, 270, 273, 281, 282

Intermediate Value Theorem, **102**, 128, 130, 251, 278, 291
International Mathematical Olympiad, 74, 87, 121, 144, 158, 179, 190, 220, 265, 293
isometric embedding, **211**

Jacobi sum, **150**, 151
Jensen's Inequality, **55**, 56
Jordan canonical form, 63

Kummer's Theorem, **238**, 290, 293

L'Hôpital's Rule, **152**, 156, 166, 269
Lagrange interpolation, 259, **260**, 289
Lagrange multipliers, **217**
Lagrange's Theorem, **57**, **268**
Lambert equal-area cylindrical projection, 255
Laplace transform, **236**
lattice polygon, 118
lattices, 91, 111, 118–120, 208, 209
Laurent polynomial, 35, **59**
Laurent series, 259
Law of Cosines, 186
Law of Sines, 269, **286**
Legendre polynomial, 270
Legendre symbol, 86, **149**
Leibniz's formula, 43, **182**, 183
lexical scanner, 123
lg, *see* logarithm
Lie group, **143**
Limit Comparison Test, **89**
linear algebra, 166
linear operator, 43, 184
linear recursion, 38, 92, **93**, 116, 117, 172, 199, 207, 208, 246, 252, 265, 267
linear recursive relation, 199
linear recursive sequence, 38, 92, **93**, 116, 117, 172, 246, 252, 265, 267
linear transformation, 191, 192, 277
little-o notation, **89**, 90, 92, 96, 103, 279
ln, *see* logarithm
Local Central Limit Theorem, 208
log, *see* logarithm
logarithm, **xi**
logarithmic derivative, 239
Lucas numbers, **212**
Lucas' Theorem, **137**, 270

Macdonald's function, **63**
Markov equation, **74**
Markov process, 207
Mean Value Theorem, 117, **198**
measure theory, 120, 160, 270
Monotone Convergence Theorem, 183, 234

Newton polygon, **128**
Newton's method, 72
Newton-Puiseux series, **260**
Nim, **213**
notation, **xi**
number theory, 72, 120, 148, 183

O, *see* big-O notation
o, *see* big-o notation
Olympiad, 74, 87, 92, 101, 121, 144, 158, 161, 179, 190, 214, 220, 247, 265, 270, 293
orthogonal polynomials, **270**
orthogonality relations for characters, **64**
outdegree, **126**

p-adic absolute value, **128**
p-adic gamma function, 147
p-adic logarithm function, **xi**
p-adic numbers, xi, **128**, 147, 148
p-adic valuation, **245**, 257
p-group, **169**
parity, 40, 45, 85, 87, 118, 141, 212, 237, 256, 258
partition, 24, 140, 141, 179, 209, 214, 216
path, **126**
Pell's equation, **279**
permutation matrix, **68**
Perron-Frobenius Theorem, 184, 207, 241, **241**, 242
Pick's Theorem, 40, 118, **119**
Pigeonhole Principle, 43, 50, **96**, 118, 175, 193, 195, 215, 248, 267, 289, 294
Pochhammer symbol, **284**
polynomial ring, **xi**
Postage Stamp Theorem, **143**
Power Mean Inequality, **55**, 56
primitive root, 202, **247**
probability, 2, 62, 115, 159, 160, 182, 183
product rule, 9, 88, 136
Puiseux series, **259**
Pythagorean Theorem, 85, 287, 294
Pythagorean triple, **85**

quadratic form, 95
quadratic reciprocity, 86

radius of curvature, 210
Ramsey theory, **219**
random walk, 207, **208**, 209, 241
Ratio Test, **111**, 125, 235, 284

recursion, 38, 46–48, 85, 92, 93, 116, 117, 165, 172, 199, 207, 208, 223, 229, 230, 239, 246, 252, 265, 267, 270
reduced totient function, *see* Carmichael's lambda function
reflection trick, 48, **254**
règle des nombres moyens, 180
representation theory, 64, 99, 168, 169
residue theorem, 59
Riemann sum, 12, 114, **115**, 134
Riemann zeta function, **183**
ring, **xi**
ring of integers, **xi**
Rolle's Theorem, 42, 50, **136**, 137, 155, 198, 291
Rouché's Theorem, **102**, 231

set theory, 173
Siegel zero, 95
sine law, *see* Law of Sines
sketching, 9, 43, 88
spherical cap, 255
Sprague-Grundy theory, **214**
stationary phase approximation, **285**
Stirling's approximation, 208, **225**
Stokes' Theorem, 37, **79**
strongly connected, **126**, 242
sum of two squares, 33, 150, 263, 278, 279
Sylow subgroup, 238
symmetry, 35, 37, 38, 42, 45, 71, 80, 88, 109, 112, 117, 118, 136, 160, 162, 169, 192, 208, 281

Taylor series, 38, 45, 48, 89, 103, 104, **107**, 156, 204, 257, 275
Taylor's Formula, **155**, 257, 273
Taylor's Theorem, **133**
Thue-Morse sequence, **158**
Thue-Siegel Theorem, **261**
transcendental number, **121**, 158
Trapezoid Rule, 132, 243
triangle inequality, 103, 218, 252
trigonometric substitution, 37, 46, 50, 85, 192, 217, 292
Turing machine, 95
Twin Prime Conjecture, **279**

ultraspherical polynomials, 270
unique factorization domain, 59, 259
USA Mathematical Olympiad, 214, 247

Vandermonde determinant, **70**, 98, 276
Vandermonde's identity, **82**, 83, 145
vectors, 46, 93, 159, 186, 187, 218, 230, 231, 233, 276
Venn diagram, 53

weakly connected, **126**
Weierstrass M-test, **111**
Weierstrass's Theorem, **111**
Weil Conjectures, vii, **151**, 261
Well Ordering Principle, **112**
well-ordered set, **112**, 194, 195
Weyl's Equidistribution Theorem, **96**, **216**
winding number, 230, 231
Wolstenholme's Theorem, **147**, 245
Wythoff's game, **180**

Zorn's Lemma, 111, **112**, 195

Kiran S. Kedlaya is from Silver Spring, MD. He received an AB in mathematics and physics from Harvard (where he was a Putnam Fellow three times), an MA in mathematics from Princeton, and a PhD in mathematics from MIT, and currently holds a National Science Foundation postdoctoral fellowship at the University of California, Berkeley. Other affiliations have included the Clay Mathematics Institute and the Mathematical Sciences Research Institute. His research interests are in number theory and algebraic geometry.

He has been extensively involved with mathematics competitions and problem solving. He has taught at the Math Olympiad Summer Program, served on the USA Mathematical Olympiad committee and on the executive committee of the 2001 International Mathematical Olympiad (held in Washington, DC), served as a collaborating editor for the *American Mathematical Monthly* problems section, maintained problem information on the World Wide Web for the American Mathematics Competitions, and edited Olympiad compilations for the Mathematical Association of America.

Bjorn Poonen is from Boston. He received the AB degree *summa cum laude* in mathematics and physics from Harvard University, and the PhD degree in mathematics from the University of California at Berkeley, where he now holds the title of associate professor of mathematics. Other affiliations have included the Mathematical Sciences Research Institute, Princeton University, the Isaac Newton Institute, and the Université de Paris-Sud.

He is a Packard Fellow, a four-time Putnam Competition winner, and the author of over 50 articles. His main research interests lie in number theory and algebraic geometry, but he has published also in combinatorics and probability. Journals for which he serves on the editorial board include the *Journal of the American Mathematical Society* and the *Journal de Théorie des Nombres de Bordeaux.* He has helped to create the problems for the USA Mathematical Olympiad every year since 1989.

Ravi Vakil is from Toronto, Canada. He received his undergraduate degree at the University of Toronto, where he was a four-time Putnam Competition winner. After completing a PhD at Harvard, he taught at Princeton and MIT before moving to Stanford, where he is a tenure-track assistant professor. He is currently an American Mathematical Society Centennial Fellow and an Alfred P. Sloan Research Fellow. His field of research is algebraic geometry, with connections to nearby fields, including combinatorics, topology, number theory, and physics. He has long been interested in teaching mathematics through problem solving; he coached the Canadian team to the International Mathematical Olympiad from 1989 to 1996, and one of his other books is titled *A Mathematical Mosaic: Patterns and Problem Solving.*